694POL

Carpentry and Joinery

Work Activities

Carpentry and Joinery

Work Activities

Brian Porter LCG, FIOC

Reg Rose MCIOB, DMS, DASTE, FIOC
formerly Assistant Principal, Leeds College of Building, UK

ELSEVIER
BUTTERWORTH
HEINEMANN

AMSTERDAM • BOSTON • HEIDELBERG • LONDON • NEW YORK • OXFORD
PARIS • SAN DIEGO • SAN FRANCISCO • SINGAPORE • SYDNEY • TOKYO

Elsevier Butterworth-Heinemann
Linacre House, Jordan Hill, Oxford OX2 8DP
30 Corporate Drive, Burlington, MA 01803

First published 2000
Reprinted 2001, 2002, 2003, 2004, 2006

British Library Cataloguing in Publication Data
A catalogue record for this book is available from the British Library

Library of Congress Cataloguing in Publication Data
A catalogue record for this book is available from the Library of Congress

ISBN 0 340 69241 3

For information on all Elsevier Butterworth-Heinemann
publications visit our website at www.bh.com

Working together to grow
libraries in developing countries

www.elsevier.com | www.bookaid.org | www.sabre.org

ELSEVIER   BOOK AID International   Sabre Foundation

Printed and bound in Great Britain by MPG Books Ltd., Bodmin, Cornwall

Contents

Preface

This book has been written as a companion volume to *Carpentry and Joinery Bench and Site Skills*, and its purpose is to cover the manufacture, fixing and maintenance of wood and wood-based components used in buildings and associated with the work of the carpenter and joiner. In general terms the well-illustrated text seeks in the simplest way possible to cover the theoretical knowledge which underpins craft skills recognised by NVQ at level III. Each area of work is dealt with in general terms, as to do otherwise would require each chapter to be developed into a book of its own. This book should be used in conjunction with *Carpentry and Joinery Bench and Site Skills*, and if a more detailed study of the subject matter of the individual chapters is required, reference should be made to more specialist books and/or to the standards indicated in the text.

One of the biggest changes in the latter half of the twentieth century has been the codifying of materials and craft practices into standard codes. The final changeover to the ISO metric system, and the adoption of International and European standards covering the materials, manufacture and fixing of carpentry and joinery structures and components. The reader should not underestimate the importance of the standards set by the regulating bodies, or the influence they have on the work you are involved with.

It is important to recognise that structural components are specifically designed to meet their in-use requirements, and if the designer requires nails or screws of a specific number, length, or diameter to connect components together, they should be used. Many failures have occurred because the designer's specifications for fixings were ignored.

In the study and practice of your craft, knowledge and understanding are the added value to your skill. We wish you well in your studies and development of craft skills in your chosen career as a carpenter and joiner, and that this book will be helpful to you in achieving your goals.

W. R. Rose FIOC, MCIOB, DMS.

Acknowledgements

We would like to acknowledge our gratitude to the following companies and organisations for their help and support in providing us with technical information, and in many cases permission to reproduce items of artwork and/or photographs. Without their help many aspects of the work would not have been possible. Allan Wilson using AutoCAD, to whom we are most grateful, produced the illustrations that head each chapter, and the house plan in the appendix.

Annstar Group Ltd (Architectural ironmongery)
British Gypsum Ltd
Chasmood Ltd (Architectural hardware)
Draper (The Tool Company)

Fischer Fixing Systems
Henderson Ltd (Garage doors, Sliding and folding door hardware)
Ingersoll-Rand Architectural hardware (Briton, Legge, Newman Tonks)
Optical Measuring Instruments Cowley Ltd
Reynolds (UK) Ltd (Crompton Hardware & ERA Security Products
Royde & Tucker Ltd
Rugby Joinery
Smallwood I & D Ltd
Stramit Industries (UK) Ltd
Swish Building Products
Trend Machinery & Cutting Tools Ltd
Trustspan Ltd, Murton York

1
Ground and upper floors

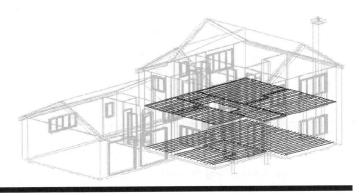

1.1 INTRODUCTION

This chapter deals with ground and upper suspended timber floors in single occupancy domestic dwellings with no more than three stories.

1.2 SUSPENDED TIMBER GROUND FLOORS

A suspended timber ground floor is an arrangement of **joists** (beams) set parallel to each other, being supported at their ends and possibly at intermediate points along their length (see Figure 1.1a). Joists are the structural part of the floor supporting a platform of boards or sheet material which carry those things we wish to place on a floor (Figure 1.1b). As the name implies this type of floor is only just above ground level, and requires care in its construction if it is to give trouble-free service during the life of the building. Protecting the timbers from the ingress of moisture is of prime importance when constructing ground floors (see Figure 1.6).

1.3 JOISTS

Joists are load-bearing members which transfer the loads on the floor to the supporting structure (Figure 1.1c). Floor joists are designed to carry two types of loading:

• Dead loads. The self-weight of the floor structure [*Building Regulations Approved Document "A" and "D"*] (Figure 1.2a).

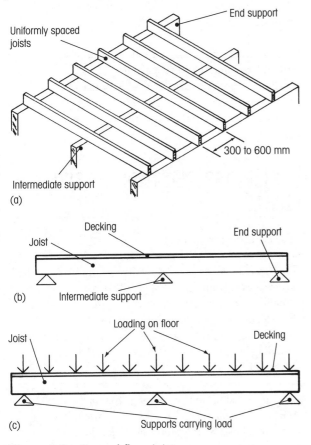

(a)

(b)

(c)

Figure 1.1 Ground floor joists

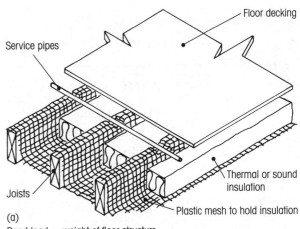

(a)
Dead load = weight of floor structure

NB. Building Regulations Approved Documents Table A1 takes into account the weight of all floor elements except the joists

Figure 1.2 (a) Dead loading on floors; (b) continued overleaf

Figure 1.2 (b) Live loading on floors

- Imposed or live loads. People and other items that the floor deck supports [*Building Regulations Approved Document "A" and "D"*] (Figure 1.2b).

1.4 JOIST SPAN

The span of a joist is normally considered the distance between the centres of its bearing supports, and is known as the **effective** span [*Building Regulations Approved Document "A" and "D"*].

The term **clear** span is sometimes used, and this refers to the clear distance between supports (see Figure 1.3a).

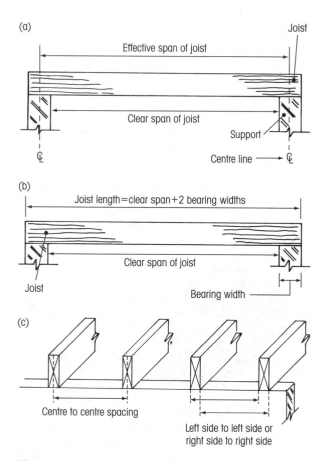

Figure 1.3 (a) Clear and effective span; (b) joist length; (c) joist spacing

1.5 JOIST LENGTH

The length of the joist is the clear span plus the two bearing widths (Figure 1.3b).

Note. The width of the bearing must not be less than 90 mm [*Building Regulation Approved Document "A" and "D"*].

It is recognised that the length and cross-sectional size of timber is limited, and it is not normal for timber floor joists to have a clear span of more than 5 m. Otherwise the cost of the joists increase out of all proportion to their size.

1.6 JOIST SPACING

Refers to the distance, usually measured centre to centre between one joist and the next. The distance between the centres of joists may normally vary between 300 and 600 mm (Figure 1.1b) depending upon:

- The imposed (live) and dead loads the floor has to carry.
- The size of the joist.
- The span of the joist.
- Strength of the joist.
- The type, strength, and thickness of the decking material.

Note. When sheet materials are used as decking they normal determine the joist spacing.

When spacing out joists it is more practical to measure from one right face to the next, or from one left face to the next, rather than to their centres (Figure 1.3c).

1.7 CROSS-SECTIONAL SIZE OF JOISTS

This size is determined by:

- The way the joists are laid relative to the loads applied (geometrical properties) (Figure 1.4a).
- The distance between supports (Figure 1.4b).
- The loads to be supported (Figure 1.4c).
- The strength of the joists. Determined by "Strength Grading" [BENCH AND SITE SKILLS, SECTION 1.8.4, PAGE 37].

Using the tables in the Building Regulations Approved Document "A" and knowing the following factors it is possible to determine given the cross-sectional size of joists from the following information.

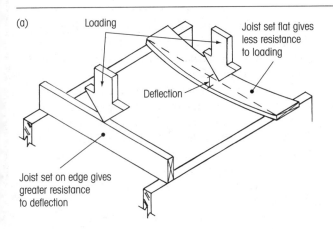

(a)

Loading

Joist set flat gives less resistance to loading

Deflection

Joist set on edge gives greater resistance to deflection

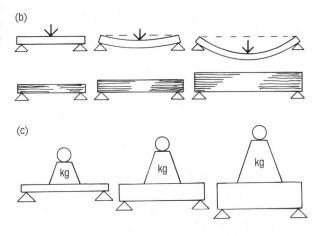

(b)

(c)

Figure 1.4 Influence on span and loading on joist size: (a) reasons for setting joists on edge; (b) the greater the span the deeper the joist; (c) the larger the loading the deeper the joist

- For example, the strength classification of timber [BENCH AND SITE SKILLS, TABLE 1.13]. Select Hemlock with a strength class of 3. Read from table for timber of strength class 3 (SC3) (C16 new classification).
- For example the dead load the floor will carry is 0.5 kN/m².
- The joists are to be spaced at 600 mm centres.
- The clear span between supports is taken as 2 m.
- Joist width 50 mm.

Note. For other than single occupancy domestic dwellings a structural engineer would design the floor.

Given these five pieces of information and using Table 1.1 a suitable joist depth can be determined as follows.

- Look at the top of the table for the given dead load (not more than 0.5 kN/m²).
- Look for the column in this group of three columns marked 600 mm.
- Looking down this column identify the value nearest to two (2.19), but not less than two, and horizontally across to the group of sizes starting with a thickness of 50 mm.
- Trace across to the left from the figure greater, but nearest to two (2.19) and you will see in the first column of the table that the size of joist you require is 50 thick × 122 mm deep.

Note. The joist thickness also influences the span, however. If you wish to keep all the joists the same depth, and the span in some areas is reduced you can reduce the joist thickness (see Table 1.2).

Table 1.1 Span tables for joists (Building Regulations Approved Document "A")

Maximum clear span of joist (m). Timber of strength class SC3 (C16 new classification)

Size of joist mm × mm	Dead Load (kN/m²) excluding self-weight of joist:								
	Not more than 0.25			More than 0.25 but not more than 0.50			More than 0.50 but not more than 1.25		
				Spacing of joist (mm)					
	400	450	600	400	450	**600**	400	450	600
47 × 122						2.09			
47 × 195									
47 × 220									
50 × 97						1.54			
50 × 122						2.19			
50 × 147						**2.61**			
50 × 170						2.99			
50 × 195						3.39			
50 × 220						3.79			
63 × 97									
63 × 122						2.45			

Table 1.2 Joist span related to joist thickness. Building Regulations Approved Document "A"

Section in mm		Max clear span in m at 600 mm centres
Variable thickness	Constant depth	
38	122	1.76
47	122	2.09
50	122	2.19
63	122	2.45
75	122	2.60

1.8 FLOORING (DECKING) MATERIALS

The three most common forms of floor decking are tongued and grooved:

(1) Flooring grade chipboard [BENCH AND SITE SKILLS, TABLE 2.8, PAGE 77].
(2) Softwood floorboards [*BS 1297: 87*].
(3) Plywood floor panels.

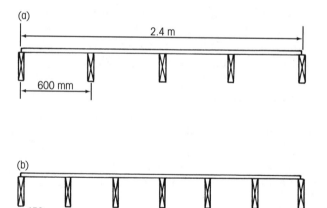

(a)

2.4 m

600 mm

(b)

450 mm

Positioning of joists to carry decking – joint spacing for 2.4 m long chipboard floor panels

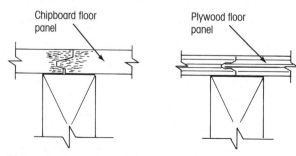

Chipboard floor panel

Plywood floor panel

(c) Tongue and grooved edge joints on chipboard and plywood

When using chipboard and plywood flooring panels, the spacing of the joists must be such that cross-tongued and grooved ends of the panels are supported on a joist (Figure 1.5a–c).

Provided the joists are spaced at the recommended centres for softwood floorboards they can be cut and butt jointed on the nearest joist, and if the short end (offcut) is sufficiently long to span over at least three joists, it can be used in the next row of boards. If it is shorter it may be used for other things. Usually there is very little waste when cutting floorboards.

It is necessary that the thickness of the decking material is such that undue deflection does not occur as it spans between joists (see Table 1.3 and Figure 1.5d).

1.9 JOIST SUPPORTS

Joists require supporting at their ends and in the case of ground floors at intermediate points along their length. The use of intermediate supports reduces the clear span and consequently the cross-sectional size of the joists required. In the case of ground floors the distance between supports is usually 1.5–2.0 m. The bearing point areas at any position along the joist must not be less than the width of the joist by 90 mm (see Figure 1.6a). If joists are jointed at intermediate bearing points, the joist ends should not project more than 100 mm beyond the bearing (Figure 1.6b) as the effect of deflection at the passing ends of the joists creates an uneven surface for fixing the decking material (Figure 1.6c).

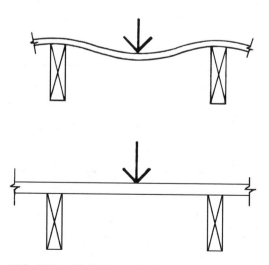

(d) Positioning of joists to carry decking

Figure 1.5   Floor decking to joists

Table 1.3 Guide to decking material thickness

Decking material	Finished thickness (mm)	Maximum joist spacing on centres (mm)	Reference
Tongued and	16	450	
grooved softwood flooring	19	600	Building Regulations. Approved Document 'A'
Flooring grade chipboard	18/19	450	National House Building
Tongued and grooved on all edges	22	600	Council (NHBC) BS5669
Tongued and grooved plywood decking	12.5	300	
	16	400	
	19–20.5	600	

Ground floor joists should not be built into the inner leaf of external cavity walls unless the ends are protected against the ingress of moisture and the joists are treated with preservative (Figure 1.6d). It is better practice to support the end of ground floor joists on honeycomb sleeper walls (half brick wall with ventilation gaps) with their ends clear of the outer wall (Figure 1.6e), or on mild steel galvanised hangers (Figure 1.6f).

1.10 LAYOUT OF JOISTS FOR PART OF A GROUND FLOOR

Figure 1.7 shows the plan for the layout of ground floor joists spaced at 600 mm centres to carry 2400 × 600 × 22 mm tongued and grooved chipboard flooring panels laid at right angles to the line of the joists.

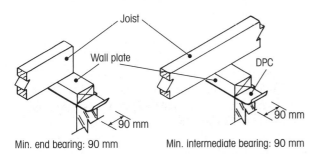

Min. end bearing: 90 mm   Min. intermediate bearing: 90 mm

(a) Joist bearing

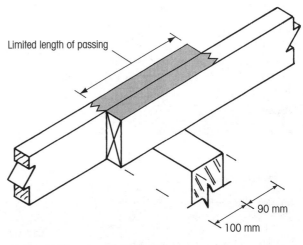

(b) End of joist overhang limited to 100 mm

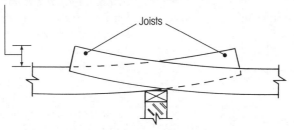

(c) Effects of deflection on joists with excessive overhang

Figure 1.6 Joist ends supported by joist hangers: (a)–(c) shown above; (d) and (e) continued overleaf

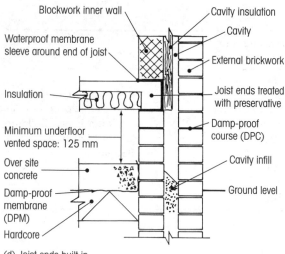

(d) Joist ends built in

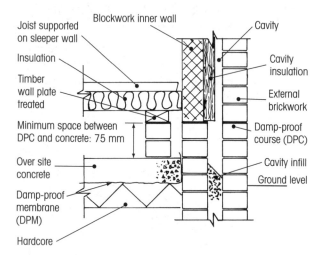

(e) Joists supported off sleeper walls

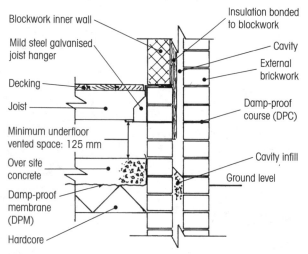

NB. Insulation not shown under floor

(f) Joist ends supported by metal hangers

Figure 1.6 End and intermediate support of ground-floor joists

1.11 CALCULATING THE NUMBER OF JOISTS REQUIRED FOR A GIVEN ROOM

Measure the length of the room at right angles to the line of the joists and divide [BENCH AND SITE SKILLS, SECTION 13.2.5] this length by the joist spacing. Round up any fraction to the next whole number and add one.

Note. The divisor and dividend *must be* in the same units. All in millimetres or metres.

Example 1

Number of joists required for the larger room (lounge) in Figure 1.7

$$4845 \div 600 = 8.075$$

rounded up to the nearest whole number gives 9. Then add 1 to give the number of joists as 10.

Example 2

Number of joists required for the smaller room (dining-room) in Figure 1.7

$$2500 \div 600 = 4.167$$

rounded up to the nearest whole number gives 5. Then add 1 to give the number of joists as **6**.

Total number of joists required for the floors in the two rooms is 10 + 6 = **16**.

Note. Because of the wide opening between the two rooms they have been treated as a single area for the purpose of setting out the joists to give correct spacing of joists for decking sheets.

1.12 LAYOUT OF THE GROUND FLOOR JOISTS

Looking at Figure 1.7(b) you will see that joist 2 is set 620 mm to centre from the wall to give a slight gap between this wall and the end of the first flooring sheet, and in Figure 1.7 joists 2, 3, 4, 5, 8, 9, 10, 11, 12, 13, 14 and 15 are set out at 600 mm centres from each other. The remaining joists, 1, 6, 7 and 16, are set adjacent to the walls to carry the ends of the decking material (see Figure 1.7a–c). The typical layout of the ground floor joists is illustrated in Figure 1.7(d).

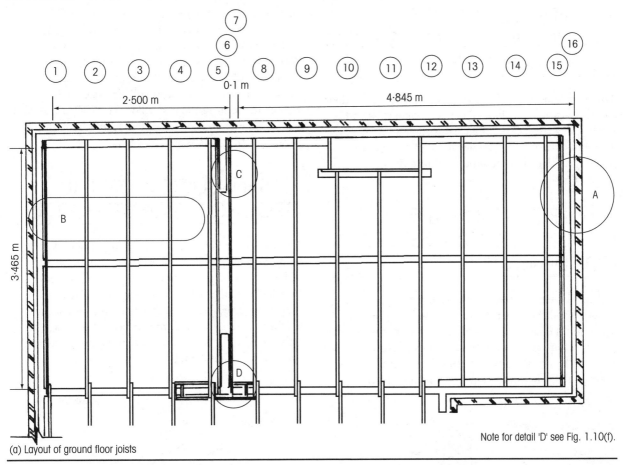

(a) Layout of ground floor joists

Note for detail 'D' see Fig. 1.10(f).

Enlarged views of (A)–(D) are shown below or in Fig. 1.10(f).

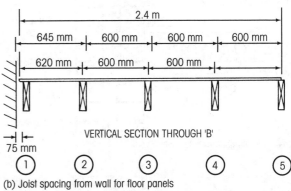

(b) Joist spacing from wall for floor panels

VERTICAL SECTION THROUGH 'B'

75 mm

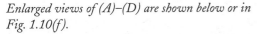

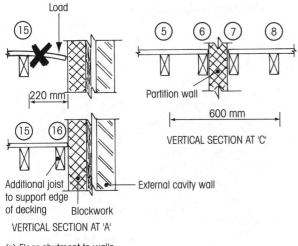

Load

Partition wall

VERTICAL SECTION AT 'C'

600 mm

220 mm

Additional joist to support edge of decking — External cavity wall

Blockwork

VERTICAL SECTION AT 'A'

(c) Floor abutment to walls

(d) Typical layout of ground floor joists

Figure 1.7   Joist layout (lounge and dining-room)

1.13 PROTECTING THE GROUND FLOOR JOISTS FROM RISING MOISTURE

The relative position of the various elements in the floor and the position of DPCs (damp-proof courses) and DPMs (damp-proof membranes) is shown Figure 1.6. The timber members must always be above the damp proofing so that they are protected from rising moisture.

1.14 PROTECTING THE GROUND FLOOR JOISTS FROM CONDENSATION

If there is a temperature difference between the floor structure and the under floor void, condensation will occur and the joists will absorb the moisture thus raising their moisture content above safe limits [BENCH AND SITE SKILLS, SECTION 3.3.5], unless the under floor area is ventilated. The ventilation is provided by placing air

vents in the perimeter walls of the building. The air vents are usually placed just below ground floor level and are of such a size and spacing that they provide 3000 mm^2 of venting per m run of wall. The vents (air bricks) should be positioned in such away that there are no dead air pockets (areas of still air) where condensation could occur in the under floor void (Figure 1.8a). If possible the main flow of ventilating air should move parallel to the joists (see Figure 1.8b).

1.15 INSULATION OF THE GROUND FLOOR STRUCTURE

Because of the flow of cool air through the underfloor void drawing heat from the room, thermal insulation of blanket or ridged sheet type can be used to reduce heat loss through the floor. The insulating material being supported below the floor decking by plastic netting (see Figure 1.2a), timber bearers or a cage of light-weight plastic-coated metal mesh hung from the joists (see Figures 1.9a, b).

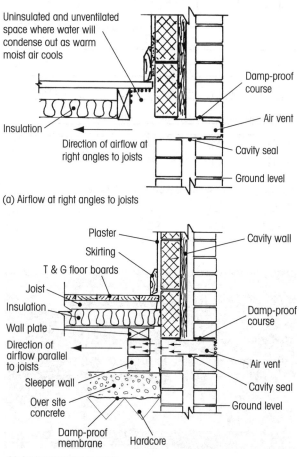

(a) Airflow at right angles to joists

(b) Air flow parallel to joists

Figure 1.8 Under floor ventilation

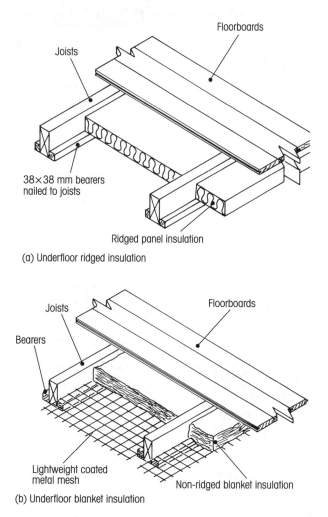

(a) Underfloor ridged insulation

(b) Underfloor blanket insulation

Figure 1.9 Underfloor thermal insulation

1.16 PLACING AND LEVELLING GROUND FLOOR JOISTS

To make the best economic use of timber, joists are designed to be placed across the shortest span.

To lay and level ground floor joists you will require the following tools and equipment.

- Straight edge (this can be a straight and parallel piece of floorboard).
- Spirit level (900 mm or longer). Pencil and rule or tape.
- Hammer and nails
- Fixing lath and packings.

Operations involved in fixing the joists as shown in Figure 1.7.

- Cut the wall plates to length, and cut any necessary halving joints for lengthening, or at a change of direction [BENCH AND SITE SKILLS, SECTION 7.5.4]. The bricklayer will assist in bedding the wall plates level and on top of the DPCs. Mark out the spacing of the joists on the perimeter wall plates as indicated in Figure 1.7.
- Select 16 joists of uniform depth (the joists may be regularised) [BENCH AND SITE SKILLS, SECTION 1.9.1]. If the joists are not regularised you may require packings which should be of the correct thickness, with an area equal to that of the bearing and securely fixed to the wall plates.
- Cut joists 13, 14, 15 and 16 to fit between the walls with a 20 mm gap at each end, and joists 1, 2, 3, 4, 5, 6, 7, 8, 9 and 12 to the width of the room plus 200 mm. Joists 10 and 11 should be cut 400 mm shorter than joist 12 so as not to encroach into the hearth space.
- Place joist 1 cambered (crown) side up 50 mm in from the wall, and using the spirit level, level in the joist. The end that rests on the inner partition wall may need packing. Secure the joist by nailing to the wall plates (Figure 1.10a).
- Place joist 6 in position and level in (as shown in Figure 1.10b) and nail to wall plates.
- Position joists 2, 3, 4 and 5 at the positions marked on the wall plates. Level by packings if necessary until the tops of the joists are level and in line, this can be checked by the use of the straight edge (see Figure 1.10c) and nail joists to wall plates.
- If joists 1–6 are to be built into the inner partition wall, mark out the joist spacing on a lath and temporarily tack it to the top of the joists, but clear of the partition wall (Figure 1.10d) until the joist ends are built in.

- In the larger room (lounge) place in position joists 7, 12 and 16 (as described in Figure 1.10b–e). Then fix in position joists 8, 9, 13, 14 and 15 (as in Figure 1.7f).
- Using short lengths of joist material form a kerb by framing round the fender wall to the hearth and levelling in from joists 9 and 12 (Figures 1.7a and 1.10e). Finally level and fix joists 10 and 11 by securing to the kerb and wall plates. Joists 6–12 will again require holding in place by a lath, temporarily nailed to the top edge of the joists, until their ends are built in (Figures 1.7a and 1.10d).
- Short ends of joist material can now be used to form noggins around door openings to support the ends of the floor decking material (see Figures 1.7 and 1.10f).

1.17 LAYING SHEET FLOOR DECKING

For this operation you will require:

- Flooring nails (ring shanked) and hammer [BENCH AND SITE SKILLS, SECTION 1.1.3].
- Hand saw, or power saw.
- String line, square, pencil, rule or measuring tape.
- Adhesive and timber blocks from short ends of timber.

(a) After checking the thermal insulation is in place mark a line across the top of the joists running the whole length of the two rooms, one or a multiple of sheet widths plus 20 mm from the wall (to allow for moisture movement) running at right angles to the joists. In the case of Figure 1.11(a) the line is established two sheet widths away from the wall.

(b) Position the first sheet (as shown in Figure 1.11a) making sure the end of the sheet is positioned in the centre of joist 5 and the long edge is parallel to the line marked along the top of the joists; and secure with two or three floor nails. Cut sheet 2 to fit between the end of sheet 1 and the projecting partition wall. With some adjustment of its length the remainder of sheet 2 will be cut round the hearth and finish with its edge in the centre of joist 11.

(c) Apply adhesive to the cross-tongued and grooved edge of the sheets 1 and 2 and place into position, mating the tongue and groove together, and nail in position.

(d) Sheet 3 will also be cut round the concrete hearth slab, and finish on joist 15. Cut a piece off sheet 4 to make up the space between sheet 3 and the end wall, and fix as described above. Finally nail down this first row of sheets before the adhesive cures. Nails are spaced at 150 mm centres.

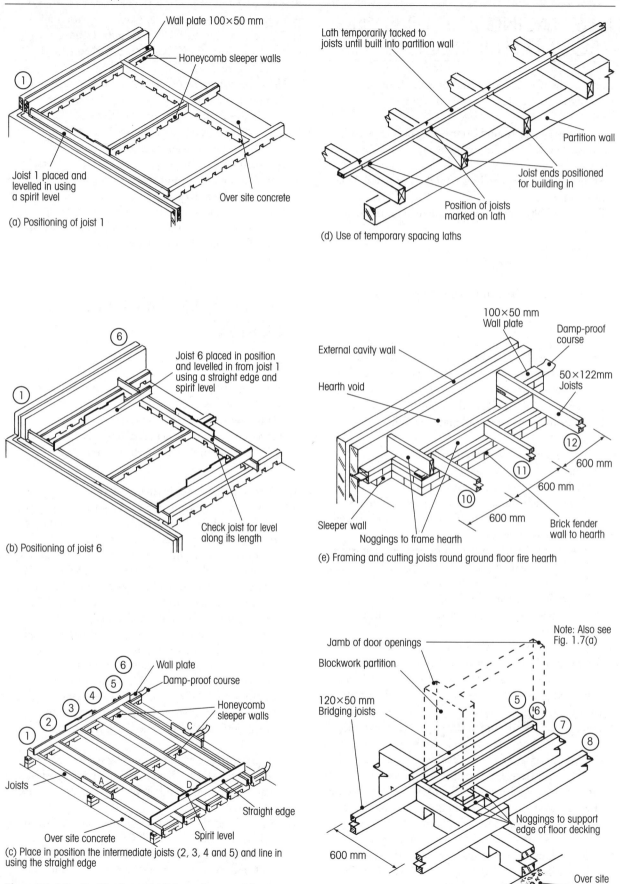

(a) Positioning of joist 1

(b) Positioning of joist 6

(c) Place in position the intermediate joists (2, 3, 4 and 5) and line in using the straight edge

(d) Use of temporary spacing laths

(e) Framing and cutting joists round ground floor fire hearth

(f) Trimming round door openings to support floor decking

Figure 1.10   Positioning and fixing of ground floor joists

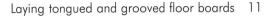
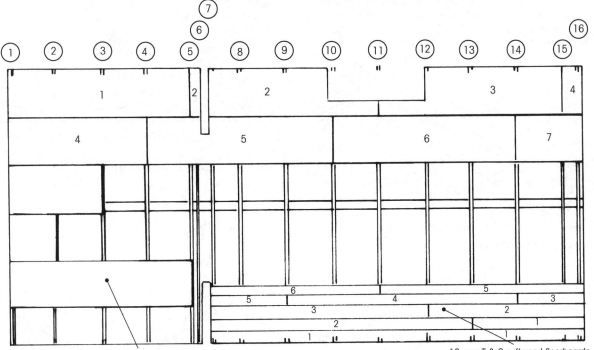

22 mm flooring grade T & G chipboard flooring panels: 2440×600 mm

19 mm T & G softwood floorboards

(a) Layout of floor decking

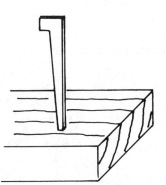

(b) Cut floor brad used as an alternative to lost head nails

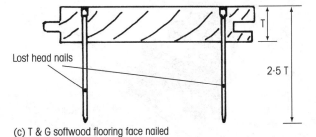

Lost head nails

T

2·5 T

(c) T & G softwood flooring face nailed

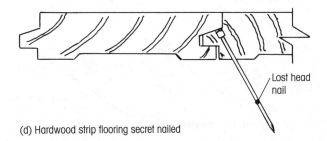

Lost head nail

(d) Hardwood strip flooring secret nailed

Figure 1.11   Economic cutting and fixing of floor decking

(e) The remainder of sheet 4 with some adjustment to its length is used to start the second row with its edge finishing on joist 4. This ensures the header joints are staggered across the floor.

(f) Repeat the process (a)–(e) making sure the adhesive is applied to all board edges that are in contact. The sixth and last row of sheets will require cutting along their length. The width of the sheet being reduced to the remaining gap less 20 mm (Figure 1.11).

1.18 LAYING TONGUED AND GROOVED FLOOR BOARDS

For this work you will require.

- Flooring nails, hammer and nail punch [BENCH AND SITE SKILLS, SECTION 5.8.1 & 2].
- Panel saw, adjustable square, pencil, measuring tape or rule.

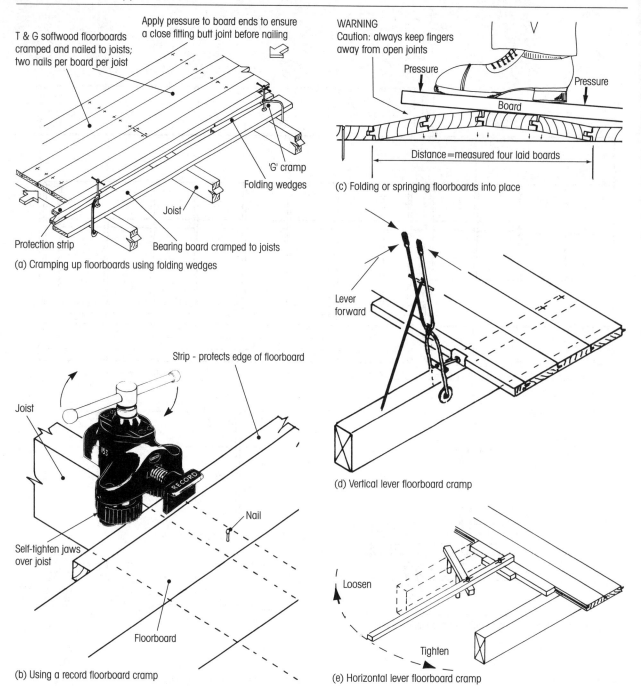

Figure 1.12   Methods of cramping floor boards: (a) cramping up floor boards using folding wedges; (b) using a record floor cramp; (c) cramping floor boards by springing into place; (d) vertical lever cramp; (e) horizontal lever cramp

- String line [BENCH AND SITE SKILLS, CHAPTER 6, SECTION 6.2.1].
- Floorboard cramps.

(a)   As in the case of laying floor panels, because of the width of the opening between the two rooms (Figures 1.7 and 1.11), a line should be established a number of floorboard widths away from the long wall. In the case of Figure 1.11 with a board

covering width of 105 mm six board widths plus 12 mm will be sufficient.

(b)   Lay the first floor board with the off centre tongue closest to the joist (Figure 1.11a, less allowance) and cut to fit between the two walls and nail in position making sure that it is parallel to the established line with the distance away from the line equal to five board widths. Using the short end from the first board mark its position on joist 13 (this allows it to span over three joists) and cut off square to the centre of the joist. Mark to length, and cut the end of board 2 to fit against the end of the first board, making sure the ends are butted tightly together (header joint).

(c) Cut a further two or three rows of boards and lay together.

Note. It is normal to work with the tongues of the boards facing away from the working edge.

(d) Cramp the loose boards to the first fixed row, making sure all the end joints are in close contact and spot nail to hold them in place (Figure 1.12a–e).

(e) Continue as described in (b)–(d) until the whole area is boarded; then lining up with the spot nails drive in the remaining nails. Two nails at every point where a board crosses a joist, at butt joints four nails would be used. If service pipes and cables run below the decking and are supported by the floor joists their position should be marked on the top surface of the floor decking, and care taken not to damage them when driving in the nails.

Note. If hardwood strip flooring is used (that is boards 100 mm wide or less) the boards would be secret nailed a row at a time (Figure 1.11d). Hardwood strip flooring is cross tongued across the end of each strip so that butt joints need not be made on a joist, provided they are staggered (Figure 1.15).

1.19 FINISH OF FLOOR BOARDS AT THE EDGE OF A HEARTH

To finish off the ends of the floor boards butting to the hearth 50 mm margin strips are mitred in.

The arrangement of the margin strips will depend on which way the boards run against the hearth (Figure 1.13).

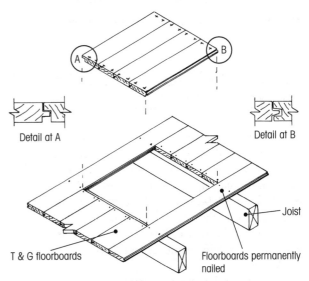

(a) Forming an access panel (trap) in T & G floorboards

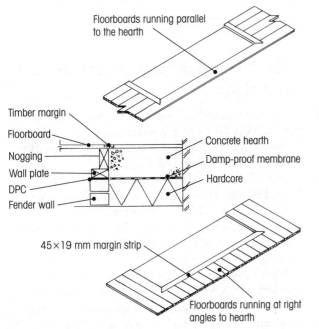

Figure 1.13 Finish to ends of floor boards at a hearth

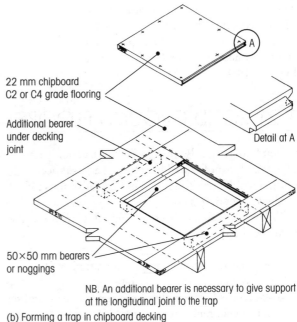

(b) Forming a trap in chipboard decking

Figure 1.14 Forming access traps in hollow floors: (a) and (b) shown above; (c) see the next page

(c) Forming traps in floor decking using a Trend Routabout jig and inserts for forming a circular trap in sheet floors

Figure 1.14   Forming access traps in hollow floors

1.20 FORMING ACCESS TRAPS IN FLOOR DECKING

Occasionally it is necessary to obtain access to services that run in the floor void, and to do this traps are formed as the floor decking is laid, so that when needed they can be taken up. Traps can be formed in both floor boards (Figure 1.14a) and in sheet flooring panels (Figure 1.14b). If possible the traps should be positioned in the areas of least traffic. It may be necessary to modify the underfloor thermal insulation.

For forming circular access traps in sheet flooring after it is laid, "Trend" produce a Routabout jig for use with a portable router for cutting traps up to 250 mm in diameter, and to form an insert. Which when used with a spacer ring will exactly fit the circular hole cut to form the trap. It can be used on chipboard flooring up to 22 mm thick (see Figure 1.14c).

1.21 SOLID GROUND FLOORS

Ground floors can also be of solid construction with timber and manufactured boards used as a surface finish, providing resilience and or decorative finish if hardwood flooring is used.

Timber floor finishes to solid ground floors are of the following basic types.

- Board, sheet, or strip flooring laid on timber fillets (battens) attached to the concrete.
- Wood blocks laid directly on the concrete base and set in hot or cold bitumen mastic
- Sheet floor panels over ridged insulation, which can be obtained as composite panels with the insulation bonded to the back of the decking.

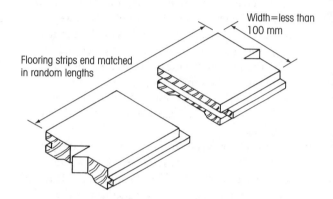
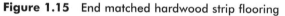

Figure 1.15   End matched hardwood strip flooring

In floors of this type the timber used must be protected from moisture and careful incorporation of an effective damp proofing system is necessary.

1.21.1 Board Strip or Sheet Finishes to Solid Floors

Hardwood tongued and grooved end matched strip flooring (less than 100 mm wide) (see Figure 1.15). Softwood floorboards, plywood, or chipboard flooring grade sheets are nailed to timber fillets.

Note. The width of hardwood strip flooring is limited to 100 mm to reduce the effect of shrinkage, and for the best finish it is recommended that the strips should not be wider than 75 mm. Table 1.4 gives a list of some suitable hardwood for use as strip flooring.

The timber fillets treated with preservative [BENCH AND SITE SKILLS, SECTION 4.3] are fixed to the concrete subfloor by folded galvanised, or sheradised mild steel clips, which when opened out form a cradle into which the fillets sit, being secured to them by nails (see Figure 1.16a).

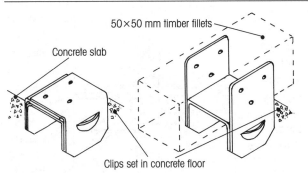

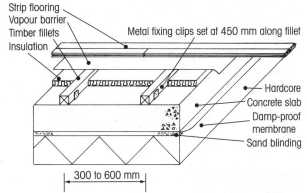

(a) Metal floor clips in open and folded positions

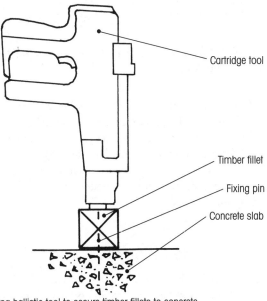

(b) Using ballistic tool to secure timber fillets to concrete

(c) Section through a solid ground floor with hardwood strip flooring finish

Figure 1.16 Solid ground floor with hardwood strip flooring

As an alternative the fillets can be secured directly to the concrete by the use of ballistic driven fixing pins [BENCH AND SITE SKILLS, SECTION 8.19]. A vapour barrier is used between the underside of the strip flooring and the fillets to stop water vapour passing through and condensing out in the air space between the underside of the strip flooring and the top of the rigid sheet insulation placed between the fillets on top of the concrete (see Figure 1.16c).

When using plywood or chipboard flooring panels they can be laid as a floating floor directly on top of a ridged layer of sheet insulating material of varying density and thickness from 15 to 75 mm without direct fixing to the subfloor (see Figure 1.17a). Some manufacturers will supply flooring grade chipboard with the insulation bonded to it, so that both sheet and insulation can be laid simultaneously. Under normal loading in domestic properties, compression of the insulation will

Table 1.4 A guide to hardwoods suitable for flooring

Name	Wearing properties	Moisture movement	Cost
Abura	Low	Small	Medium to high
Afrormosia	Moderate to low	Small	Low
Agba	Moderate to low	Small	Low
Beech	High	Large	Medium
Iroko	Moderate	Small	Medium
Mahogany (African)	Moderate to low	Small	Medium
Makore	High	Small	Medium
Maple	High	Medium	Medium
Oak (European)	High	Medium	Medium to high
Olive (East African)	Very high	Medium	Medium to high
Opepe	High	Small	Medium
Sapele	Moderate to high	Medium	Medium
Teak	Moderate to high	Small	High

Note. Timber can be grouped into three classes by shrinking and swelling movement:
1. Small movement class. When the sum of the radial and tangential movement is less than 3.0%.
2. Medium movement class. When the sum of the radial and tangential movement is between 3.0 and 4.5%.
3. Large movement class. When the sum of the radial and tangential movement is greater than 4.5%.

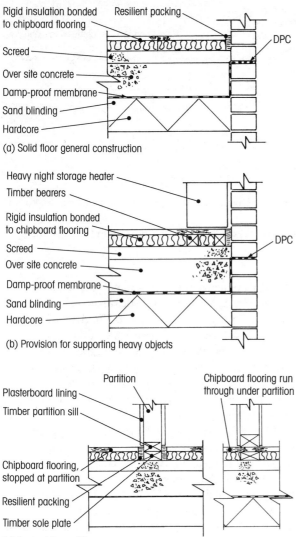

(a) Solid floor general construction

(b) Provision for supporting heavy objects

(c) Support for partitions

Figure 1.17 Insulated decking to solid floors

be limited to about 3%. If areas of the floor are subject to heavy loading, such as night storage heaters, then bearers set in the insulation must be used to carry the weight (see Figure 1.17b). Partitions should not be fixed on top of the insulated decking, but be fixed via a sole plate to the subfloor (Figure 1.17c).

1.21.2 Adjacent Suspended and Solid Ground Floors

When both types of floor are incorporated in one building it is necessary to make sure:

- That the suspended floor is adequately ventilated. This may require air ducts to be formed in the solid floor to provide ventilation of the under floor void (see Figure 1.18).
- That the damp-proof course and membranes have continuity forming a completely waterproof barrier to rising moisture (Figures 1.18 and 1.19a).

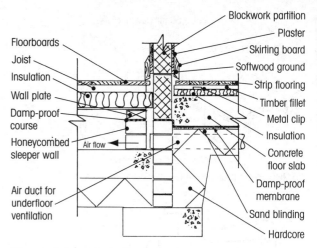

Figure 1.18 Underfloor ventilation at the junction of hollow and solid floors

- That differential movement at the junction of solid and suspended floors in door openings are allowed for by the use of thresholds (cover strips) (Figure 1.19b). In certain cases the floor decking can be continued from a suspended to solid floor construction if board, strip, or sheet materials are used (Figure 1.19a).

1.21.3 Wood Block Flooring

Figure 1.20(a) shows details of a solid concrete floor with wood blocks laid in a herring bone pattern with a two block margin round the room. Laying wood block flooring is a specialist function as the blocks have to be laid in hot or cold bitumen mastic (which acts as an adhesive and retains some elasticity when set) on a sand cement screed which must be flat and level, perfectly dry and coated with a bitumen primer. As wood blocks expand and contract under the changes in moisture content, a gap of about 20 mm is left between the perimeter edges of the blocks and the walls of the room. this gap is filled with cork strip and then covered by the skirting board. Wood blocks are usually made from hardwoods which have reasonable dimensional stability under changes of moisture content (see Table 1.4) although it is still possible to obtain them in softwood. The blocks vary in length from 150 to 300 mm and in widths up to 75 mm. Their thickness can vary between 15 and 30 mm and the length being a multiple of the width (Figure 1.20b). A dovetailed rebate is worked round the bottom edge of the block to give a key to and take up the surplus bitumen as the block is bedded. When all the blocks are laid the surface of the floor is dressed off flat using a floor sander, and then treated with a surface coating to protect the blocks and emphasise the natural beauty of the timber.

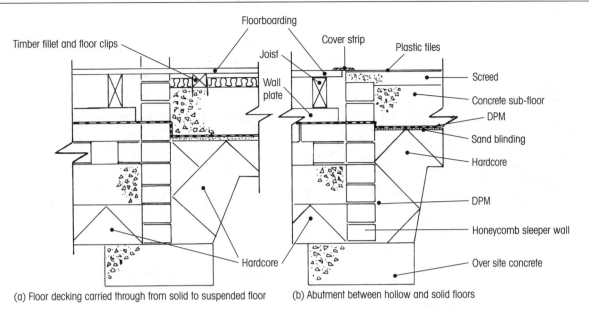

(a) Floor decking carried through from solid to suspended floor

(b) Abutment between hollow and solid floors

Figure 1.19   Decking abulments through openings at the junction of solid and hollow floors

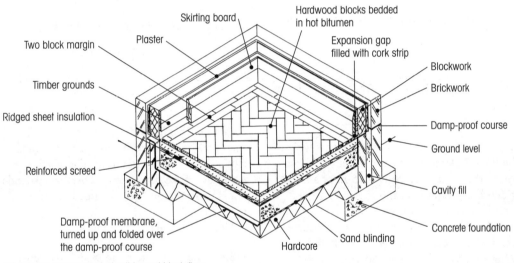

(a) General arrangement of solid wood block floor

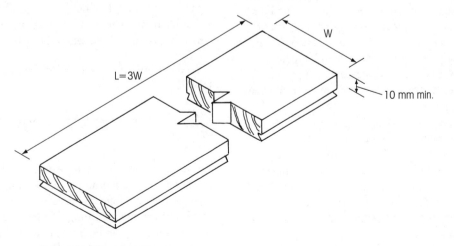

(b) Hardwood flooring block

Figure 1.20   Solid ground floor with wood block finish; hardwood flooring block

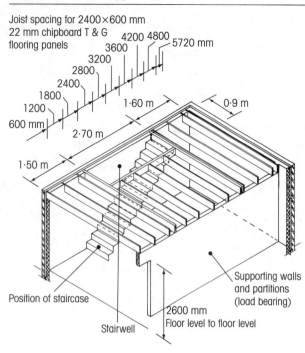

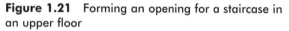

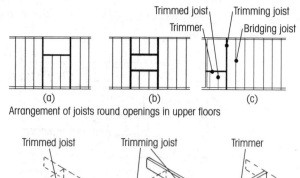

Arrangement of joists round openings in upper floors

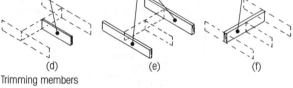

Trimming members

Figure 1.22   Trimmings to openings in upper floors

Figure 1.21   Forming an opening for a staircase in an upper floor

1.22 SUSPENDED TIMBER UPPER FLOORS

Joists in upper floors are the structural elements which support the floor decking above and the ceiling below. As the span of these joists can be up to 5 m or a little over, their depth is greater than those used in ground floors where the span is never more than 2 m. Upper floors are reached via a staircase and to accommodate this an opening (stairwell) is formed in the floor structure (Figure 1.21). If masonry carrying a flue projects from the perimeter wall into the floor space the joists must be supported round rather than being built into the projection as a precaution against the risk of fire [*Building Regulations Approved Document "A"*]. The inner leaf of the perimeter cavity wall and any load-bearing inner walls or partitions below the upper floor will support the ends of, and provide some intermediate support for the joists.

On upper floors the working heights are greater than 2 m above the ground and you must at all times work from a safe platform which is so constructed that you are safe, and anyone working below is not in danger from falling objects and tools that you are using [*Health and Safety at Work Act. Construction Regulations*].

Upper floor joists have a greater width-to-depth ratio than ground floor joists.

For example: ground floor ratio. Depth divided by width

$$122 \div 50 = \mathbf{2.4}$$

or

$$1 : 2.4$$

Upper floor ratio. Depth divided by width

$$200 \div 50 = \mathbf{4.0}$$

or

$$1 : 4.0$$

because of this increase in ratio of upper floor joists and their greater span they will require stiffening along their length by the use of strutting. The spacing of joists in upper floors are the same as those for suspended ground floors to accommodate the same type of decking materials.

Upper floors are grouped into:

- **Single floors** where there is no intermediate support other than that provided by load-bearing walls and partitions below the floor.
- **Double floors** where an intermediate supporting member is introduced when the span exceeds 5 m.

To determine the size of joists see Section 1.7 and Table 1.1.

1.22.1 Forming Openings in Upper Floors

Figure 1.21 shows how an opening (stairwell) is formed in a floor, and the method of supporting the joists at the edges of the opening. The process of framing the opening is termed trimming. Figure 1.22(a–c) shows how the opening can be framed depending on its size, position in the floor and the direction in which the joists run. The members forming the opening and the joists supported by this framing are termed.

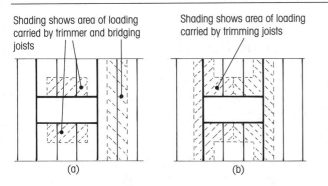

Shading shows area of loading carried by trimmer and bridging joists

Shading shows area of loading carried by trimming joists

(a)   (b)

Figure 1.23 Area of loading carried by trimmed members

- **Trimmed joists**. A joist that terminates at and is supported by the trimmer (Figure 1.22d).
- **Trimming joists**. The member that supports the trimmer (Figure 1.22e).
- **Trimmers**. That member which gives support to the trimmed joists (Figure 1.22f).

Note. Bridging joists are those joists that span across the floor (Figure 1.22c).

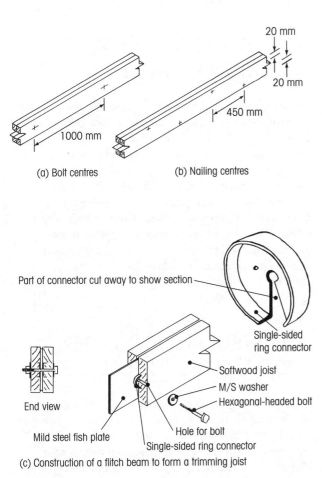

20 mm

20 mm

450 mm

1000 mm

(a) Bolt centres   (b) Nailing centres

Part of connector cut away to show section

Single-sided ring connector

Softwood joist

M/S washer

Hexagonal-headed bolt

End view

Mild steel fish plate

Hole for bolt

Single-sided ring connector

(c) Construction of a flitch beam to form a trimming joist

Figure 1.24 Jointing bridging joists to form trimming members

1.23 SIZE OF TRIMMERS AND TRIMMING JOISTS

As these members have to carry additional loads transferred via the trimmed joists, their width has to be increased, or two bridging joists connected together can be used. Figure 1.23(a) shows the area of loading supported by the trimmer and Figure 1.23(b) shows the area of loading supported by the trimming joist. As a rule of thumb the width of the trimmer and trimming joist would be increased in width 3 mm for every trimmed joist carried. If six trimmed joists are supported by the trimmer the width would be increased to 50 + (6 × 4) = 74 mm. It is now required that tables giving sizes of trimmers be used, or the size of these members are calculated by a structural engineer.

If the trimmers and trimming joists are formed by connecting two bridging joists together. The bolting or nailing of the members together must be to the designers specification [*Building Regulations Approved Document "A"*]. The surfaces of the abutting joists must be in contact and the bolt or nail points must be able to transfer the loads without failure [BENCH AND SITE SKILLS, SECTION 11.1.4]. Figure 1.24(a, b) shows nail and bolt spacing as recommended by the NHBC (National House Builders Council). If exceptional loads have to be carried then a flitched beam can be formed to act as a trimmer, by sandwiching a flat metal plate between two bridging joists and connecting them together with bolts and single-sided timber connectors (Figure 1.24c).

1.24 JOINTING OF TRIMMERS, TRIMMING AND TRIMMED JOISTS

There are two ways of connecting these members together.

- Cutting joints into the timber members in the form of **tusk tenons** (Figure 1.25) and housing joints [BENCH AND SITE SKILLS, SECTION 7.5.5] connecting the trimmed to the trimming joists (Figure 1.26).
- The use of pressed or fabricated galvanised or sherardised mild steel joist hangers (Figure 1.27a).

1.25 USE OF METAL HANGERS

When using metal hangers to transfer loads (forces) between structural floor members, care must be taken to make sure that:

- Hangers are housed in flush with the top and bottom edges of the member to maintain a flat surface on which to fix the floor decking and ceiling (Figure 1.27a, b).

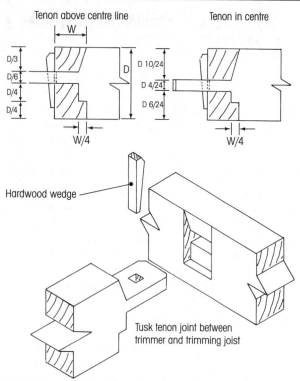

Figure 1.25   Traditional tusk tenon joints

- Hangers fit closely to the surface of the members being connected (Figure 1.27d).
- Shoe part of the hangers provide sufficient bearing area to carry the end of the joist without compression failure of the timber (Figure 1.27e).
- Gauge of the nails holding the hangers in place are the correct size to fit snugly into the holes drilled in the hangers to receive them (Figure 1.27f).

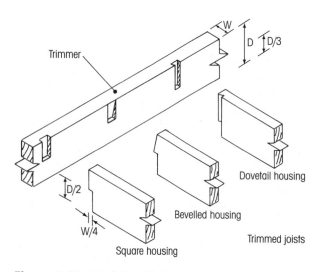

Figure 1.26   Traditional joins between trimmer and trimmed joists

1.26 SETTING OUT UPPER FLOOR JOISTS

The spacing setting out and levelling upper floor joists is the same as for suspended ground floors (see Section 1.12).

1.27 TRIMMING ROUND A PROJECTING FLUE (CHIMNEY BREAST)

Timber joists must be kept a safe distance away from the sources of heat produced by the hot gasses of combustion rising up through a flue (chimney) [*Building Regulations Approved Document "A"*]. The flue must be surrounded by at least 100 mm thickness of masonry, and no timber is to be built into the structure closer than 200 mm to the flue (Figure 1.28). If a fireplace opening and hearth is to be formed at the upper floor level, a hearth of incombustible material 125 mm thick must project 500 mm in front and 200 mm to each side of the fireplace opening (Figure 1.29).

1.28 SECURING FLOOR JOISTS TO EXTERNAL WALLS

It is required that upper floors are tied into the external walls to give them lateral support. If the joist ends are built in, or joist hangers without a hook over are used, or the joists run parallel to the wall, then galvanised mild steel straps should be used at intervals no greater than 2 m [*Building Regulations Approved Document "A"*]. All metal strapping must fit correctly and packings at least two thirds the depth of the joists must be used between the joist, if the joists run parallel to the wall. The strap must run at right angles across at least three joists, and be supported and fixed to packings in addition to the joists (Figure 1.30).

1.29 STRUTTING TO UPPER FLOOR JOISTS (FIGURE 1.31)

As upper floor joists have a longer span and a greater width to depth ratio (see Section 1.22) they are prone to lateral movement (sideways movement) and buckling if they are not stiffened at intervals along their length (Figure 1.31a). Although the floor decking and ceiling panels will provide some resistance to lateral movement of the joists, reliance is not placed on them and strutting at right angles to the line of the joists is introduced into the floor structure, usually not more than 1.5 m apart [*BS5268 Part 2. 1996*].

Strutting is of two types:

(1) Solid timber at least two thirds of the depth of the joists (usually cut from the short ends of joists) and fixed slightly out of line to allow for nailing (Figure 1.31b).

(2) Timber herring bone, or cross strutting using joist short ends of sufficient length ripped down to 50 × 38 mm. An alternative is to use proprietary galvanised pressed metal strutting. Pressed metal struts should have a profile that does not buckle as they are in compression when fixed (Figure 1.31c).

1.30 CUTTING AND FITTING HERRING BONE STRUTTING

Two parallel lines using a pencil and straight edge, or chalk line are marked out at right angles across the

(c) Joist hangers used in trimming to a stairwell

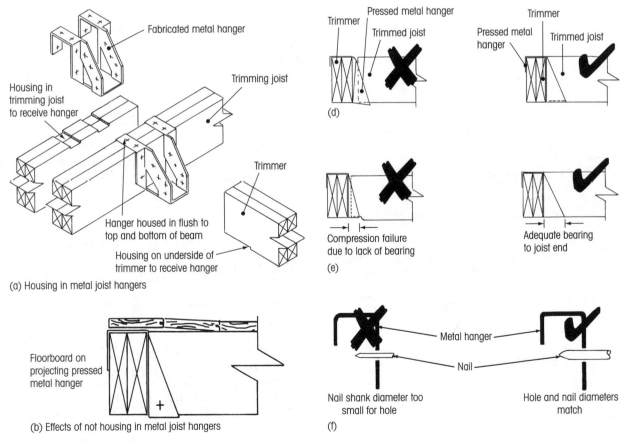

Figure 1.27   Use of metal joist hangers and their correct fixing

NB. No timber to be built into walls unless 200 mm away from flue

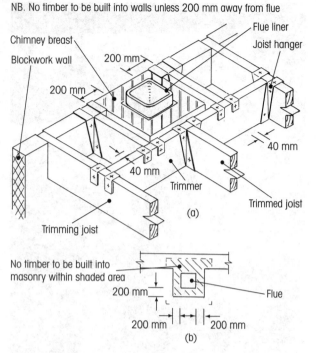

Figure 1.28 Trimming round a chimney breast using joist hangers

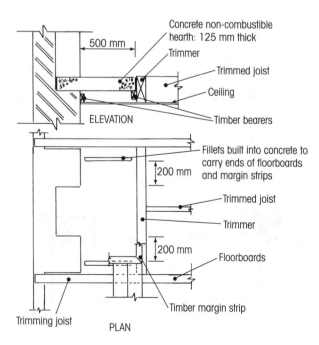

Figure 1.29 Trimming to a hearth at first floor level

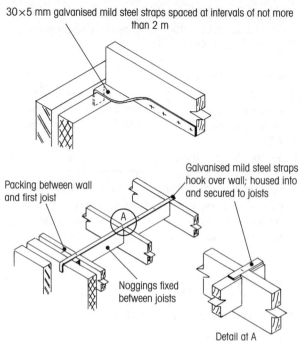

Figure 1.30 Anchoring perimeter walls to floors

joists, the depth of the joist less 10 mm apart. The strutting is laid diagonally across the joists between the parallel lines, marked off with a pencil and cut, a pair for each space. As there may be some variance in the spacing between joists the pairs of struts are not interchangeable (see Figure 1.31d).

1.31 FLOOR DECKING

The types of decking material used and the techniques of laying and fixing are the same as for hollow suspended timber ground floors (Sections 1.17–1.19).

1.32 TRAPS IN UPPER FLOOR DECKING

These are formed as described in Section 1.20.

1.33 NOTCHING AND HOLES IN UPPER FLOOR JOISTS (FIGURE 1.32)

Service pipes and cables are run in the thickness of an upper floor, rather than under the floor as is the case in hollow ground floors. These pipes and cables could be run at ceiling level, but they would be exposed to view. If they run parallel to the joists there is no problem as they are simply fixed to the sides of the joists for support. It is when they run at right angles to the joists that there is a need to notch and bore the joists to accommodate them, and this has the effect of

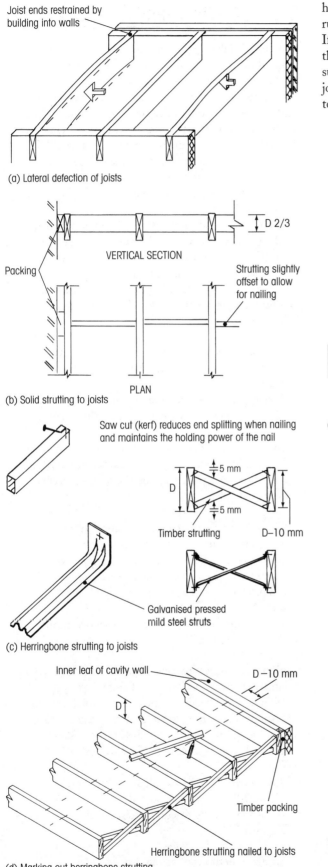

(a) Lateral defection of joists

VERTICAL SECTION

Packing

Strutting slightly offset to allow for nailing

PLAN

(b) Solid strutting to joists

Saw cut (kerf) reduces end splitting when nailing and maintains the holding power of the nail

Timber strutting

D–10 mm

Galvanised pressed mild steel struts

(c) Herringbone strutting to joists

Inner leaf of cavity wall

D –10 mm

Timber packing

Herringbone strutting nailed to joists

(d) Marking out herringbone strutting

Figure 1.31 Lateral restraint to deepfloor joists

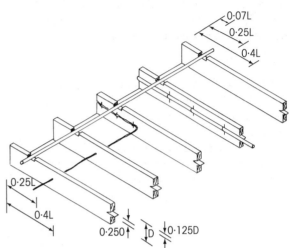

(a) Service pipes and cables running in thickness of the floor

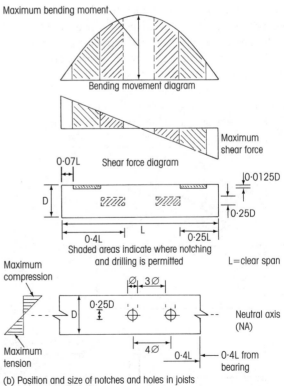

Maximum bending moment

Bending movement diagram

Maximum shear force

Shear force diagram

Shaded areas indicate where notching and drilling is permitted

L=clear span

Maximum compression

Neutral axis (NA)

Maximum tension

(b) Position and size of notches and holes in joists

Figure 1.32 Notching and drilling joists for services within the floor structure

weakening the structural strength of the joists (Figure 1.32a). To limit this weakening effect and to maintain the structural integrity of the joists account should be taken of any notching and boring that is to occur when the size of the floor joists are being determined. Joists can be made deeper or placed closer together.

Notches are formed in the top or compression edge of the joist where the bending stress is at a minimum and holes drilled along the neutral axis of the joist at proscribed positions along the length of the joist "L" (span). This has the effect of limiting the position where notches and holes are cut to the least stressed areas of the joists.

- The depth of notches should not be greater than 0.125 (one eighth) of the joist depth and only cut in the shaded areas shown in Figure 1.32(b).
- Hole diameters must not exceeding 0.25 of the depth of the joist can be drilled in limited areas along the neutral axis (NA) provided they are at least three diameters apart (see Figure 1.32b).

1.34 THERMAL INSULATION

In the upper floors of single occupancy dwellings it is not usual to insulate against heat loss, unless the part of the under side of the floor is exposed to the external atmosphere in the case of a floor overhanging to form the ceiling of an open porch (Figure 1.33). If this occurs then the floor must be insulated and constructed to comply with the Building Regulations Approved Documents "B" Fire safety and "L" Conservation of fuel and power.

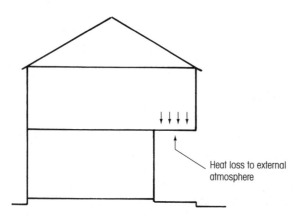

Figure 1.33 Exposed area of floor requires thermal insulation and protection against fire

1.35 SOUND INSULATION

When a stimulus applied to the inner ear produces a sense of hearing, we call it sound. Sound is generated by the vibration of a body causing changes in pressure of the surrounding air forming waves which travel outwards through the air from the source. The unit of sound intensity is the **Bel** and each Bel unit is divided into ten subunits which are called **deci Bels** (dBs). Sound waves are created by speaking, using amplifiers, playing musical instruments, or by directly striking the walls, floors, or ceilings of a building (Figure 1.34a).

Sound generated in one room of a building without insulation may be considered a nuisance in another room (Figure 1.34b) and one of the functions of sound insulation is to dissipate the energy of the sound waves as shown in Figure 1.34(c) with sound insulation. Sound reduction can be achieved in the following ways:

- By making the structure as solid and as heavy as possible so that the sound waves do not vibrate the structure.
- Inserting into the structure materials that will absorb the sound wave energy without vibration (see Figures 1.36b, 1.36c).
- Isolating elements of the structure from each other. Figure 1.35 illustrates how this can be achieved in composite floor panels by the use of visco-elastic strips.

If sound proofing is required in upper timber floors the reduction in airborne sound should be 51 dBs or better [*Building Regulations Approved Document Part E*]. (This is the intensity of sound produced by shouting as loudly as possible.)

Sound transmission through a suspended timber floor can be reduced by:

- Filling the space between the joists with high density slag wool to a depth of from 50 to 100 mm laid directly on the ceiling, which should be sufficiently strong to support the weight, rather than on pugging boards (Figure 1.36a) or sound absorbent mineral fibre panels (Figure 1.36b).
- Floating the floor on a quilt above the joists (Figure 1.36c).
- Use of acoustically designed floor panels (Figure 1.35).

All these methods of sound proofing can be incorporated in upper timber floor construction.

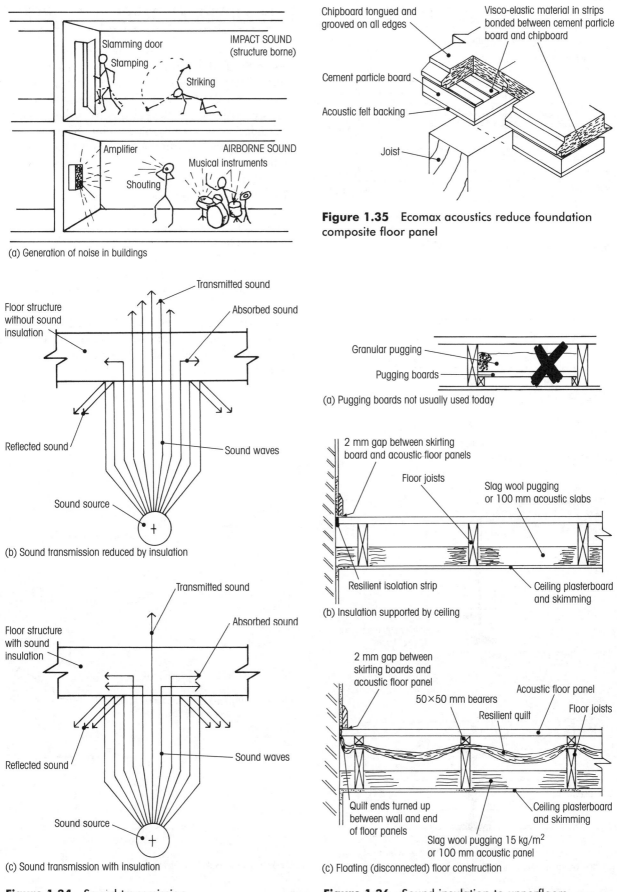

(a) Generation of noise in buildings

Figure 1.35 Ecomax acoustics reduce foundation composite floor panel

(b) Sound transmission reduced by insulation

(a) Pugging boards not usually used today

(b) Insulation supported by ceiling

(c) Sound transmission with insulation

Figure 1.34 Sound transmission

(c) Floating (disconnected) floor construction

Figure 1.36 Sound insulation to upperfloors

Section through floor at AA

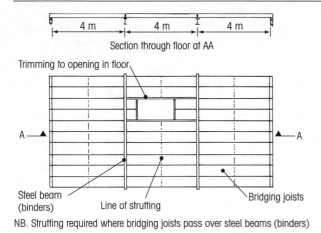

NB. Strutting required where bridging joists pass over steel beams (binders)

Figure 1.37   Layout of a double floor

1.36 UPPER FLOORS (DOUBLE)

If the span of upper floor joists is in excess of 5 m the cross-sectional of the joist becomes larger than the commonly supplied sizes of timber and consequently the cost becomes uneconomic. To keep a reasonable cross-sectional size for the joists the span is broken up by introducing steel beams (binders) into the floor structure (Figure 1.37). In this type of floor the beams can be:

- Completely below the ceiling line (Figure 1.38a).
- Partially below the ceiling line (Figure 1.38b).
- Within the thickness of the floor structure (Figure 1.38c).

Figure 1.38(d) shows the use of a template for marking out joist ends for notching on to the flange of girders.

Steel beams that project below the ceiling line are encased to improve their atheistic appearance and to protect them for a period of time against the effects of fire see Section on casing in.

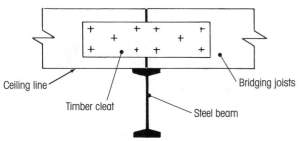

(a) Steel beam fully below ceiling line

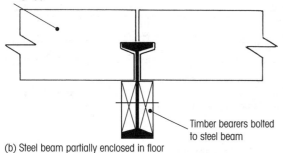

(b) Steel beam partially enclosed in floor

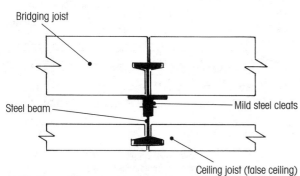

(c) Steel beam fully enclosed in floor

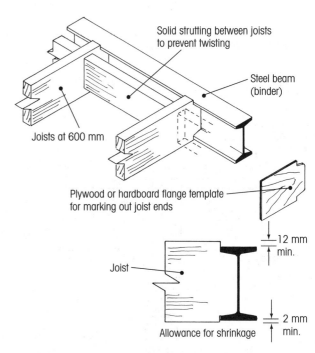

(d) Notching joists ends into steel bearings

Figure 1.38   Connection of timber joists to steel beams (binders)

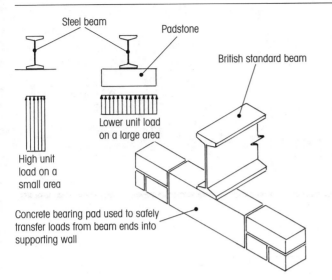

High unit
load on a
small area

Lower unit load
on a large area

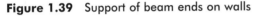

Concrete bearing pad used to safely
transfer loads from beam ends into
supporting wall

Figure 1.39 Support of beam ends on walls

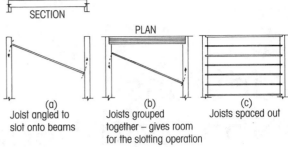

(a)
Joist angled to
slot onto beams

(b)
Joists grouped
together – gives room
for the slotting operation

(c)
Joists spaced out

Figure 1.40 Method of slotting bridging joists onto steel beams

1.37 DOUBLE FLOOR ASSEMBLY

In assembling this type of floor the steel beams (binders) are placed in position approximately 5 m apart with their ends resting on pads set in the walls (concrete blocks to help distribute the weight from the end of the beam into the wall) (Figure 1.39).

When the beams are level and in place the joists can be laid on the beams and the decking placed on the joists. If the joists have both their ends notched round the flanges of the beam, then before spacing the joists their ends will have to be threaded into the beam flanges by angling them into position as Figure 1.40.

Note. In this type of floor the joists will still require strutting, tying into the external walls, drilling and notching and the forming of service traps (see Sections 1.20, 1.28, 1.29 and 1.33).

2
Partitions (Non-load-bearing)

Partitions [*BS5234: 1&2: 1992*] are walls which divide up interior space into compartments (Figure 2.1) – doors, windows, hatch openings, etc., or their combination may be incorporated into the design (Figure 2.2).

Note. Screen partitions can be regarded as partitions, which do not extend fully from floor to ceiling.

This chapter is concerned with full height non-load-bearing partitions. In other words, a dividing wall not designed or intended to provide any structural support, or form any part of the main fabric of the structure.

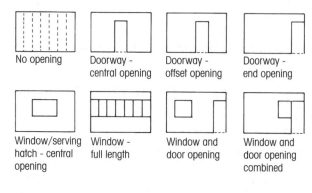

No opening | Doorway - central opening | Doorway - offset opening | Doorway - end opening

Window/serving hatch - central opening | Window - full length | Window and door opening | Window and door opening combined

Figure 2.2 Partitions openings

Even though compared to a load-bearing masonry dividing wall (shown in Figure 2.3a with its own foundation) they are lightweight in their construction, but provision must be made at floor level to carry the partition's weight (Figure 2.3b).

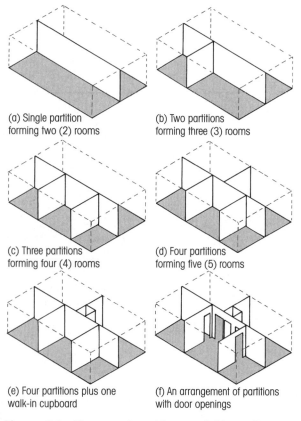

(a) Single partition forming two (2) rooms

(b) Two partitions forming three (3) rooms

(c) Three partitions forming four (4) rooms

(d) Four partitions forming five (5) rooms

(e) Four partitions plus one walk-in cupboard

(f) An arrangement of partitions with door openings

Figure 2.1 The use of partitions to divide up floor space

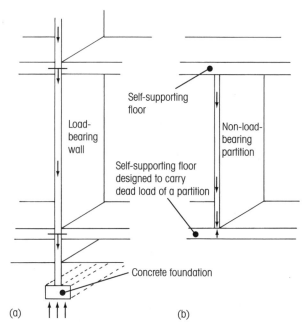

Figure 2.3 Dividing walls

Non-load-bearing partitions can be divided into two main groups:

- Timber stud partitions.
- Proprietary partitions of panel construction.

The former being based on traditional timber frame construction, whereas the latter includes many proprietary (manufactured and sold by a particular firm) systems.

2.1 TIMBER STUD AND SHEET PARTITIONS

This type of partition consists of a number of vertical members called studs (from which the partition takes its name) – as can be seen from Figure 2.4(a) they are framed by a sole plate (sometimes termed its sill) at the base, two end studs (possibly wall studs), and a head plate at the top.

The intermediate studs (or middle studs) are spaced at centres to suit both the overall size and the permitted span of the lining material (Figure 2.4b). Studs are stiffened by using one or more rows of noggings (Figure 2.4c) these consist of short lengths of timber of the same section as the studs. The number and location of the rows of noggings will depend on the sheet size and the position of any trims, such as deep skirting board, dado rails and chair-back rails, etc. as shown in Figure 2.4(d).

Construction: framework (carcass) can be made up of 75 mm × 38 mm or 75 mm × 50 mm section. Where sheets abut, 75 mm × 50 mm section should be used to ensure adequate bearing (cover for nailing purposes) as shown in Figure 2.4(b).

Pre-fabrication of a partition or framework is generally not practicable for reasons of transport and access through doorways, etc., but if access is possible – perhaps to a single storey warehouse or an industrial development with large loading bay doors, etc. – then pre-fabrication could be considered.

Partitions within domestic dwellings are therefore erected either by being built in-situ or, if the situation allows, the framework may be built-up on or off the floor, then reared up into position for fixing. The latter method will mean making the framework smaller than the opening and fixing packing pieces between the framework and all the fixing points. We can describe these two methods as:

- Built-in framework (Figure 2.5).
- Pre-assembled framework (Figure 2.6).

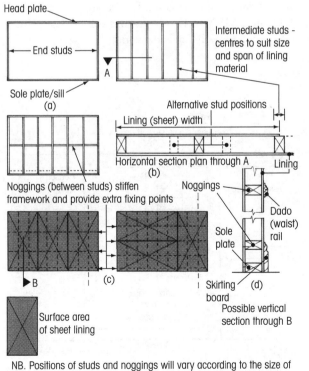

NB. Positions of studs and noggings will vary according to the size of lining sheets, amount of stiffening required and fixing points for trims, etc.

Figure 2.4   Timber stud partitioning

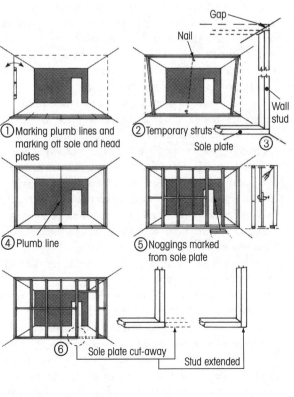

Figure 2.5   Built-in framework

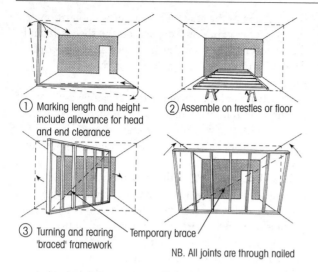

① Marking length and height – include allowance for head and end clearance

② Assemble on trestles or floor

③ Turning and rearing 'braced' framework

Temporary brace

NB. All joints are through nailed

Figure 2.6   Pre-assembled framework

2.1.1 Built-in Framework (Figure 2.5 – For Fixing Between Walls)

- Cut the sole-plate and head-plate to length from straight stock. If the walls are plumb, stud positions may be marked while both plates are placed together as shown. If the plate lengths differ, the length of the head-plate must be adjusted accordingly – if it is longer than the sole-plate, marking-off can be carried out with the timber laid across or down the rooms length. Joints between the plates and studding should be prepared at this stage. The sole-plate may now be fixed to the floor as shown in Figure 2.10, and plumb lines marked on the walls at each end.
- The head-plate can now be fixed to the ceiling (Figure 2.10). Temporary struts may be used until fixings are made – if a mid-strut is required, use a nail driven into the head-plate to steady the strut.
- Alternatively – and perhaps the safer method – the end studs may be fixed first, leaving a gap at the top to accommodate the head-plate. Provided there is no cornice (moulded fillet – usually plaster – between the wall and ceiling), the head-plate can then be pushed into position from one side, then fixed to the ceiling.
- Once the outer framework is completed, check that the intermediate stud marks are plumb. If satisfactory, complete fixing to the floor and ceiling.
- Mark the length of each stud by placing one end onto the sole-plate and marking the underside of the head-plate or housing. Alternatively, use a pinch rod [BENCH AND SITE SKILLS, SECTION 6.1.3, PAGE 160]. Each stud should be marked separately, because floor-to-ceiling heights will vary slightly across the room. After the studs have been fixed (Figures 2.7 and 2.8), the height of the noggings are levelled across the

partition. Nogging lengths are taken off the sole-plate. In this way, bent studs will be straightened as the noggings are fixed into position (Figure 2.7).
- If a door or other opening is required, studs will either be left out or cut short, in the case of door openings, the sole-plate will require cutting away. Alternatively, as shown in Figures 2.5 and 2.8(e, f) studs forming the opening may be allowed to extend over the ends of the sole-plate.

2.1.2. Pre-assembled Framework (Figure 2.6 – For Fixing Between Walls)

- The lengths of the sole-plate, head-plate, and end studs should be shorter than the lengths required to fit between the floor, ceiling, and walls, so that the completed framework can be turned or reared into its final position – otherwise the arc it would follow would cause the ends or top to bind on the ceiling and or walls as it was being reared into position.
- Once the required lengths and height are known, the whole framework can be assembled either on the floor or on a trestle staging (the latter method being less troublesome to one's back and knees).
- Once assembled, the framework can be reared and swung into its final position – either on the turn or, where floor space allows by rearing straight up. A temporary brace may be used to keep the framework rigid during the operation. End fixings are shown in Figure 2.11. *Note.* Door openings will be dealt with as stated for built-in framework.

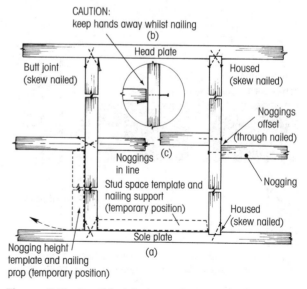

CAUTION: keep hands away whilst nailing (b)

Head plate

Butt joint (skew nailed)

Housed (skew nailed)

Noggings offset (through nailed)

Noggings in line   (c)

Nogging

Stud space template and nailing support (temporary position)

Housed (skew nailed)

Nogging height template and nailing prop (temporary position)

Sole plate   (a)

Figure 2.7   Possible jointing techniques for built-in framework

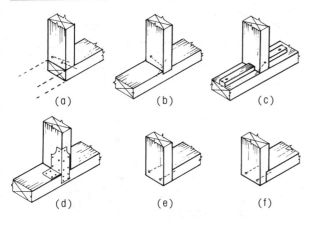

Figure 2.8 Alternative jointing arrangements between stud and sole-plate

2.1.3 Joints (Figures 2.7 and 2.8)

Members may be joined either by a butt joint (Figure 2.8a), housed joint [BENCH AND SITE SKILLS, SECTION 7.5.5] (Figure 2.8b), or notched joint (Figure 2.8c), and depending on access, through or skew nailed (driven at an angle) together. In some situations framing anchors (Figure 2.8d) can be useful. With the exception of pre-framed structures members will require restraint while the nails are being driven – height and distance templates can be a useful aid for this purpose, one end of the template should be slightly under cut (sloping) to aid its removal. A partly driven nail below or at the back off the member is also very effective (see insert in Figure 2.7). **A word of warning here** – always keep your fingers away from the joint whilst nailing if there is a gap (no matter how small) between the members. Many a trapped finger (blood blister) has resulted from this practice.

- **Studs to sole-plate** (Figures 2.7 and 2.8). Direct nailing is not possible except where frames are pre-assembled, and at door openings where, as shown in Figure 2.8(e, f), studs may be cut long to allow the stud to run over the ends of the sole plate. Although time consuming, stud placement can be more easily achieved by nailing a lath along the middle of the sole-plate and notching the ends of the studs over it (Figure 2.8c), or by housing (Figure 2.8b) the sole-plate to receive the stud ends.
- **Studs to head-plate** (Figure 2.7). Except where the partition is not full storey height, or above or below openings, similar fixing arrangements can be made as for the sole-plate (Figure 2.8). With in-situ work notching the head-plate or using temporary template support can help resolve any overhead nailing difficulties.
- Noggings to studs (Figure 2.7), where noggings are used for stiffening purposes only, they can be offset –

in which case, with the exception of end (wall) studs, through-stud nailing is possible. If, on the other hand, noggings are to provide an edge bearing for the lining material at some point, nails will have to be driven at an angle or skew-nailed into the studs as shown.

Note. Noggings should always be provided where support is required for fittings, such as kitchen units or radiators, etc.

2.1.4 Positioning and Fixing Framework

Details of the partition will either be as according to the working drawings, or by on site agreement. You should be given directional positions and distances to meet the requirements, and should there be any openings their size and position. In all cases you must verify the details and make a special note of the position and size of any openings.

When setting out the floor position always double check the layout, pay particular attention to any square [BENCH AND SITE SKILLS, SECTION 6.3] or angular [BENCH AND SITE SKILLS, SECTION 6.4] corners. Don't forget to allow for the thickness of the partition (studs, linings, and trim) when determining the finished room shape, or the net floor size.

Once the positioning details and method of fabrication is established, fixing to the main building fabric can commence. We will now consider some of the methods that may (providing the base material and surrounding fabric is suitable) be employed.

Fixing framework at corners and junctions (Figure 2.9)

Nailing or screwing a stud to a stud, or a stud to a nogging, should present little difficulty (Figure 2.9a), but providing edge support for lining materials at internal corners might. There are, as shown, two possible solutions:

- Providing an extra studs, or
- Proprietary steel junction clips.

On the other hand, if one side of the framework is lined before abutment takes place, then as shown in Figure 2.9(b) – edge support is not necessary.

Fixing framework to floors and ceilings (Figure 2.10)

If the partition forms part of the original building designed, provision for fixing may be made during the building construction. If the partition is being added to an existing building, ceiling joist centres must be located. This can be done by gently tapping the ceiling

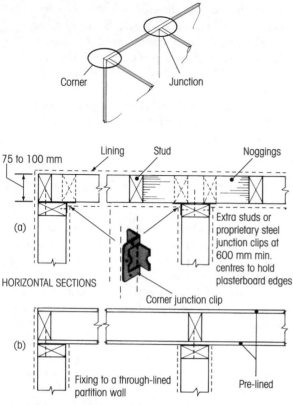

Figure 2.9   Alternative corner and junction details

lining – the joist will make a solid sound, compared with the hollow sound of the areas between joists, and their location is confirmed by inserting a bradawl at these points – alternatively a battery-operated stud detector may prove useful (Figure 2.14).

Figure 2.10(a) shows how the plates of a built-in framework are positioned at right angles to joists and fixed before or after the decking and ceiling linings are in place.

Figure 2.10(b) shows how a pre-assembled framework is wedged and packed-up off the floor before fixing. Where the head-plate falls between two ceiling joists, noggings are used as restraints and fixings.

In Figure 2.10(c), a pre-assembled framework falls within two floor joists: depending on the design loading – cross-bridging or an extra floor joist positioned directly under the partition may be used as support. At the head, lining material will need support, and this can be offered either by nailing bearers to the sides of the ceiling joist or by junction clips nailed to the head-plate.

Figure 2.10(d) shows how a pre-assembled framework meets up to a concrete floor and ceiling slab. Various types of proprietary fixing devices are available for making satisfactory fixings here: alternatively, preservative-treated wood fillets or timber runners may have been inserted into the concrete screed (top finishing layer).

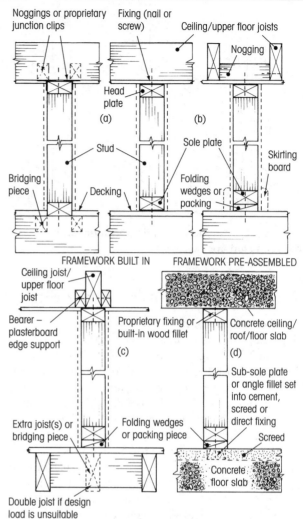

Figure 2.10 Vertical sections through built-in and pre-framed framework

Fixing framework to walls (Figure 2.11)

Many types of building blocks permit direct nailing; whereas brick and concrete [BENCH AND SITE SKILLS, SECTION 8.19] will require plugging [BENCH AND SITE SKILLS, SECTION 11.6–11.7] for nails or screws. Where packings are required between the framework and the wall, they should be positioned at the point of fixing as shown.

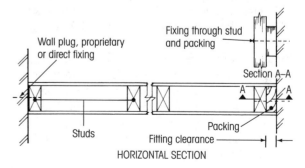

Figure 2.11   Wall stud fixing details

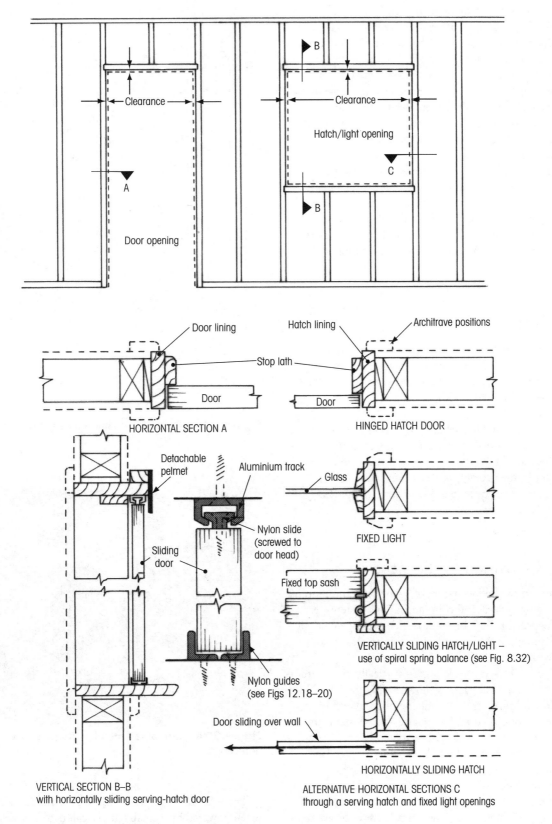

Clearance

Door opening

Clearance

Hatch/light opening

Door lining

Hatch lining

Architrave positions

Stop lath

Door

Door

HORIZONTAL SECTION A

HINGED HATCH DOOR

Detachable pelmet

Aluminium track

Glass

Nylon slide (screwed to door head)

Sliding door

FIXED LIGHT

Fixed top sash

VERTICALLY SLIDING HATCH/LIGHT –
use of spiral spring balance (see Fig. 8.32)

Nylon guides (see Figs 12.18–20)

Door sliding over wall

HORIZONTALLY SLIDING HATCH

VERTICAL SECTION B–B
with horizontally sliding serving-hatch door

ALTERNATIVE HORIZONTAL SECTIONS C
through a serving hatch and fixed light openings

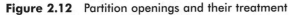

Figure 2.12 Partition openings and their treatment

2.1.5 Provision at Openings

Openings such as doorways, windows (to provide borrowed light from another room), serving hatches, etc. should where practicable fall within the room run off studwork and within a single sheet of lining material. Openings should be made oversize – minor adjustments can then be made at the time of fixing a door or window frame. Adjustments are made with the aid of folding wedges and packings at fixing points – gaps are made good with trim (architrave).

Figure 2.12 shows typical horizontal and vertical sections through door and hatch lining arrangements. Closing a hatch opening can be done in several ways, for example, by using a single or double hinged, or horizontal sliding doors – vertical sliding hatch have also been popular. A beaded fixed light of toughened glass is also shown.

2.1.6 Lining Materials

Plasterboard is the most common lining material – suitable types, and edge jointing techniques are described in *Bench & Site Skills* [BENCH AND SITE SKILLS, SECTION 2.8] together with suitable sheet sizes.

There are situations where some types of wood-based manufactured boards can be used, but this will depend of the location of the partition and whether it must comply with current building regulations – see Section 2.4 [*Building Regulation Approved Document "B"*].

2.1.7 Provision for Services

Because of the hollow nature of a partition, the running of cables (electrical and communications systems, etc.) usually presents little or no difficulty as they are usually installed before the lining material is attached to the studwork.

Where cables, etc. have to past vertically or longitudinally through a partition as shown in Figure 2.13. Holes should be drilled central to the members' width of a diameter not exceeding one-quarter of the members width or thickness or closer one to another than four times the diameter of the hole.

Proprietary steel cover plates are available to help protect cables, etc. from indiscriminate nailing.

Note. Carpenters and joiners must always be aware of the possibility of hidden services – particularly electrical, behind plasterwork, ceilings, under floors, behind and within existing partitions.

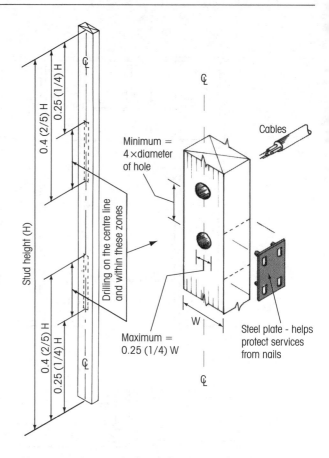

Figure 2.13   Guide for drilling holes for services

Figure 2.14   Combined metal, voltage, and stud detector

There are now several devices on the market which when powered by a small battery can detect electrical cables and pipework. Some of these, like the one illustration in Figure 2.14 can also detect timber studs and joists from behind plasterwork.

2.2 PROPRIETARY PARTITIONS OF PANEL CONSTRUCTION

The trade often knows these systems by their commercial product name. For the purpose of this book we shall classify them into two groups:

- Demountable partitions (Monobloc panel) – prefabricated units capable of being dismantled and reassembled elsewhere with little disturbance between interconnecting components, or the fabric to which it was attached.
- Non-demountable partitions – a permanent fixture.

2.2.1 Demountable Partition Systems

These are ideally suited to dividing up large floor areas for commercial or industrial use, to provide divisions or separate office accommodation.

A modular framework of steel, aluminium, or timber is purpose-built to house panels. These panels may be fabricated from composite plasterboard, strawboard, particle board, toughened glass, etc. or their combination. Panels may also be surface-fixed or hung to the framework, possibly with an insulation material sandwiched between them.

Figure 2.15 shows how panel modules can be designed to suit various linings, openings, and their combination, for example:

- solid
- solid and glass
- door and solid
- door and glass (fanlight)
- glass.

Note. Door storey frames with or without a window (fanlight) over the door may be of a narrower module.

2.2.2 Non-demountable Partition Systems

Mainly used in domestic housing to provide a lightweight partition system with either a plastered finish or dry decorative finish.

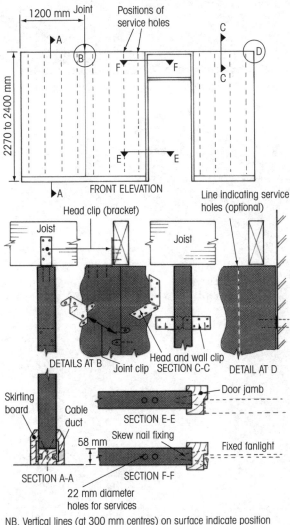

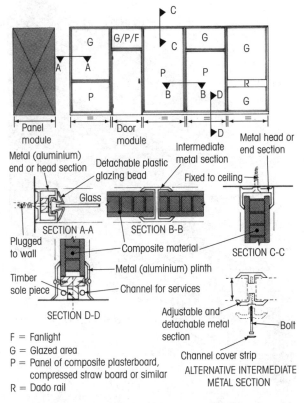

F = Fanlight
G = Glazed area
P = Panel of composite plasterboard, compressed straw board or similar
R = Dado rail

Figure 2.15 Possible arrangement of a demountable partition system incorporating solid and glazed panels

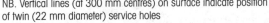

NB. Vertical lines (at 300 mm centres) on surface indicate position of twin (22 mm diameter) service holes

Figure 2.16 Assembly and fixing details for a 'Stramit and Easiwall System'

Figure 2.17 Installing a Stramit 'Easiwall' panel

Because of the rigidity of the panels used in these systems, framing is usually only necessary as a means of anchorage and fixing medium at floors, walls, and ceiling. Purpose-made clips, fixing devices, or wood block inserts are generally all that are necessary to fasten panel to panel.

Figure 2.16 shows joint and fixing techniques used with 'Stramit Easiwall Partitioning System'. Easiwall panels consist of a core of highly compressed straw, encased with a plastering quality lining over a fibreglass reinforced mesh. Panels have tapered long edges to allow for a scrim to be applied over the joints before plastering. Pre-formed twin holes (22 mm diameter) are at 300 mm centres for running cables – a green line on the surface indicates their position. Fixing packs including clips and nails are supplied by the manufacturer to correspond with the number of panels required.

Figure 2.17 shows an 'Stramit Easiwall Partitioning System' being installed as a first fixing operation (work prior to plastering).

Figure 2.18 shows an exposed view of a 'Gyproc Paramount Dry Partition' together with some fixing details – each pre-fabricated panel consists of two tapered edge Gyproc wallboards separated by a cellular lattice which is bonded to them.

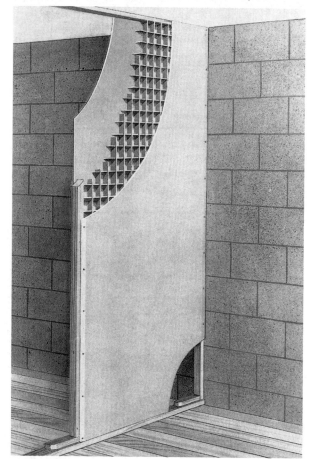

Figure 2.18 Exposed view of a Gyproc Paramount panel and fixing details

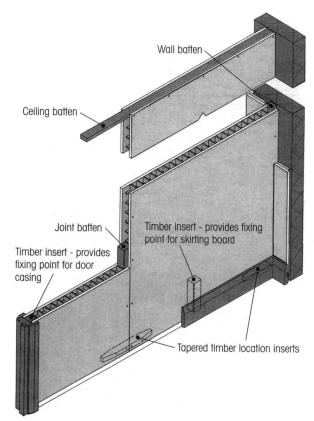

Wall batten

Ceiling batten

Joint batten

Timber insert - provides fixing point for skirting board

Timber insert - provides fixing point for door casing

Tapered timber location inserts

Figure 2.19 Isometric cut-away details showing hidden fixings of a Gyproc Paramount Dry Partition

Table 2.1 Gyproc Paramount Dry Partition panel sizes

Width (mm)	Edge profile of board	Thickness (mm)	Length (mm)
900	Tapered	57* 63†	2350, 2400, 2700
1200	Tapered	57*	2350, 2400

*Incorporating 9.5 mm thick Gyproc wallboard.
†Incorporating 12.5 mm thick Gyproc wallboard.

As can be seen in Table 2.1 several combinations of panel size are available. Figures 2.19 and 2.20 detail the construction at various junctions and inter connections.

A 'Gyproc Laminated Partition' is shown in Figure 2.21 with portions cut away to expose its method of construction. It is made up of a framework of timber (38 × 25 mm) fixed around the perimeter and openings – this in turn forms an integral part of the three layer sandwich of plasterboard – each layer being bonded by vertical bands of adhesive at not less than 300 mm centres. Board joints are staggered to produce a strong bond. Figure 2.22 shows how joints are made at junctions, corners, and openings. If services are to run

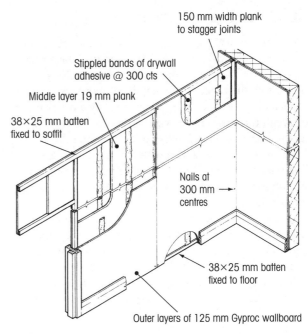

Typical construction, 50 and 65 mm partitions

Figure 2.21   Isometric cut-away view of a 'Gyproc Laminated Partition'

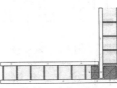

Figure 2.20 'Gyproc Paramount Dry Partition' construction details

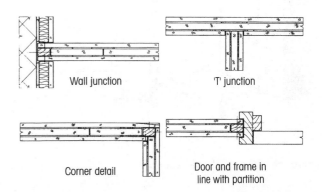

Figure 2.22 'Gyproc Laminated Partition' construction details

within the panel, provision for a duct must be made within the middle core during its construction.

2.3 INSULATION

2.3.1 Thermal Insulation

Thermal performance or heat loss for a building will not under normal conditions applied to a non-load-bearing partition, unless a partition separates a heated room or area from one which is not.

2.3.2 Sound Insulation

What we are really guarding against is unwanted sound – i.e. noise! Sound travels in waves through the air (airborne sound) as a result of conversation, etc. Noise is very often the result of overamplified sound from radios, TVs and music systems – it may travel by a direct route through the structure or by an indirect route via gaps around openings or the edges of the partition. In which case, gaps should be sealed with a sound-absorbent material or mastic. Flanking may be another route by which sound can travel to adjoining rooms, as shown in Figure 2.23.

Noise may, as shown in Figure 2.24 also materialise as a result of impact (impact sound), when the building structure is used as a vehicle to transmit noise which may have been transmitted via a door closing (slamming), a striking object, or vibration from various sources. Such noise is also known as 'structure-borne' noise.

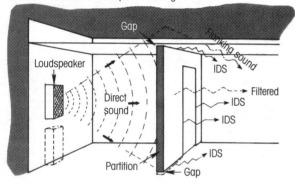

IDS = Indirect sound - via gaps in walls and around openings, etc.

Figure 2.23 **Airborne sound**

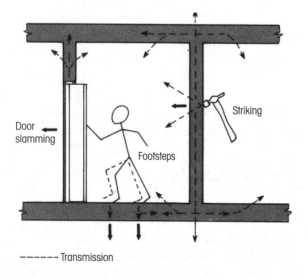

------- Transmission

Figure 2.24 **Impact sound (structure-borne)**

When airborne and impact (structure-borne) sound waves strike a partition, the resulting vibration will be lessened only according to the amount of sound insulation offered by the partition.

The changes in sound pressure as a result of a noise can be measured in what are known as decibels (see Section 1.35) (dB) – at the bottom of the scale with a reading of 0 dB would be silence, where as normal conversation could be expected to register between 50 and 60 dB. As for a partition, let's say as an example that a partition had to provide a means of reducing an airborne noise of 80 dB from a loud radio on one side to 40 dB on the other, then the partition would have to be constructed in such a way to give a sound reduction of 40 dB.

When a partition separates a room containing a water closet from a living room, dining room, study, or bedroom (excluding an en-suite situation) within a dwelling under the control of the National House Building Council (NHBC) [*NHBC Standards Sec., 6.3*] it must be built to satisfy a minimum sound-reduction index of not less than 35 dB (the higher the decibel value, the better be insulation). Examples of how these levels of insulation can be achieved with stud partitions are shown in Figure 2.25. Some proprietary systems can meet these requirements.

As shown in Figure 2.26 transmission of sound may be reduced by up to 50 dB or more by incorporating one or more of the following methods of construction:

* Having alternate studs set proud to leave a separate staggered gap.
* Using insulation board within or as part of the lining.
* Increasing the mass of insulation material.
* Using a resilient backing to linings and the joints between the panels or partitions, and the floor, walls, and ceiling. [*BS. 5234: 1975 Appendix "A"*].

2.4 PROTECTION FROM FIRE

2.4.1 Fire Resistance

We have to consider here the ability of the partition (as an element in the building) to resist collapse, passage of flame, or heat liable to ignite material on its opposite side, for a given period of time [for example, 30 min, or 1 hour (see Figure 2.27)], when under attack by fire from one side. The required legal amount in time of fire resistance will depend on the location and purpose of the partition [*BS476:22:1987*], [*Building Regulation Approved Document "B"*].

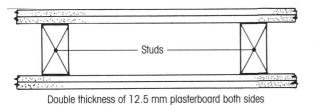

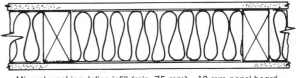

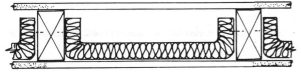

Double thickness of 12.5 mm plasterboard both sides

Mineral wool insulation infill (min. 75 mm) - 12 mm panel board (fibreboard) both sides

25 mm mineral wool insulation quilt - 12.5 mm plasterboard both sides

25 mm mineral wool insulation quilt - 9.5 mm plasterboard and 5 mm plaster coating both sides

Figure 2.25   Alternative arrangements of providing sound insulation – to satisfy the National House Buildings Council (NHBC) standards

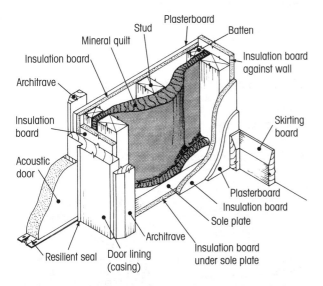

Figure 2.26   An effective method of providing sound insulation

2.4.2 Surface Spread of Flame

Surface finishes and lining materials are usually required to be classified as to the extent and rate at which flame would spread across them.

Materials are tested according to the methods set down in British Standards – they are then given a number from 1 to 4. The higher the number, the greater the risk of flame spread. Class '0' may be regarded as non-combustible. The Building Regulations stipulate what classification will be required for a given situation, such as the building type and floor area which the partition or partitions will enclose [*BS476:7:1987 (1993)*], [*Building Regulation Approved document "B"*].

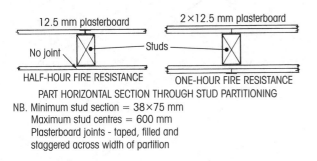

12.5 mm plasterboard   2 × 12.5 mm plasterboard

No joint   Studs

HALF-HOUR FIRE RESISTANCE   ONE-HOUR FIRE RESISTANCE

PART HORIZONTAL SECTION THROUGH STUD PARTITIONING

NB. Minimum stud section = 38 × 75 mm
Maximum stud centres = 600 mm
Plasterboard joints - taped, filled and staggered across width of partition

Figure 2.27   Examples of how fire resistance can be achieved

3
Flat roofs and finishes

3.1 INTRODUCTION

The term flat roof refers to any roof which has its upper surface incline at an angle (slope) not exceeding 10° as shown in Figure 3.1(a).

The illustration at the start of this chapter shows how the methods of construction are not dissimilar to the independent timber upper floor featured in Section 1.22 – where means of support, anchorage, stiffening by strutting and the decking all follow a similar procedure.

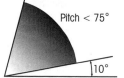

(a) FLAT ROOF – roofs with a slope within shaded area

(b) PITCHED ROOF – roofs with a slope within the shaded area

Figure 3.1   Roof slope (pitch) in relation to terminology

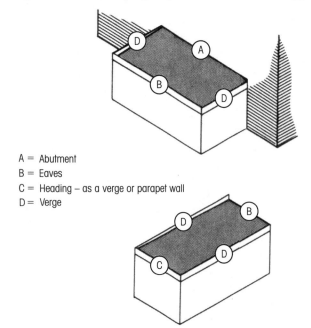

A = Abutment
B = Eaves
C = Heading – as a verge or parapet wall
D = Verge

Figure 3.2   Key to roof edge details

Figure 3.2 shows how a flat roof may form part of either an attached or detached structure. The illustration also serves as a useful key for identifying different types of edge treatments, we will be dealing with these separately later in this chapter.

3.2 ROOF FALLS

The direction of the fall (slope) can depend on:

- The type and location of the building.
- The type of roof construction.
- The position of the surface water outlet.
- The use of an integral or external gutter.

1. Single fall – via external gutter and fallpipe

2. Single fall – via integral box gutter and built-in funnel outlet (see Fig. 3.20)

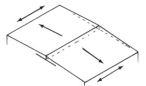
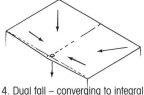

3. Dual fall – via opposite external gutter and fallpipe

4. Dual fall – converging to integral valley funnel outlet

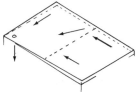

5. Dual fall – converging to integral corner funnel outlet

6. Four falls – converging to central integral funnel outlet

Figure 3.3   Methods of dispersing surface roof water to outlets by using different directional changes of fall

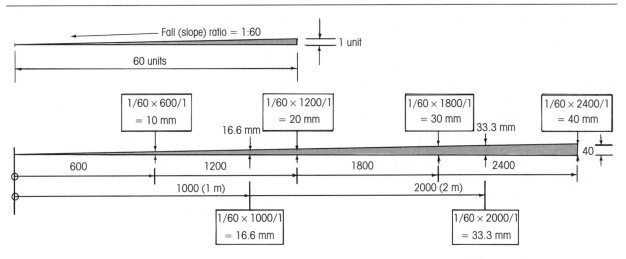

Figure 3.4 Roof rise over 600 mm (0.6 m) and 1000 mm (1.0 m) increments with a fall ratio of 1:60

The amount of fall should be sufficient to clear water away to the outlet pipe(s) as quickly as possible across the whole roof surface – this may involve several directional changes of fall as shown in Figure 3.3. Failure to dispose of surface water effectively could result in ponding (forming pools of water) which may also increase the load on the roof, and also provide a catchment area for enough silt to encourage and support plant life – which could have a harmful effect on the roof covering and become powerful enough to break open water seals [*Building Regulation Approved document "A1/2" (Sec 1.3)*].

Table 3.1 shows how a comparison can be made between the ratio of the roof's fall and the degree of slope shown as an angle. The percentage slope has been shown as a useful means for measuring the height of fall of a given ratio over a 1 m run of joist. For example, if we are given that the roof requires a slope of a 1:60 ratio then for every metre run of joist it will have a rise from the horizontal of approximately 17 mm. This is calculated as follows:

1:60 over 1 m (1000):
$1/60 \times 1000 = 16.66$ mm lift for every metre run of joist.
In other words for every 60 mm of run we would

have to lift the joist 1 mm above horizontal (Figure 3.4). As can be seen from Figure 3.5 fall can be achieved by one or a combination of the following:

(a) *Sloping joists* (Figure 3.5a)– decking and ceiling run parallel. This type of construction is suitable for garages, outbuildings, or where there is to be no ceiling or a level ceiling is not required.
(b) *Tapered firring pieces* (Figure 3.5b) – tapered pieces of timber firrings nailed along the top edge of horizontal joists.
(c) *Deepened joists* (Figure 3.5c) – lengths of timber of decreasing section which are positioned at right angles to the fall of the roof and nailed to top edges of the joists.

Table 3.1 Cross-reference between falls, angle of slope, and percentage of slope.

| | Fall of roof | |
Ratio	Roof slope in degrees (°)	Roof slope in percentage (%)
1:80	0.7	1.3 (1.25)
1:60	*1.0*	*1.7 (1.66)*
1:40	1.4	2.5
1:20	3.0 (2.8)	5.0
1:10	6.0 (5.9)	10.0

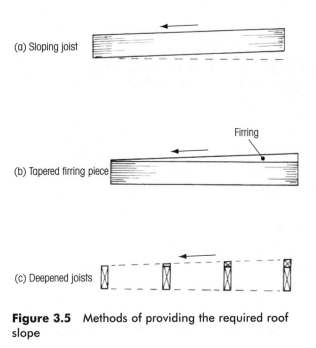

Figure 3.5 Methods of providing the required roof slope

3.3 CARCASS CONSTRUCTION (FRAMEWORK)

Figure 3.6 shows a skeletonised plan of a large flat roof featured on the chapter headed page. The roof area in this case has been divided into two, by introducing a central laminated timber beam. This has enabled the effective span of each joist and firring to be reduced by half, this will also mean that the edge finishes of the roof will also be lessened in depth, and that the roof water will run to both eaves.

3.3.1 Roof Joists

All roof joists must be in accordance with the requirements of current building regulations [*Building Regulations Approved document "A"*]. These regulations relate to sectional sizes grades of timber, dead loads to be placed upon them, their maximum clear span (Figure 3.7) and spacing (fixing centres). Whenever

practicable joist centres should be sub-multiples of the dimensions (length and width) of the decking material. [BENCH AND SITE SKILLS, TABLE 2.21]. The size of ceiling materials (plasterboard, etc.) may also be taken into account.

Note. All unsupported edges of both decking and ceiling materials will require noggin support.

Depending on the joist span they may also require strutting as with upper timber suspended floors (see Section 1.29) to stiffen the roof structure as a whole.

3.3.2 Joist Bearing

At the eaves and or verge depending on the direction of joist run, joist ends will be supported either directly onto the blockwork or via a timber wall plate. They *must be anchored* to the sub-structure by the means stated in the specifications [BENCH AND SITE SKILLS, SECTION 12.9.2], which may be by galvanised steel straps, and joist clips (see Figure 3.16), or skew nailing, or a combination of these methods.

Where joists abut a wall as shown in Figure 3.2(a) various bearing methods can be employed as shown in Figure 3.8, they are:

(a) *built-in* (Figure 3.8a) – joist ends should be treated with wood preservative.
(b) *flange of a steel beam or joist* (Figure 3.8b) – where the roof adjoins an opening in the abutting wall.
(c) *steel joist hanger* (Figure 3.8c) – built into and tight up against the wall face.
(d) *timber wall piece* (Figure 3.8d) – supported by steel corbel brackets built into the wall. The joist bearing should be reinforced with steel-framing anchor plates.
(e) *steel angle* (Figure 3.8e) – bolted to the wall with rawlbolts or similar, depending on the base material of the wall.
(f) *framed* (Figure 3.8f) – headers can bear on steel corbel brackets or angle. The whole roof framework may be constructed at ground level then hoisted into position – with or without roof decking. This method is suitable for roofs, or roof sections, of a small surface area.

Figure 3.9 shows how the abutment can be dealt with against a parapet wall. A parapet wall is a wall that protrudes above either the verge or eaves, and terminated by a weathered coping.

3.3.3 Openings

Forming an opening for a roof light, ventilation turret, or piping, etc. passing through the roof may simply require utilising the space between two joists as shown in Figure 3.10.

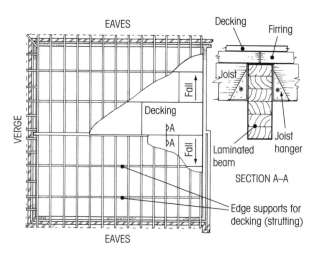

Figure 3.6 Skeletonised plan of a large flat roof (double roof) area – see the illustration at the start of this chapter

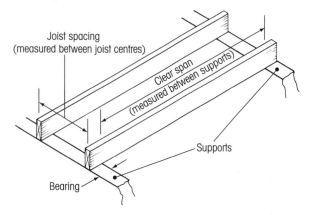

Figure 3.7 Dimensions required for sizing roof joists

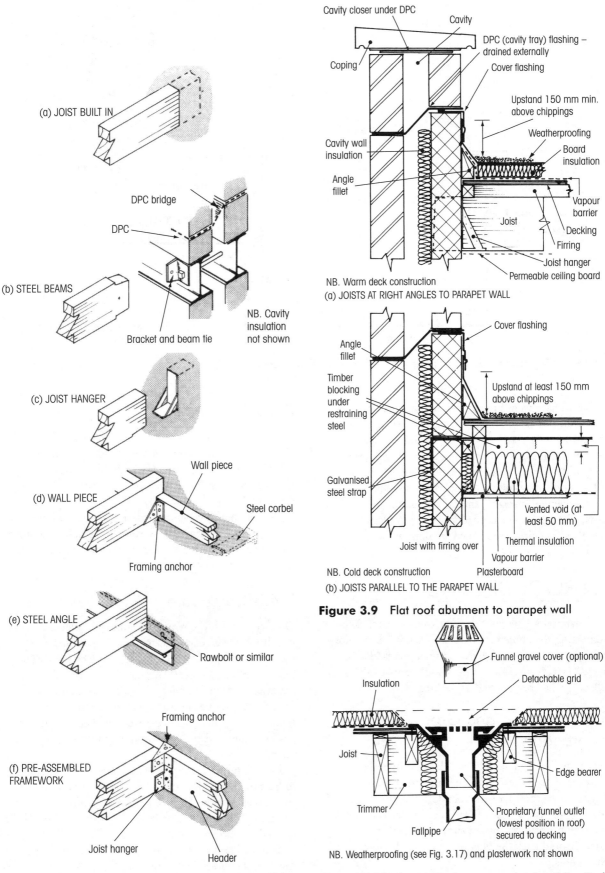

(a) JOIST BUILT IN

(b) STEEL BEAMS

DPC bridge

DPC

Bracket and beam tie

NB. Cavity insulation not shown

(c) JOIST HANGER

(d) WALL PIECE

Wall piece

Steel corbel

Framing anchor

(e) STEEL ANGLE

Rawbolt or similar

(f) PRE-ASSEMBLED FRAMEWORK

Framing anchor

Joist hanger

Header

Figure 3.8 Joist bearings – abutting a house wall (firring pieces not shown)

Cavity closer under DPC

Cavity

Coping

DPC (cavity tray) flashing – drained externally

Cover flashing

Upstand 150 mm min. above chippings

Weatherproofing

Board insulation

Cavity wall insulation

Angle fillet

Vapour barrier

Joist

Decking

Firring

Joist hanger

Permeable ceiling board

NB. Warm deck construction

(a) JOISTS AT RIGHT ANGLES TO PARAPET WALL

Cover flashing

Angle fillet

Timber blocking under restraining steel

Upstand at least 150 mm above chippings

Galvanised steel strap

Vented void (at least 50 mm)

Thermal insulation

Vapour barrier

Plasterboard

Joist with firring over

NB. Cold deck construction

(b) JOISTS PARALLEL TO THE PARAPET WALL

Figure 3.9 Flat roof abutment to parapet wall

Funnel gravel cover (optional)

Detachable grid

Insulation

Joist

Edge bearer

Trimmer

Fallpipe

Proprietary funnel outlet (lowest position in roof) secured to decking

NB. Weatherproofing (see Fig. 3.17) and plasterwork not shown

Figure 3.10 Integral rainwater outlet detail (vertical section)

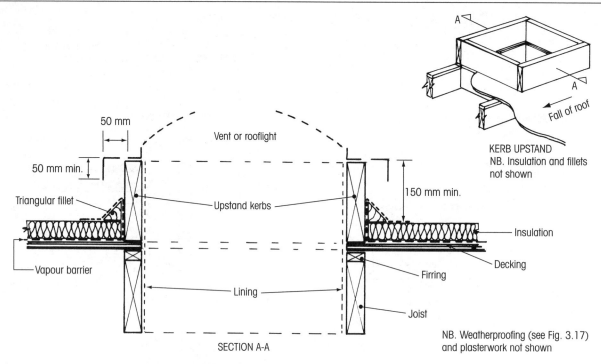

Figure 3.11 Upstand details for kerb to roof vent (vertical section)

You will see from Figure 3.11 that an opening in a roof will need extending above its upper surface to prevent any ingress of moisture from water splashing and snow. This is achieved by incorporating a kerb upstand to accommodate a weatherproof apron.

Larger openings on the other hand will require an arrangement involving the use of trimming joists, trimmers, and trimmed joists – examples of how these openings are formed are illustrated in Section 1.23.

3.4 ROOF DECKING

The material for the roof decking must be capable of supporting the chosen roof system (examples are shown in Figure 3.17) of dead, live, and wind loads [*Building Regulations Approved document "A1/2"*]. Any of the following may be specified as decking.

3.4.1 Tongued and Grooved Floorboards

Very rarely used these days as a decking material, however, if used they should be laid either with or diagonally to the fall of the roof. Cupping of tangential sawn boards laid across the fall could cause the roof covering to form hollows of sufficient depth to partially hold back the flow of water off the roof and leave pools.

The methods of laying and fixing details are the same as those described in Section 1.18 for floor boarding.

3.4.2 Plywood

Only roofing grades must be used – plywood is classified according to its veneer bond performance (WBP), and veneer durability. Suitable grades are listed in *Bench and Site Skills* [BENCH AND SITE SKILLS, SECTION 2.1.3].

Boards should be supported on all edges, leaving a joint gap of 1 mm per metre run of board, be it along its length or width. This allowance is made in case of any moisture movement.

3.4.3 Chipboard

Only types with the required water resistance and which have been classified for this purpose must be used. Suitable grades are listed in *Bench and Site Skills* [BENCH AND SITE SKILLS, SECTION 2.3.3]. Boards are available which have a covering of bituminous felt bonded to one surface – this gives the board temporary protection against wet weather once it has been laid and the edges and joints have been sealed.

Edge support, laying and fixing are similar to floors in Section 1.17. Moisture movement will be greater than plywood therefore a 2 mm provision per metre run should be allowed with at least a 10 mm gap around the roof edges and abutments.

Tongued and grooved boards should be laid as per the manufacturer's instructions.

3.4.4 Oriented Strand Board (OSB)

Generally more stable than chipboard, but again only roofing grades are suitable [BENCH AND SITE SKILLS, SECTION 2.41].

Provision for moisture movement should be made as with chipboard.

3.4.5 Cement Bonded Chipboard

Strong and durable, with high density (much heavier and greater moisture resistance than standard chipboard). For full property details refer to *Bench and Site Skills* [BENCH AND SITE SKILLS, SECTION 2.72].

Moisture movement provision should be as chipboard.

3.4.6 Metal Decking

Profiled sheets of aluminium or galvanised steel are used and a variety of factory-applied coloured coatings are available. This form of decking is more usually associated with large steel sub-structures and fixed by specialist installers. However they may be used on small span roofs to some effect − sheets can be rolled to different profiles and cut to almost any reasonable length to suit individual requirements. The special colour co-ordinated fixings are available from the manufacturers.

Note. Translucent sheeting both corrugated and flat (e.g. Polycarbonate Twinwall, etc.) are required to be installed following the manufacturer's instructions.

3.5 EAVES TREATMENT

Eaves treatment is illustrated in Figures 3.12 and 3.13.

The eaves are where the roof water flows over the roof edge (known as weir flow) into a gutter. The gutter may be a separate component fixed externally to the fascia board, or formed within the roof structure as a box. This box gutter may be formed within the confines of the external walls, or as shown in Figure 3.13(c) outside the external wall (overhanging box gutter). The eaves arrangement may be either 'flush' or 'overhanging'.

3.5.1 Flush Eaves (Figure 3.12)

In this case the fascia board stands just proud of the outer face wall − this allows for any face-wall unevenness and also provides a 'drip'.

In Figure 3.12(a) joists have been laid with the fall of the roof and allowed to protrude thereby providing a fixing for the fascia board. Because this portion of joist is enclosed and in direct contact with outer brick or blockwork, if pre-treated timber has not been used it would be advisable to treat this section with a wood preservative.

Figure 3.12(b) on the other hand shows the joists set at right angles to the fall − the fascia board in this case will need to be fixed to an out-rigger (Figure 3.15) or a framework of treated timber grounds as shown later in Figure 3.14(b).

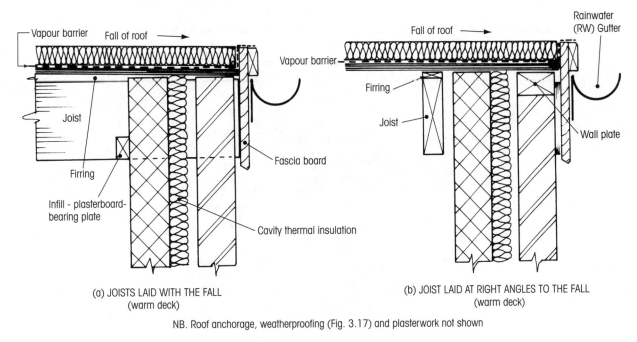

(a) JOISTS LAID WITH THE FALL
(warm deck)

(b) JOIST LAID AT RIGHT ANGLES TO THE FALL
(warm deck)

NB. Roof anchorage, weatherproofing (Fig. 3.17) and plasterwork not shown

Figure 3.12   Flush eaves details (vertical section)

3.5.2 Overhanging Eaves (Figure 3.13)

As the name implies this type of eaves arrangement projects beyond the face wall to provide a short form of canopy. The other side of this overhang can either be left open (open eaves) or as shown, closed (although vented) by using a soffit board.

Figure 3.13(a) shows details of how the joists run over and through the outer wall to provide a fixing for the fascia that supports the external gutter. A fly-proof mesh must cover vents in soffit boards. Figure 3.13(b) differs in that the inner structural wall is of timber-framed construction. It is important to note that each joist should be positioned onto the head plate directly above a vertical stud. Also that a gap is left between the

under-side of the soffit and head of the outer brick leaf – this gap makes provision for the differential thermal and moisture movement between the timber framework and masonry construction.

Figure 3.13(c) is similarly constructed to Figure 3.13(a). Except that a gutter has been formed within the overhang – this is known as a boxed gutter (see also Figures 3.20 and 3.21).

The boxed gutter forms an integral part of the roof structure, it is more expensive than an externally fixed eaves gutter, but it is very efficient and gives a neater finish to the eaves. The gutter outlet (funnel) can be sited at that either end or mid-length of the gutter. A purpose-made funnel cover should be fitted over the outlet to help prevent the rainwater fallpipe becoming blocked at a

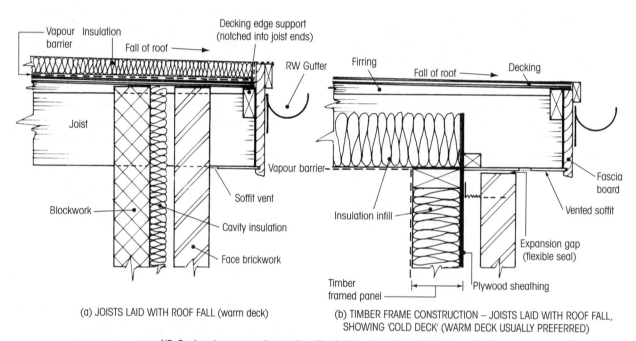

(a) JOISTS LAID WITH ROOF FALL (warm deck)

(b) TIMBER FRAME CONSTRUCTION – JOISTS LAID WITH ROOF FALL, SHOWING 'COLD DECK' (WARM DECK USUALLY PREFERRED)

NB. Roof anchorage, weatherproofing (Fig. 3.17) and plasterwork not shown

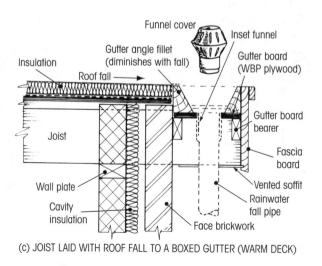

(c) JOIST LAID WITH ROOF FALL TO A BOXED GUTTER (WARM DECK)

Figure 3.13 Overhanging eaves details (vertical section)

bend possibly caused by a buildup of silt, leaves or loose stone chippings. Stone chippings are used in conjunction with some roof coverings (white stone chippings help keep the roof surface cool by reflecting solar heat and they also add to its fire resistance).

3.6 VERGE TREATMENT (FIGURES 3.14 AND 3.15)

A verge is an edge of the roof that lies parallel to the fall of the roof. The arrangement at this point can (as with the eaves) be either flush with the outer wall or overhanging. Both methods of construction will require a kerb edge (upstand) to prevent roof water running off the edge.

Figures 3.14(a) and (b) show details of how the verge for a warm roof deck is formed flush with the outer walls. Similarly, Figures 3.15(a) and (b) detail overhanging verges.

The verge fascia board on short roof runs can be fixed level (but need not be), in which case the inner upstand kerb will need to be tapered to suit the fall. This is achieved by using a tapered kerb angled fillet, or by

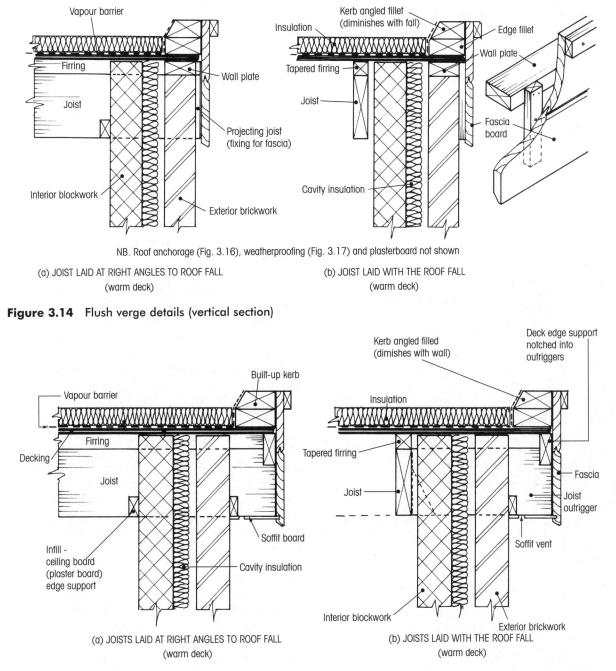

NB. Roof anchorage (Fig. 3.16), weatherproofing (Fig. 3.17) and plasterboard not shown

(a) JOIST LAID AT RIGHT ANGLES TO ROOF FALL
(warm deck)

(b) JOIST LAID WITH THE ROOF FALL
(warm deck)

Figure 3.14 Flush verge details (vertical section)

(a) JOISTS LAID AT RIGHT ANGLES TO ROOF FALL
(warm deck)

(b) JOISTS LAID WITH THE ROOF FALL
(warm deck)

Figure 3.15 Overhanging verge details (vertical section)

leaving a tapered margin between the angled fillet and fascia board top.

Soffit boards are either fixed to the protruding joist end, or outrigger end (Figure 3.15b) or by using treated timber grounds fixed to the wall plate and face wall as shown in Figure 3.14(b).

3.7 ROOF ANCHORAGE

Except where roof joists are fully built into a wall such as the arrangement shown in Figure 3.8(a), the joist will require anchoring to the sub-structure to resist upward (lift) thrust induced by strong winds.

Galvanised steel anchor straps of the required sectional size, length and shape, together with the specified fixings can provide suitable anchorage [*Building Regulations Approved Document "A"*]. Examples of how anchorage can be achieved are shown in Figure 3.16.

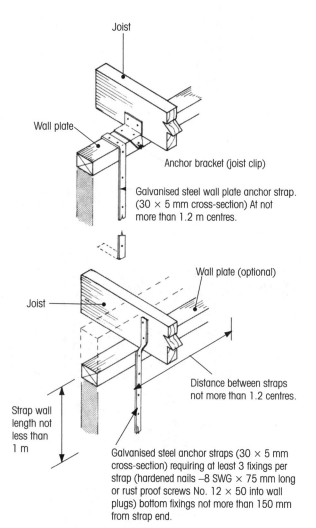

Joist

Wall plate

Anchor bracket (joist clip)

Galvanised steel wall plate anchor strap. (30 × 5 mm cross-section) At not more than 1.2 m centres.

Wall plate (optional)

Joist

Distance between straps not more than 1.2 centres.

Strap wall length not less than 1 m

Galvanised steel anchor straps (30 × 5 mm cross-section) requiring at least 3 fixings per strap (hardened nails −8 SWG × 75 mm long or rust proof screws No. 12 × 50 into wall plugs) bottom fixings not more than 150 mm from strap end.

Figure 3.16 Roof anchorage to sub-structure

A similar arrangement can also offer lateral support to outer walls.

3.8 THERMAL INSULATION

Unless the roof serves only as a shelter to a shed, garage, or porch, etc., one of the main functions of roof insulation is to help reduce heat energy from entering or leaving the structure.

Materials possessing good thermal insulation quantities are cellular in their make-up. The types used to insulate flat roofs are either, in the form of a quilt or semi-rigid slab comprising of a mineral wool derived from rock (rockwool) or glass (glasswool) bonded with a resin binder, to certain degrees of stiffness to be installed directly above the ceiling. The more rigid forms of thermal insulation of varying densities are usually derived from boards of expanded polystyrene or urethane foams. These types of board are installed above the roof deck. Where added stiffness is required composite boards are used, for example insulating foam pre-bonded to a backing of WBP plywood, or bitumen-impregnated fibreboard.

The position of the thermal insulation will categorise the type of roof, for example, in the former case where the insulation material was at ceiling level, we call this type of roof a 'cold roof' or 'cold deck roof construction'. If on the other hand the insulation material is in contact with the roofs under surface of its weathering, then we term this type of roof a 'warm roof' or of 'warm roof construction'. However, if the insulation material is placed on top of the weathering instead of under it, we refer to this type of roof as an 'upside-down roof' or as 'inverted roof construction'. All three methods are shown in Figure 3.17.

3.8.1 Cold Roof Construction (Figure 3.17a)

By placing the insulation material at ceiling level heat is retained at this point. However, unless a vapour barrier (an impervious membrane, such as polythene sheeting) is positioned at the warm side of the insulation water vapour will pass through the insulation and, on contact with the cold impervious weathering, will turn into droplets of water (condensation).

Dampness in this void can mean that there is a danger from fungal attack and that the insulation will become wet and eventually compacted and therefore less effective – not to mention the presence of damp patches on the ceiling below. To prevent the risk of condensation, voids between joists must be thoroughly ventilated.

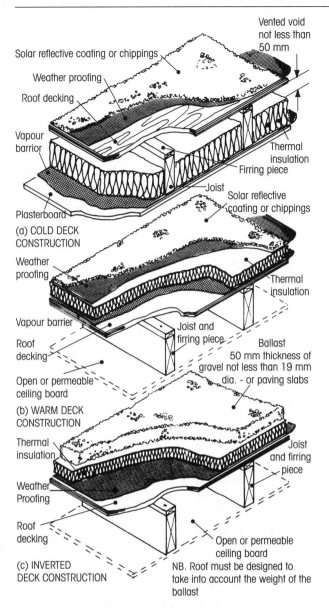

Solar reflective coating or chippings

Weather proofing

Roof decking

Vapour barrier

Plasterboard

(a) COLD DECK CONSTRUCTION

Vented void not less than 50 mm

Thermal insulation

Firring piece

Joist

Weather proofing

Solar reflective coating or chippings

Thermal insulation

Vapour barrier

Roof decking

Open or permeable ceiling board

(b) WARM DECK CONSTRUCTION

Joist and firring piece

Ballast 50 mm thickness of gravel not less than 19 mm dia. - or paving slabs

Thermal insulation

Weather Proofing

Roof decking

Joist and firring piece

Open or permeable ceiling board

(c) INVERTED DECK CONSTRUCTION

NB. Roof must be designed to take into account the weight of the ballast

Figure 3.17 Types of roof deck construction and methods of providing thermal insulation

A gap of at least 50 mm should be left between the decking and the thermal insulation to allow for free air space.

Note. 'Through-' or cross-ventilation can be improved by the use of counter battens set at right angles to the joists as shown in Figure 3.18(c).

As shown in Figure 3.18 the options for providing cross-ventilation include:

- Figure 3.18(a) – cross-ventilation is provided by leaving gaps equivalent to a 25 mm wide continuous strip along each of two opposition sides. The inner void should be such as a 50 mm gap between the underside of the roof decking and upper side of the insulation is left open to allow free movement of air.

- Figure 3.18(b) – where cross-ventilation is not possible (due to an abutting wall), a proprietary head venting system may be installed such as the one shown.

- Figure 3.18(c) – an alternative to Figure 3.18(b), is to use counter battens – these must be strong enough to support and provide a satisfactory base for fixing the decking too, and security anchored to the sub-structure.

3.8.2. Warm Deck Construction (Figure 3.17b)

In this case the thermal insulation is positioned at decking level, with a vapour barrier to its warm under side. Ceiling voids will not need ventilating. Apart from the reduced risk of fungal attack due to interstitial condensation, another advantage of this method is that, if access to the roof void is required for services (electric cables, etc.), this can be achieved without damaging the vapour barrier.

3.8.3. Inverted Deck Construction (Figure 3.17c)

This is another form of warm roof more usually associated with a reinforced concrete roof slab. In this case the insulation is positioned on top of the weatherproof covering held in place either by a gravel ballast or paving slabs. It goes without saying these roofs must be specifically designed to carry this extra dead load.

The temperature within the roofs components should remain above dew point and therefore reduce any risk of interstitial condensation – there is however a risk of surface condensation to the underside of the deck if cold rain water flows under the insulation.

Note. Reference should be made to current Building Regulations with regard to insulation type and relevant thickness for both cold and warm deck construction. [*Building Regulation Approved Documents "F & L"*].

3.9 ROOF COVERING (WEATHERPROOFING)

Built-up roofing consists of two or more layers of a bitumen-based membrane (reinforced with either a glass or polyester fibre). The first layer is stuck directly onto the decking and subsequent layers to each other, with a top layer of bitumen-bedded stone chippings.

Mastic asphalt; or metals such as aluminium, copper, or lead may be also used as a weather-proof covering.

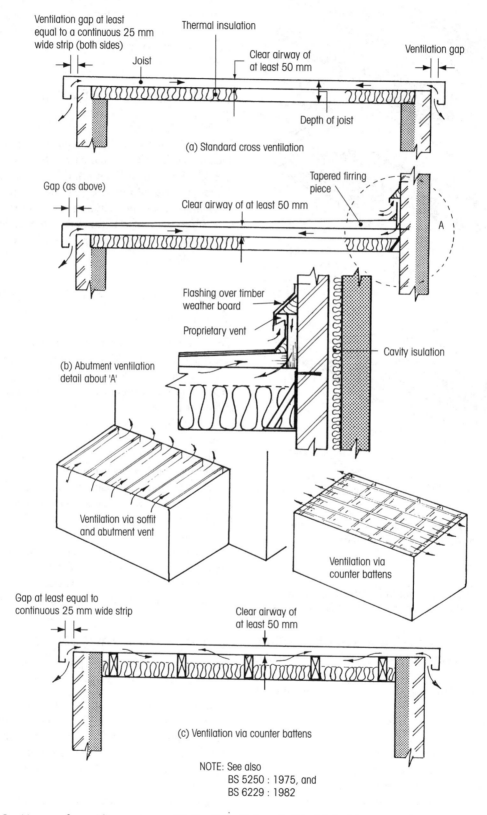

Ventilation gap at least equal to a continuous 25 mm wide strip (both sides)

Thermal insulation

Joist

Clear airway of at least 50 mm

Ventilation gap

Depth of joist

(a) Standard cross ventilation

Gap (as above)

Tapered firring piece

Clear airway of at least 50 mm

A

(b) Abutment ventilation detail about 'A'

Flashing over timber weather board

Proprietary vent

Cavity isulation

Ventilation via soffit and abutment vent

Ventilation via counter battens

Gap at least equal to continuous 25 mm wide strip

Clear airway of at least 50 mm

(c) Ventilation via counter battens

NOTE: See also
        BS 5250 : 1975, and
        BS 6229 : 1982

Figure 3.18 Means of providing cross-ventilation to a flat roof of 'cold deck' construction

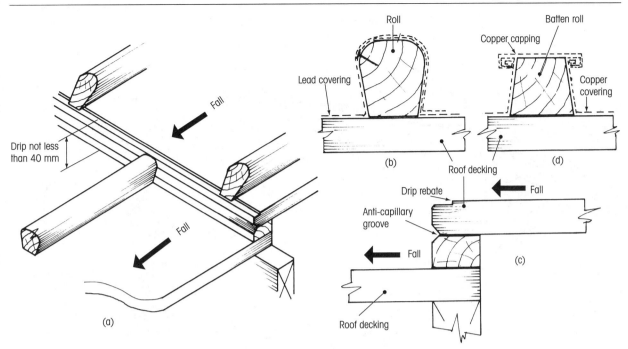

Figure 3.19 Provision for thermal movement of a metal covered flat roof

With metals there is the problem of thermal movement (expansion and contraction according to environmental temperature change), and provision must be made on the decking to allow movement to take place. This is achieved by the introduction of timber upstands known as 'rolls' (lead covering) or 'batten rolls' (copper covering).

Figure 3.19(a) shows how these rolls are positioned at intervals along the fall of the roof – the distance apart varying with the type of metal. A roll suitable for lead is shown in Figure 3.19(b) – notice that only one edge of the sheet is fixed – these rolls facilitate the formation of waterproof and expansion joints between the longitudinal edges of the lead sheets. Joints across the fall of the roof (width of sheet covering) could form a water check, in which case as shown in Figure 3.19(c) a step in the decking, known as a 'drip' would be formed as part of the roof's construction. Figure 3.19(d) shows a typical batten roll used with a copper covering – it has a different function to the lead roll in that its flat top is covered with a copper capping.

3.10 STAGES OF CONSTRUCTION

Once all the requirements are in place, for example:

- Direction of falls.
- Position of roof water outlets.
- Direction of joists (and firrings if required).
- Method of supporting joist ends (and mid supports if required).

- Position of any roof openings.
- Provisions for trimming any roof openings.
- Sectional sizes of all timber components.
- Type of decking material and methods of fixing.
- Type and position angle fillets.
- Arrangement at the eaves.
- Any provision for integral gutter.
- Arrangement at the verges.
- Position and types of roof anchorage.
- Provision for thermal insulation.
- Any requirements for cross-ventilation.
- Type of roof covering (weatherproofing) – any thermal movement provision.

And providing that all the structural details comply with current building regulations and full-certificated local authority approval has been granted – on-site skills can then be put into production.

Figure 3.20 shows an exposed isometric view of a single flat roof over a small extension to a building that abuts an outer wall. The fall runs away from the wall to a front-eaves gutter (an integral boxed gutter is shown but an external gutter fixed to the fascia board may be used). Both forms of thermal insulation may be considered – the drawing shows provision for warm deck construction. If the cold deck method is used full cross-ventilation **must** be provided to meet the requirements of the current building regulations [*Building Regulations Approved Documents "F & L"*]. Figure 3.18 shows methods of providing roof space cross-ventilation.

Figure 3.21 illustrates the stages of construction used in Figure 3.20.

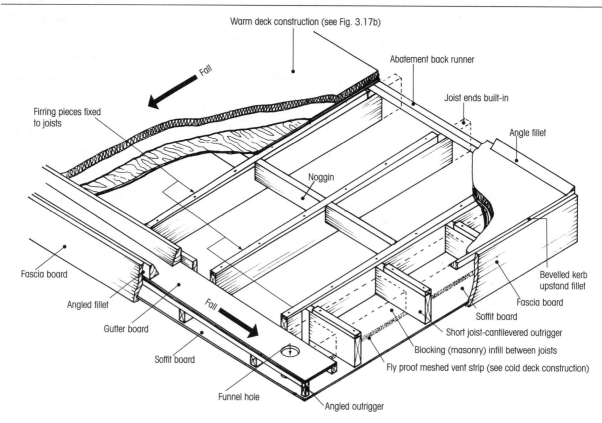

Figure 3.20   Exposed isometric view of a small single-span flat roof with one-edge abutment

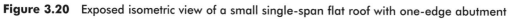

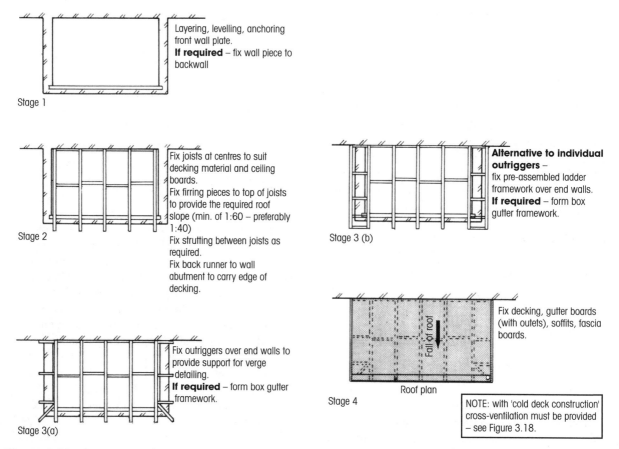

Figure 3.21   Basic stages for constructing a single-span flat roof with one-edge abutment

4
Pitched roofs

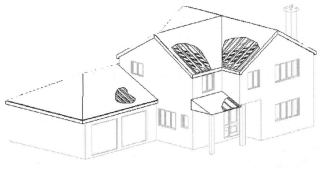

4.1 INTRODUCTION

The purpose of a roof is to cover and protect an enclosed space and its surrounding walls from the damaging effects of the elements (rain, wind, snow, sunlight). A pitched roof has sloping surfaces, which usually overhang the walls on which it sits (Figure 4.1a).

The slope or **pitch** of the roof will vary depending on the type of weather-proofing material used, and its degree of exposure (a windy or sheltered position). A pitched roof is defined as having a slope steeper than 10° (Figure 4.1b).

A roof consists of an external covering of weather-proof material [*Building Regulations Approved Document "C"*] and a layer of thermal-insulating material, [*Building Regulations Approved Document "D"*] both of which are supported by a triangular framework of timber, forming the structural part of the roof [*Building Regulations Approved Document "A"*] (Figure 4.1c).

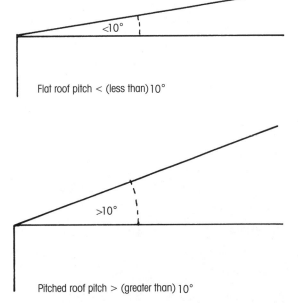

Flat roof pitch < (less than) 10°

Pitched roof pitch > (greater than) 10°

(b) Definition of flat and pitched roofs (see also Figure 3.1)

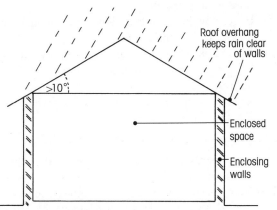

(a) Protection of an enclosed space

Roof overhang keeps rain clear of walls

Enclosed space

Enclosing walls

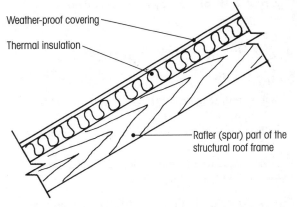

Weather-proof covering

Thermal insulation

Rafter (spar) part of the structural roof frame

(c) Roof covering and structure

Figure 4.1 Pitched roofs (see also Figure 3.1)

4.2 ROOF TYPES

The different roof types are illustrated in Figure 4.2

4.2.1 By Pictorial Outline (Figure 4.2)

(a) Mono pitch.
(b) Lean-to.
(c) Gable.
(d) Hipped.
(e) Dutch gable.
(f) Gable hip – suppressed gable (Gablet).

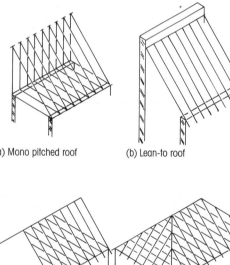

(a) Mono pitched roof   (b) Lean-to roof

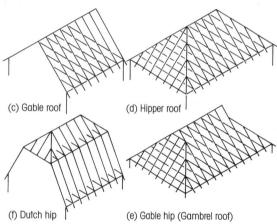

(c) Gable roof   (d) Hipper roof

(f) Dutch hip   (e) Gable hip (Gambrel roof)

Figure 4.2 Roof types by pictorial outline

4.2.2 By Structure (Figure 4.3)

Single Roofs

- Couple roof (Figure 4.3a). Clear span not greater than 3.00 m, pitch 40°+
- Collar roof (Figure 4.3b). Clear span not greater than 4.00 m
- Close couple roof (Figure 4.3c). Clear span not greater than 5.50 m pitch 25°+.

Note. As shown in Figure 4.3(d) the tie member (ceiling joist) may require support midway along its length.

Double Roofs

In this type of roof a second structural member, the **purlin** is introduced to support the rafters (sloping members) at mid or intermediate points between wall plate and ridge. This reduces the span of the rafter which can then be of a smaller cross-section, while still carrying the same live and dead loads (Section 1.2) imposed on the roof (Figure 4.3e).

Triple Roof (Trussed Roof, Section 4.11)

A third structural element the **roof truss** is introduced into the roof to transfer the loads from the rafters and purlins through the trusses to the walls (Figure 4.3f). The trusses are placed at 3–4 m intervals along the length of the roof as shown in Figure 4.3(g). *Note.* Roof trusses of this type are not in common use today, but can still be seen in older buildings (Figure 4.27).

Trussed Rafter Roofs (Section 4.12)

Trussed rafters are the most common form of roof construction used today. Each pair of rafters is framed to form a truss (Figure 4.3h). They are then, depending on the roof covering and ceiling board size, set up at from 450 to 600 mm centres along the length of the roof (Figure 4.3i).

4.3 ROOF PARTS AND COMPONENTS

Parts or elements of a roof are shown in Figure 4.4 as:

(a) **Eaves.** The lower edge of the roof which overhangs the walls.
(b) **Gable.** The triangular end of a pitched roof, or the triangular upper part of a wall that closes the gable end of the roof.
(c) **Hip.** The oblique line running from the ridge to eaves at the junction of two sloping roof surfaces at an external intersection.
(d) **Ridge.** The line of intersection of two sloping roof surfaces at their highest point.
(e) **Valley.** The oblique line running from ridge to eaves at the internal intersection of two sloping roof surfaces.
(f) **Verge.** the sloping edges of a gable roof which overhang the gable walls.

Components of a roof as shown in Figure 4.5:

(a) **Barge boards.** Finishing timber trim to the gable end of a roof.
(b) **Ceiling beams or binders.** Horizontal members running at right angles and fixed above the ceiling joists at intermediate positions along their length.

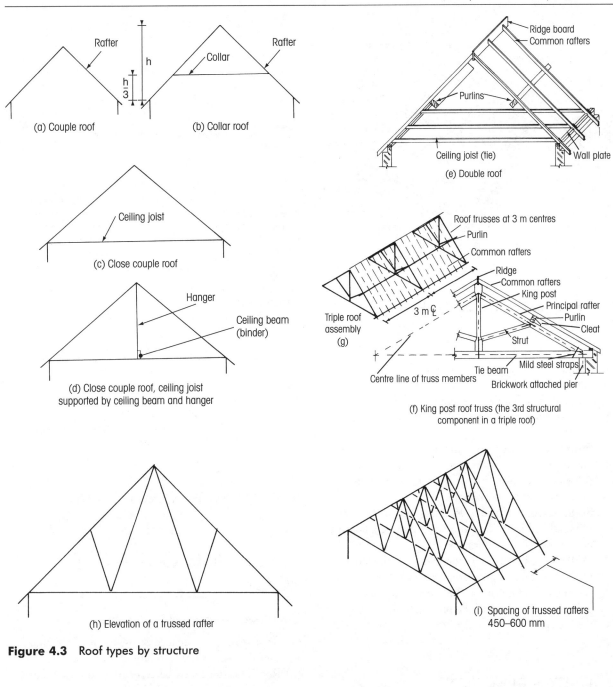

Figure 4.3 Roof types by structure

(c) **Ceiling joists.** Horizontal members running across the roof from wall plate to wall plate, to carry the ceiling and to act as a tie to the feet of the rafters.

(d) **Common rafters.** Inclined members running from wall plate to ridge to give support to the roof covering.

(e) **Fascia board.** Vertical timber trim to the lower end (feet) of the rafters.

(f) **Gable ladder.** Two pair of common rafters framed together with noggings to carry the roof framing over the gable wall.

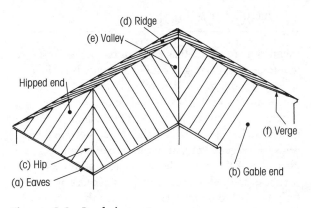

Figure 4.4 Roof elements

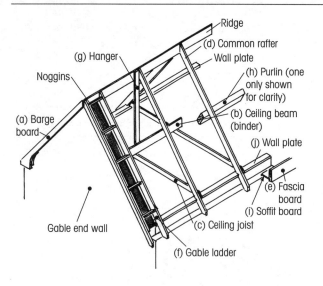

Figure 4.5   Roof components (gable roof)

(g) **Hangers**. Vertical, or near vertical members running from ridge or purlins to carry the ceiling beams (binders) in a double or single roofs.

   Note. They are not part of a roof truss, or trussed rafter.

(h) **Purlins**. A structural member running parallel to, but at mid-way or intermediate points between the ridge and wall plate and at right angles to the rafters to give them support.

(i) **Soffit board**. Horizontal trim fixed to the feet of rafters. Used in conjunction with the fascia board to box in the feet of the rafters.

(j) **Wall plates**. Horizontal members secured to the top of the walls to give fixing and support to the ends of rafters and ceiling joists, and to transfer the loads from the roof structure to the walls.

4.4 PITCHED ROOF TRIANGULATION

Inclined members such as a pair of rafters, when subject to a force (load) acting vertically downwards at their apex, will move to a horizontal position as the force causes the feet of the rafters to slide outwards, see Figure 4.6(a). Unless in the case of a lean-to roof the rafters are held securely at their top and bottom by the walls (see Figure 4.6b), or in the case of a double pitched roof the feet of the rafters are restrained by the weight of the wall, or connected at their feet by a tie (ceiling joist) to make a stable frame in the form of a triangle Figure 4.6(c).

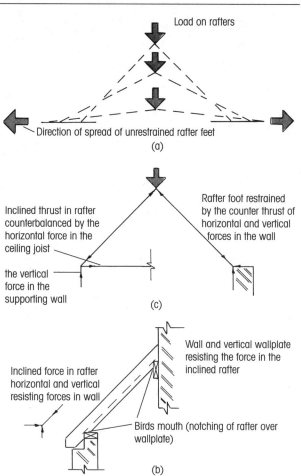

Figure 4.6   Restraining forces in a roof frame

4.5 SETTING OUT A GABLE ROOF

Terms used in setting out (Figure 4.7a).

(a) **Roof span**. The distance across the roof measured over the wall plates to their outer edges.

(b) **Roof or rafter rise**. The vertical distance measured from the top of the wall plates to the intersection of the rafters at the apex of the roof. The rafters are considered as single lines having length, but no width or thickness.

(c) **Roof pitch**. The slope of the roof expressed in degrees 0°, or as a ratio or fraction. Found by dividing [BENCH AND SITE SKILLS, SECTION 13.2.5] the rise by the span.

   Example. If the span is 4 m and the rise 2 m, what is the pitch?

$$\text{Pitch} = \frac{\text{Rise}}{\text{Span}} = \frac{2}{4} = \frac{1}{2}$$

giving a half-pitched roof. A half-pitched roof has its rafters inclined at an angle of 45°.

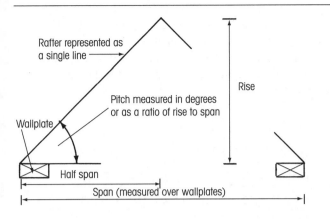

(a) Pitch of a roof

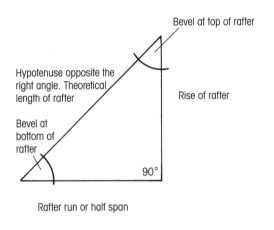

(b) The right-angled triangle

Figure 4.7 Terms used for setting out pitched roofs

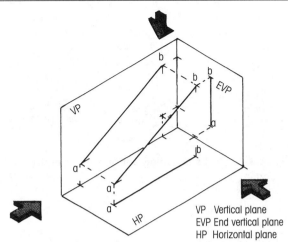

(a) Isometric sketch of rafter line ab in orthographic projection

VP   Vertical plane
EVP End vertical plane
HP   Horizontal plane

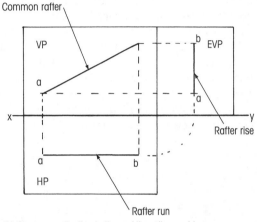

(b) Common rafter line in three of the orthographic projection views

4.5.1 Common Rafter Length and Bevels

The cross-sectional size of the timber rafters used in the roof will be determined by the loads to be carried, the strength grade [BENCH AND SITE SKILLS, SECTION 1.8.4] of the timber, and the span between the supports for the rafters.

Note. For the purpose of finding the initial lengths and bevels the rafters are considered as single lines without width or thickness.

To find the length of and bevels to be cut on a common rafter, use is made of the right-angled triangle (Figure 4.7b). Figure 4.8(a, b) shows a simple line presentation of the common rafter:

- In orthographic projection the horizontal plane (HP) plan viewed from above
- In the vertical plane (VP) front elevation viewed from the side.
- In the end vertical plane (EVP) end elevation, viewed from the end.

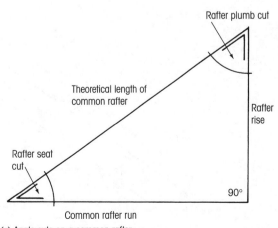

(c) Angle cuts on a common rafter

Figure 4.8 Common rafter length and bevels

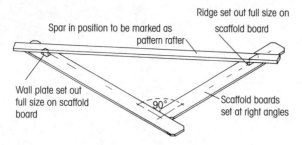

Figure 4.9 Full size setting out of a rafter on the ground

It is the plan of the rafter and the rise of the rafter in the end elevation, combined together as a right angle to form the two sides of a triangle adjacent to the right angle, while the hypotenuse, the inclined and longest line seen in the front elevation gives the true slope, length and bevels for the rafter (Figure 4.18c).

The rafter can be set out full size on site by using two scaffold boards as shown in Figure 4.9. This is a cumbersome method and great care has to be taken in setting out and checking the right angle [BENCH AND SITE SKILLS, SECTION 6.3.1]. It is probably better practice to find the lengths and bevels for the rafter by drawing

to a scale [BENCH AND SITE SKILLS, SECTION 12.3] of 1:5 or 1:10 on a piece of plywood, but remember a 1 mm inaccuracy in a 1:10 scale will give a 10 mm error full size. Accurate and careful setting out is the key to success.

4.6 SETTING OUT THE RAFTERS OF THE ROOF OVER THE GARAGE OF THE HOUSE

The span over the wallplates, measured to their outer edges is 6.830 m and half the span or run of the rafter is 3.415 m and the rise is 2.100 m. Figure 4.10(a) shows the run and rise of the rafter. Figure 4.10(b) shows the hypotenuse of the triangle drawn in as the rafter. This will give the length and bevels required. Roof members are not single lines, but have width and thickness, which must be taken into account when setting out. Figure 4.10(c) shows the wall plate, ridge and common rafter positioned in relation to this right-angled triangle.

Note. The (notch) **birdsmouth** is cut one-third into the depth of the pattern rafter.

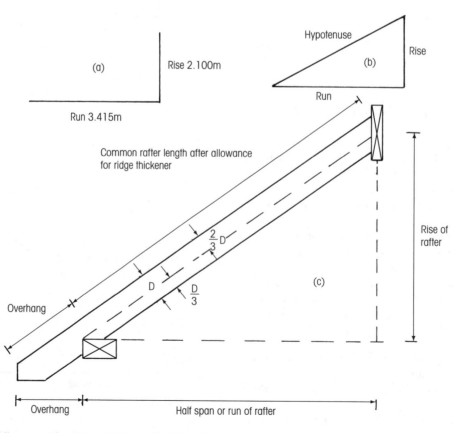

Figure 4.10 Allowance for rafter thickness and depth

Sliding bevels [BENCH AND SITE SKILLS, FIGURE 5.12] are set to the vertical, or **plumb** cut and the horizontal, or **seat** cut. Alternatively a plywood template can be cut in the form of a similar triangle to give the bevels required. Select a piece of rafter material that is reasonably straight and working from the round or crown edge mark the plumb cut (Figure 4.11a). Measure the length of the hypotenuse and scale up to full size, this should be 4.009 m and draw another plumb cut (Figure 4.11b). Draw a line two-thirds of the rafter depth and parallel to the crown edge, pitch line, and mark the seat cut bevel where the lower plumb cut crosses the pitch line. This marks out the birdsmouth (Figure 4.11c). It will be noticed that the plumb and seat cut are at right angles to each other.

To make allowance for the thickness of the ridge board use the template or bevel for the seat cut, and mark it across the plumb cut at the top of the rafter. Measure along the seat cut into the rafter, half the thickness of the ridge board and at this point draw another plumb cut. This will give the actual length of the rafter (Figure 4.11d). To cut the foot of the rafter measure the required inclined overhang along the pitch line and mark another plumb cut and then mark the seat cut at the required position (Figure 4.11e). In practice the feet of the rafters are cut after being fixed in position by marking them with a chalk line [BENCH AND SITE SKILLS, SECTION 6.2.1] to line up the rafter ends for cutting. If the wall plates are level and straight the feet of the rafters can be cut before being fixed. When the first rafter has been carefully marked out, checked and cut. It can be used as a **pattern** rafter, and must be clearly marked as such, as all the other rafters will be marked out from it (Figure 4.11f).

Note. As the rafters may be sawn stock rather than regularised [BENCH AND SITE SKILLS, SECTION 1.9.1] their depth may not be uniform and it is necessary to keep flush the upper edges of the pattern and rafter being marked out. This may slightly vary the size of the birdsmouth, but will keep the upper edges of the rafters in line when fixed.

4.6.1 Pitching a Gable Roof Over the Garage.

1. Measure the length of the wall plates and ridge. If the wall plates have to be lengthened or jointed at the corners halve together [BENCH AND SITE SKILLS, FIGURE 7.24a]. Ridges are spliced together [BENCH AND SITE SKILLS, FIGURE 7.4] if they have to be made up of more than one length of timber.
2. Pair the wall plates and mark on the position of the ceiling joists and rafters indicating the rafter

positions with a cross or some other suitable mark (Figure 4.12a).
3. From the wall plate mark onto the ridge board the position of the rafter ends (Figure 4.12b).
4. Assisted by the bricklayer, bed the wall plates to the top of the walls straight and level. Also check that they are parallel (Figure 4.12c).
5. Place the ceiling joists to their marks along the wall plate and nail in position. (Figure 4.12d). Where the ceiling joists run over the store and rear vestibule, build one end into the house wall, or secure to galvanised or sheradised mild steel hangers (Figure 4.12d) section AA [BENCH AND SITE SKILLS, SECTION 11.5.4].
6. Offer up two pairs of rafters, tack in position and temporarily brace from the wall plates (Figure 4.12e).
7. Set up a temporary scaffold in accordance with the requirements of The Construction (Working Places) Regulations 1996, the Health and Safety at Work Act 1974 and the Health and Safety Executive guidance notes GS15 and 31, to form a working platform so that the ridge board can be pushed up between the top ends of the rafters and nailed in their correct position against the ridge. Check the rafters for being plumb and adjust the temporary braces if necessary (Figure 4.12f).

  Note. A team of at least three carpenters are required for this operation. Two to check that the feet of the rafters do not slide out of place, and one to lift the ridge into position.
8. Purlins are next offered up and placed in position to be built into the gable walls (Figure 4.12g).
9. Once the purlins are built in, the remaining rafters can be pitched and nailed in position at the wall plate, purlin and ridge.
10. The gable ladders as described in Section 4.3 and Figure 4.5(f) are assembled and fixed in position.
11. Ceiling beams are built into the gable walls and supported by hangers secured to the ridge, or purlins before nailing the ceiling joists to them (Figure 4.12g) .
12. Fixing the rafters over the lean-to part of the roof will require a wall plate to be fixed by plugs and nails or metal hangers [BENCH AND SITE SKILLS, SECTION 11.5.4] to the house wall in a position to carry the top end of the rafters (Figure 4.12h). The top ends of the rafters will also have to be modified to fit over the wall plate (Figure 4.12i).
13. Finally (as shown in Figure 4.12j) the barge boards at the gable ends together with the soffit and fascia board are cut, fitted and fixed in position.

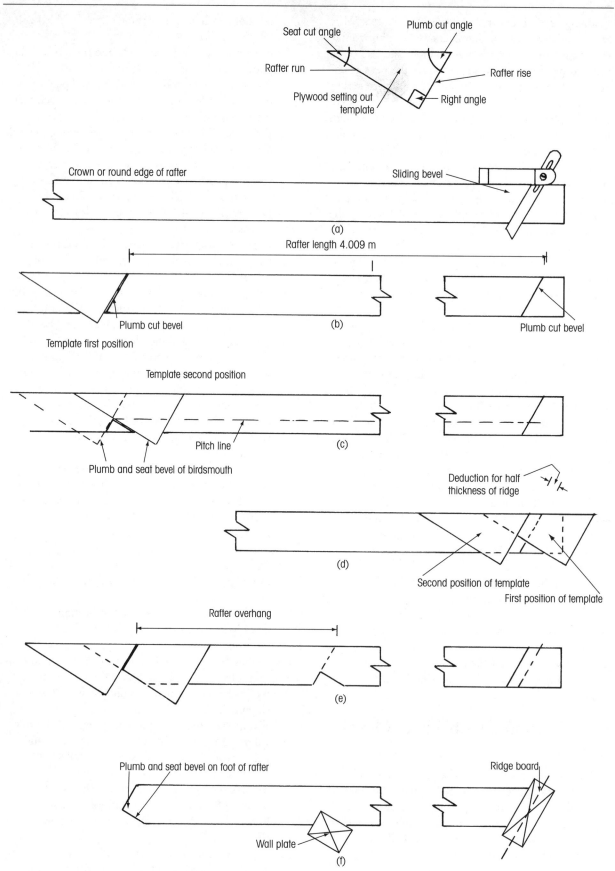

Figure 4.11 Marking out a pattern rafter

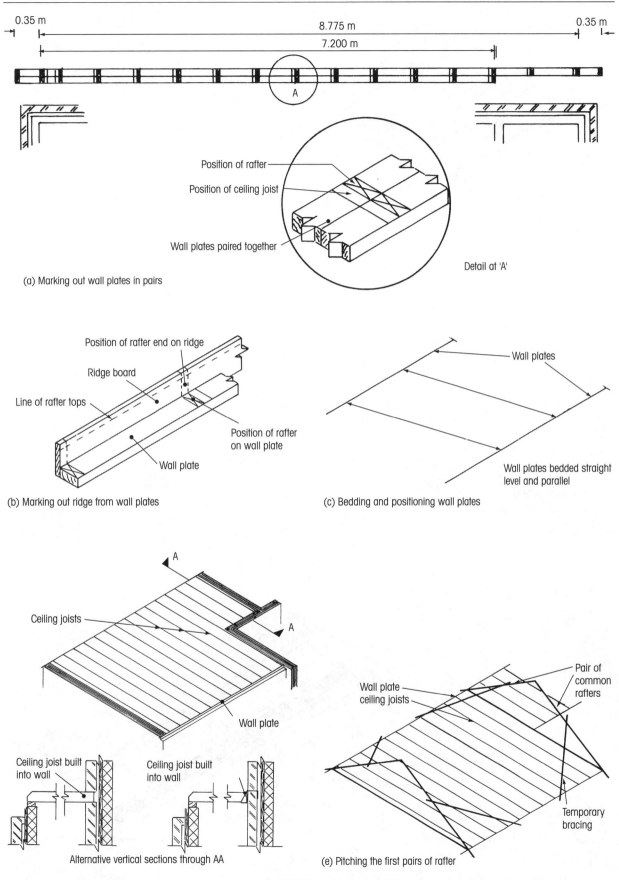

0.35 m   8.775 m   0.35 m

7.200 m

A

Position of rafter

Position of ceiling joist

Wall plates paired together

Detail at 'A'

(a) Marking out wall plates in pairs

Position of rafter end on ridge

Ridge board

Line of rafter tops

Position of rafter on wall plate

Wall plate

(b) Marking out ridge from wall plates

Wall plates

Wall plates bedded straight level and parallel

(c) Bedding and positioning wall plates

A

A

Ceiling joists

Wall plate

Ceiling joist built into wall

Ceiling joist built into wall

Alternative vertical sections through AA

Wall plate ceiling joists

Pair of common rafters

Temporary bracing

(e) Pitching the first pairs of rafter

Figure 4.12 Cutting and assembly of a traditional gable ended pitched roof: (a)–(e) shown above; (f)–(i) continued overleaf

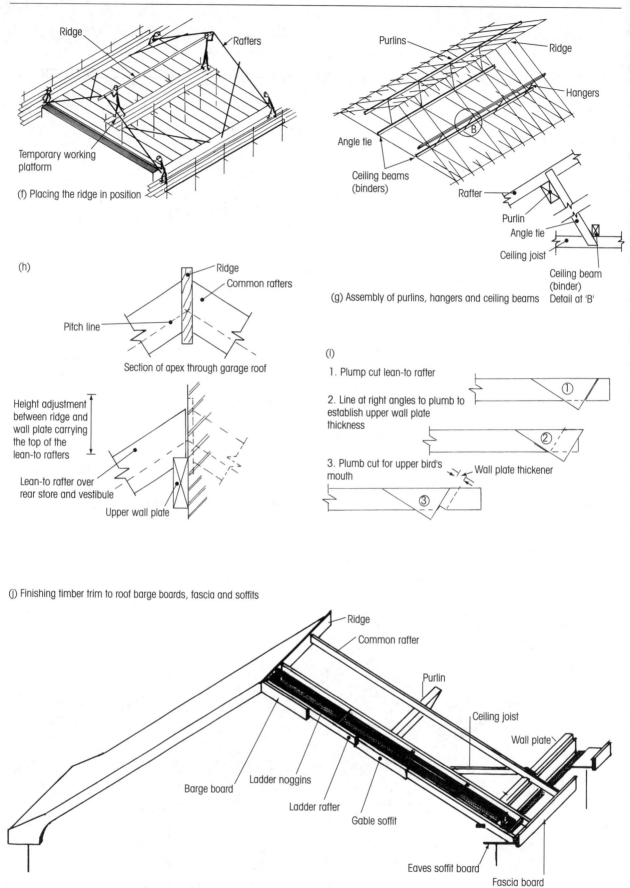

(f) Placing the ridge in position

(g) Assembly of purlins, hangers and ceiling beams   Detail at 'B'

(h) Section of apex through garage roof

(i)

1. Plump cut lean-to rafter

2. Line at right angles to plumb to establish upper wall plate thickness

3. Plumb cut for upper bird's mouth

(j) Finishing timber trim to roof barge boards, fascia and soffits

Figure 4.12   Cutting and assembly of a traditional gable ended pitched roof: (f)–(j) shown above

4.7 EAVES DETAILS

This is the finish at the lowest edge of the roof, which can be flush with, or over hang the walls as open or boxed eaves.

4.7.1 Flush Eaves (Figure 4.13a)

The rafter feet are cut plumb and project approximately 25 mm from the outer face of the wall to allow for a ventilation gap. Fascia board is nailed to the feet of the rafters to form a finish (trim) and to provide a fixing for the gutter brackets.

4.7.2 Open Eaves (Figure 4.13b)

The rafter feet are normally wrot (planed) and finished with a plumb and seat cut, or are moulded on their ends, but in all cases they project beyond the outer face of the wall. Eaves boards are fixed to the top edge of the rafters to mask the underside of the roof covering. Brackets to carry the gutters are fixed to the sides or top of the rafters.

4.7.3 Boxed or Closed Eaves (Figure 4.13c)

The feet of the projecting rafters are completely boxed in using fascia and soffit boards. To ventilate the roof space [*Building Regulations Approved Document "F"*] proprietary vermin-proof ventilation strips form part of the soffit. As an alternative the soffit can be drilled with holes into which are screwed vermin-proof circular plastic ventilators, at centres determined by the pitch and span of the roof. High-level ventilation may also be required (see Table 4.1).

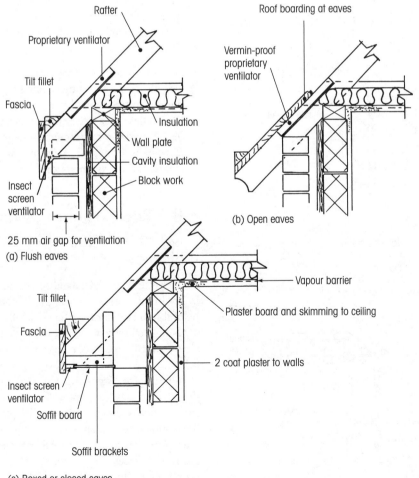

Figure 4.13   Eaves finish details

Table 4.1   Minimum ventilation for roof space voids per metre of roof

Pitch of the roof	Eaves (ceiling) level 0° to 15° (mm²/m run)	Eaves (ceiling) level above 15° (mm²/m run)	Apex (ridge) level (mm²/m run)
Ventilation at eaves level on opposite sides of the roof	25 000	10 000	
Ventilation at the apex of mono pitched roofs			5000
Ventilation at the apex of roofs pitched at an angle of more than 35° and with a span of 10 m or more.			5000

4.7.4 Sprocketed Eaves (Figure 4.14)

The pitch of a steep roof can be reduced by the use of sprockets which have the effect of reducing the speed of rain water running off the roof surface, so that washing out over the side of the gutters does not occur. Sprockets can be fixed to the top edge, or sides of the rafters (Figures 4.14a, b).

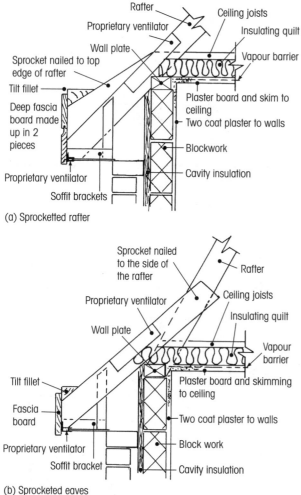

(a) Sprocketted rafter

(b) Sprocketed eaves

Figure 4.14   Roof sprockets

4.8 SETTING OUT A DOUBLE ROOF WITH HIPPED ENDS

In a fully hipped roof there are no gables and the eaves run round the perimeter of the roof. Usually at the same level, unless the overhang of the eaves varies.

Figure 4.15 shows the sketch of a hipped roof with the various parts named. All bevels cut on roof members are based on the right-angled triangle (Figure 4.16a).

The roof members can be set out using the following methods:

1. Scales drawing in orthographic projection.
2. Use of the steel roofing square (Figure 4.16b). With the roof pitch defined by:
   (i)   Linear measurement in metres.
   (ii) Angular measurement in degrees.
3. 'Cowley roof master'. A proprietary roof member length and bevel generator (Figure 4.16c).

4.8.1 Roof Member Lengths and Bevels Using Orthographic Projection

In developing an understanding of the principles involved in finding the bevels and lengths of roof members it is helpful to make cardboard models of a roof.

Using a suitable scale draw the plan, side and end elevations of a roof in orthographic projection (Figure 4.17a).

Referring to Figure 4.17(b). Place a compass point at *f* in the end elevation and rotate the distance *af* to *a'* then vertically downwards into the plan. Draw a horizontal line in plan from point *a* to meet *a''*. Join point *f* in plan to *a''*. This will give the meeting edge of the roof surfaces. Repeat for the four surfaces then cut round the outline and bend up the plan lines *cdef* until the four surfaces meet to give you the shape of the roof (see Figure 4.17c).

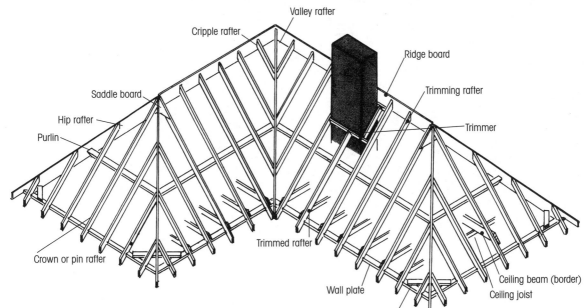

Figure 4.15 Components of a hipped roof

4.8.2 Common Rafter Length and Bevels

From the end elevation has been found (from the model (Figure 4.17b))

(a) The length of the common rafter *bc* in vertical section.
(b) The seat cut of the common rafter at 2.
(c) The plumb cut of the common rafter at 1.

4.8.3 Jack Rafter Lengths and Edge Bevel

From the model, the following have been found in the plan (Figure 4.17b):

(a) The diminishing lengths of the jack rafters in the triangle *eaf*.
(b) The edge cut of the jack rafters at 3.

Note. The plumb and seat cut of the jack rafters are the same as for the common rafter.

4.8.4 Hip and Valley Lengths and Bevels

Having developed a model of the true shape of the roof surfaces, and using the same principles, Figure 4.18(a) shows how the lengths and bevels of the hip and valley are found.

- True length of hip rafter *fk* and valley rafter *jd* in plan.
- Hip plumb cut bevel at *k*. Valley plumb cut bevel at *j*. Both in plan.
- Hip seat cut bevel at *f*. Valley seat cut bevel at *d*. Both in plan.

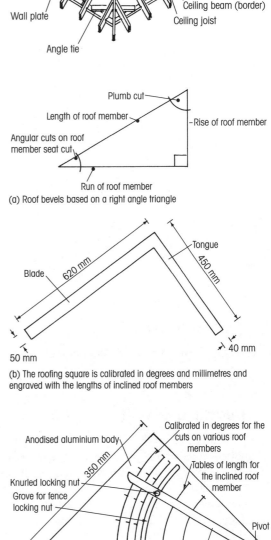

(a) Roof bevels based on a right angle triangle

(b) The roofing square is calibrated in degrees and millimetres and engraved with the lengths of inclined roof members

(c) Cowley roof master

Figure 4.16 Aids for setting out a pitched roof

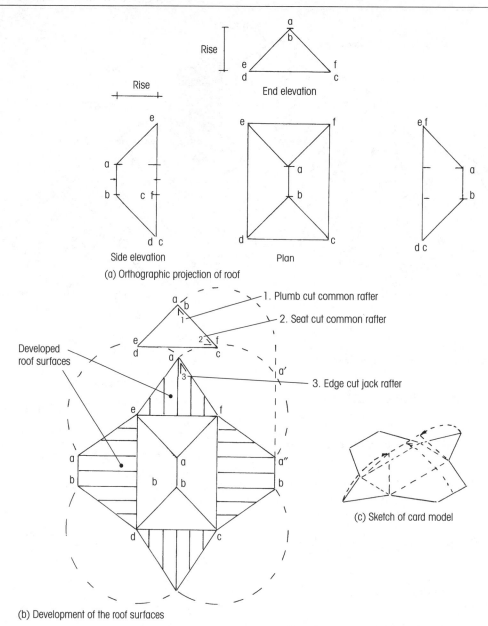

Figure 4.17   Development of a card model roof

4.8.5 Backing Bevel to the Hip

The hip rafter edge or backing angle (dihedral angle) is formed by the intersection of two adjacent roof surfaces. As the square top edge of the hip rafter does not lie in either roof surface it is necessary to bevel the top edge of the hip to give a good seating for the slate laths and slates, if a mitred rather than a bonnet finish is required at the hip.

To determine the backing bevel, draw a plan of the corner of the roof containing the hip (Figure 4.18b) and find its true length *kf* by setting up the rise of the roof *kh* at right angles to the run of the hip *fh*. Draw line *hl* from *h* and at right angles to *fk*. From *h* draw

the line *hl*. Put the compass point at *h* and rotate *l* to *l'* onto the line *fh* (plan length of the hip). Join *c* to *l'*. The backing bevel is at *l'*. Figure 4.18(c) shows a cross-section of the hip with the backing bevel applied.

4.8.6 Mitre Cut to a Hip With Backing Bevels

As the edge of a backed hip is lying in the same surface (plane) as the jack and common rafters the bevel for the mitre on the hip is the same as for the edge bevel on the jack rafter (see Figure 4.17b).

Figure 4.19 shows a practical method of forming the edge bevel on a backed hip rafter and jack rafter.

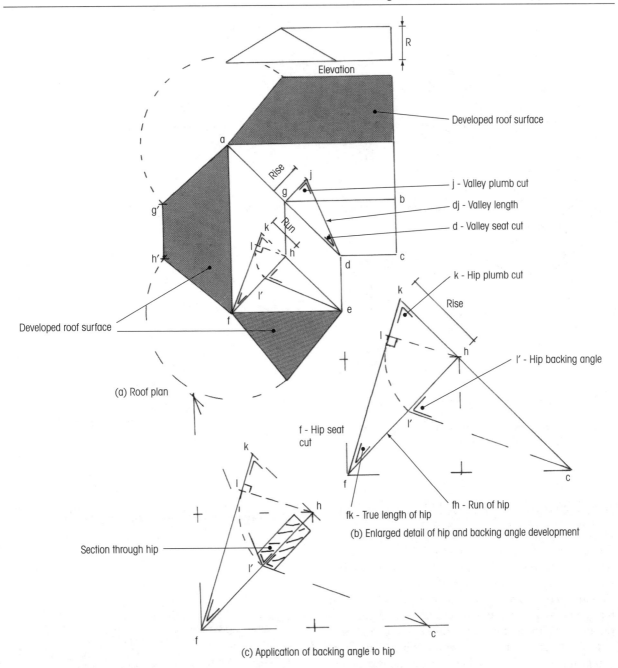

Figure 4.18 Hip and valley length and bevels

4.8.7 Mitre Cut to a Hip Without Backing Bevels

Referring to Figure 4.20(a) draw the plan of the hip *ac* and develop its true length *bc*. From point *c* draw a line at right angles to the plan of the hip *ac* to meet the extended line *ae* (the run of the rafter) at *f*. With the compass point at *c* swing point *b* downwards to meet the extended plan length of the hip *ac* at *b'*. The length *cb'* being the true length of the hip in plan. Draw a line from *f* to *b'* and the mitre bevel is at *b'*. The isometric

sketch (Figure 4.20b) shows the two intersecting roof surfaces and the surface in which the edge of the unbacked hip lies.

4.8.8 Purlin Bevels

The ends of the purlins butt against the side and lip under the bottom edge of the hip or valley rafters. If the hip or valley rafter depth is sufficient, or a bearer board is used to increase the depth of the hip or valley where they meet with the purlins, no lip cut will be required.

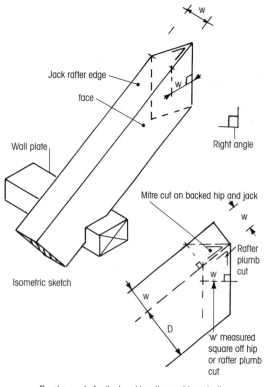

Development of mitre bevel in orthographic projection

Figure 4.19 Practical method of finding mitre bevel on backed hip and jack rafter

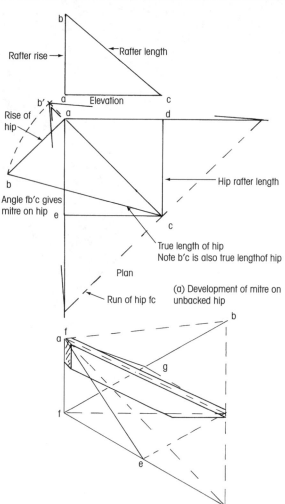

(a) Development of mitre on unbacked hip

(b) Isometric sketch of mitre cut on unbacked hip

4.8.9 Purlin Edge Bevel (Figure 4.20c)

Draw in elevation the pitch line of a rafter *ab* and draw the rectangle *cdef* at right angles to the slope of the rafter. This represents the section through a purlin. In plan draw the run of the hip *gh*. Project point *c* in elevation vertically downwards onto the plan line of the hip *gh* at *c'* and in elevation draw a horizontal line through *c*. With the compass point in *c* rotate distance *ca* to *a'*. Project *a'* vertically downwards into plan. From *g* in plan draw a horizontal line to meet the vertical line from *a* at *g'*. Join with a line *c'* to *g'* and *g'* will be the edge bevel for the purlin.

4.8.10 Purlin Side or Face Bevel (Figure 4.20d)

Draw the elevation of the rafter and section of the purlin, and the plan of the hip as in Figure 4.20(d). Project *c* vertically downwards from elevation into plan at *c'*. Draw a horizontal line through *c* in elevation and with the compass point at *c* rotate distance *ch* to *h'*. From point *h* in plan draw a horizontal line to meet the vertical line from *h'* in *h'*. Draw a line from *c'* to *h'*. The side or face bevel for the purlin will be at *c'*.

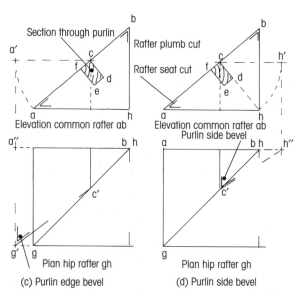

(c) Purlin edge bevel

(d) Purlin side bevel

Figure 4.20　Bevels on unbacked hips and purlins

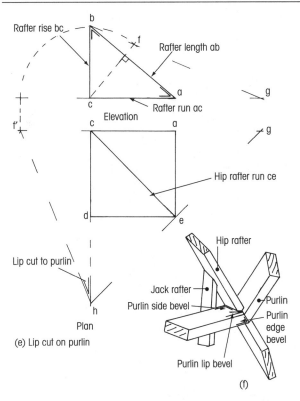

Figure 4.20 **Figure 4.20** Bevels on unbacked hips and purlins

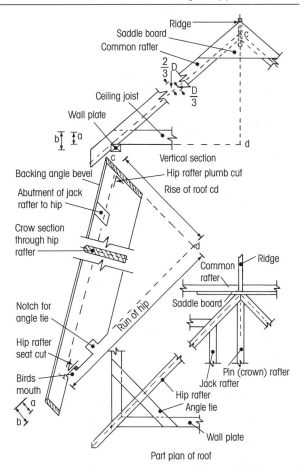

Figure 4.21 Application of bevels for a backed hip

4.8.11 Purlin Lip Cut (Figure 4.20e)

As shown in Figure 4.20(a) the square under edge of the hip lies in a different surface to that of the roof slopes. To develop the lip cut bevel, draw in elevation the rafter *ab* and in plan the run of the hip *ec*. From *e* in plan draw the line *eg* at right angles to the run of the hip *ec* to meet the extension of the line *ac* in *g* and project vertically into elevation point *g*. In plan extend the line *cd* to *h*. In elevation draw an inclined line from *g* to *b*. Draw a line through *c* and at right angles to the rafter *ab* to meet line *gb* at *f*. Placing the compass point in *c* rotate the distance *cf* to the horizontal line at *f'*. Draw a line from *h* to *f'* and the lip cut bevel will be the angle at *h*.

Figure 4.20(f) is a sketch looking up into the roof from below at the junction between hip and purlin which shows the three bevels.

4.8.12 Allowances for the Cross-sectional Size of Roof Members

So far in finding the lengths and bevels on roof members they have been considered as lines only having length. Figure 4.21 shows the part plan, and vertical section of a roof and an auxiliary elevation showing the true shape of the hip rafter, with the single setting out lines shown in relation to the width and thickness of the actual roof members, and how allowances have been made for the thickness of the actual roof members.

4.9 PITCHING A HIPPED ROOF

This will depend on how the roof members are arranged at the apex of the hips (see Figures 4.22a, b. If the arrangement is as shown in Figure 4.22(a), the first pair of rafters are pitched up at a distance along the wallplate of half the span from the corner. The pair of rafters are plumbed and braced and the crown or pin rafter is secured in place at the saddle board and wallplate. A packing piece of timber the thickness of the rafter is nailed to the saddle board below the pin rafter to form an abutment for the hips Figure 4.22(a). The distance from the apex of the roof at the saddle board to the external corner of the wallplate is measured as the hip length may vary if the roof plan is slightly out of square. After any necessary adjustment for length the plumb and mitre cuts are formed on the hips, and they are then offered up and secured in place.

Note. The plumb cut on the hip is not cut until these checking measurements of the hip rafter lengths are made (Figure 4.23).

If the hips are mitred together as in Figures 4.22(b, d), the first pair of rafters are positioned at half the span plus the thickness of the saddle board along from the

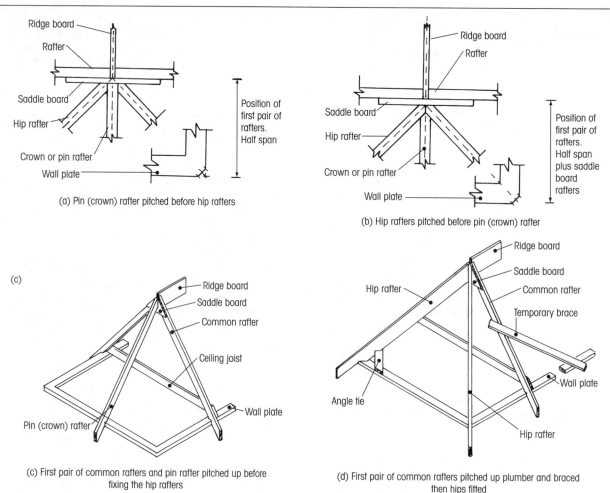

(a) Pin (crown) rafter pitched before hip rafters

(b) Hip rafters pitched before pin (crown) rafter

(c) First pair of common rafters and pin rafter pitched up before
fixing the hip rafters

(d) First pair of common rafters pitched up plumber and braced
then hips fitted

Figure 4.22 Alternative arrangements for pitching a hipped end

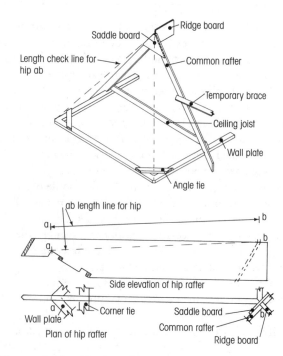

Figure 4.23 Checking hip lengths

corner of the wallplate (Figure 4.23). They are pitched up, secured at the ridge, and at their feet to the wallplate and ceiling joist, plumbed and braced. The length of the hips are checked (Figure 4.23), the plumb cut and mitre on the hips are then cut after any necessary adjustments for length are made, and the pair of hips offered into place seating against the saddle board and each other.

The purlins are then put in place and the remainder of the common and jack rafters secured in position. Ceiling beams and hangers are cut and fixed in as required. Finally fascia boards and soffits are fitted and fixed.

4.10 THE STEEL ROOFING SQUARE

As all roof bevels are based on the right-angled triangle a **steel roofing** or **framing square** is a tool to help determine the length of inclined roof members and their bevels. The roofing square shown in Figure 4.24 consists of two arms set at right angles.

The body or blade of the square is the longest arm, being 620 mm in length, while the tongue or shortest arm

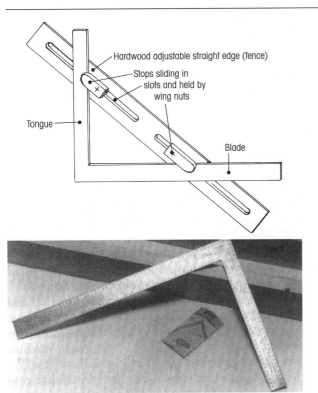

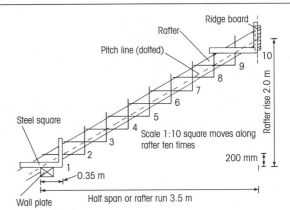

Figure 4.25 Stepping off with the steel square to find the rafter length

Once the rise and run of the common rafter is established, the lengths and bevels on all the other roof members can be found by visualising how the square lies on the roof to give the various cuts and lengths, or by remembering which lengths to take on the blade and tongue of the square.

4.10.2 Common Rafter Lengths and Bevels

Using Figures 4.26(a, b), the run of rafter on blade ab and rise of rafter be on tongue gives a rafter length ae. The plumb cut is read on the tongue at (1) and the seat cut on the blade at (2).

4.10.3 Jack and Cripple Jack Rafters

The plumb and seat cut are as for common rafters (Bevels 1 and 2).

Edge cut. Common rafter length ae on blade. Common rafter run ad^{iv} on tongue. Read edge bevel at (3) on blade (Figures 4.26a, c).

4.10.4 Hip and Valley Rafters

(a) Length, plumb and seat cut (Figures 4.26a, d)

To find the length, plumb and seat cut take the run of the hip $d'b'$ on blade and common rafter rise $b'c'$ in plan on the tongue. The seat cut is on the blade at (4) and the plumb is on the tongue at (5) The hip length is $e'd'$ (Figures 4.26a, d).

Note. The length and bevels on the valley are the same as for those on the hip.

(b) Mitre cut on a backed hip Figures 4.26a, c)

Take the rafter length ae on blade and rafter run ad^{iv} on tongue read the mitre bevel at (3) on the blade.

Note. This mitre bevel on a backed hip is the same as the edge bevel on the jack rafters.

Figure 4.24 Steel roofing square with an adjustable fence

is 450 mm long. The square is calibrated in degrees and millimetres for actual or scaled measurements. Both sides of the square are marked with various tables, the most important being the roof framing tables giving the rafter and hip lengths for the various rises measured in degrees or millimetres per unit run of the common rafter. The steel squares made by various manufacturers vary slightly in their calibration, and it is normal to use the instruction manual provided by the maker of a particular square.

4.10.1 Using the Roofing Square

To use the square we can scale down the run and rise of the common rafter to fit the dimensions marked on the blade and tongue of the square.

As an example we will use a common rafter with a run of 3.5 m and a rise of 2.0 m.

Reduce or scale down the run and rise by ten (10).

$$\text{The run will be } \frac{3.5 \text{ m}}{10} = 0.35 \text{ m or } 350 \text{ mm}$$

$$\text{and the rise } \frac{2.0 \text{ m}}{10} = 0.2 \text{ m or } 200 \text{ mm}.$$

Using the theory of Pythagoras [BENCH AND SITE SKILLS, SECTION 10.10] the length of the rafter will be $\sqrt{(350^2 + 200^2)} = 403.1$ mm. Multiplying by ten (10) will give an actual rafter length of 4031 mm or 4.031 m. Moving the square carefully ten times along the rafter will give the same result (Figure 4.25).

(c) Mitre cut on square-edged hips and valley rafters (Figures 4.26a, e)

Hip rafter length *d'e'* on blade, hip rafter run *d'b'''* on tongue. Mitre bevel (6) on blade.

(d) Backing bevel to the hip (Figures 4.26a, e)

Hip length *d'e'* on blade, hip rafter rise *e'b''* on tongue backing bevel at (7) on tongue.

4.10.5 Purlin Bevels

(a) Purlin edge cut (Figures 4.26a, g)

Common rafter length *ae* on blade, common rafter run *ad^{iv}* on tongue. Bevel for purlin edge cut on tongue at (8).

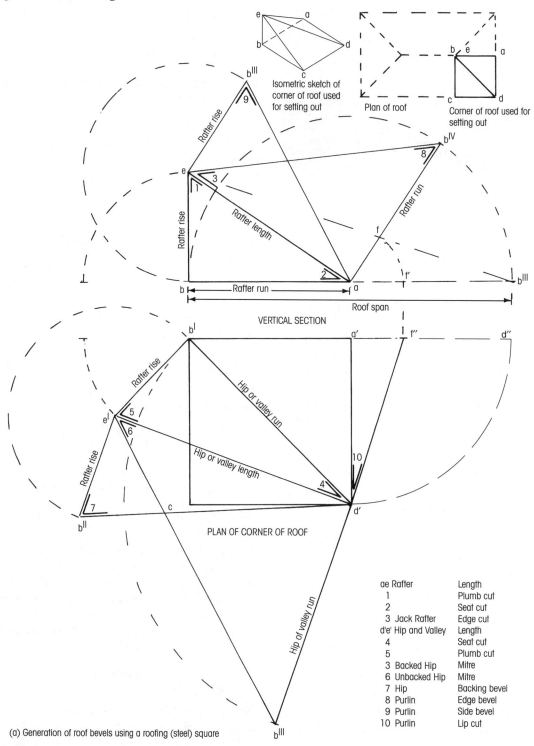

ae Rafter		Length
1		Plumb cut
2		Seat cut
3	Jack Rafter	Edge cut
d'e' Hip and Valley		Length
4		Seat cut
5		Plumb cut
3	Backed Hip	Mitre
6	Unbacked Hip	Mitre
7	Hip	Backing bevel
8	Purlin	Edge bevel
9	Purlin	Side bevel
10	Purlin	Lip cut

(a) Generation of roof bevels using a roofing (steel) square

Figure 4.26 *Continued opposite*

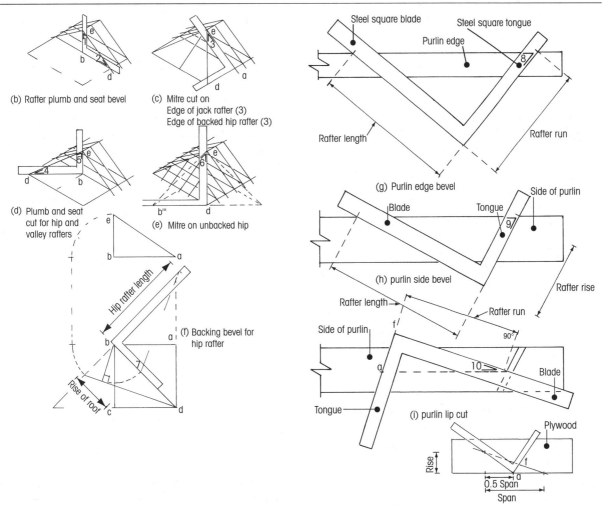

Figure 4.26 Use of the roofing square to obtain lengths and bevels on roof members

(b) Purlin face or side cut (Figures 4.26a, h)

Common rafter length *ae* on blade, common rafter rise *eb'''* on tongue. Bevel for purlin side cut on tongue at (9).

(c) Purlin lip cut (Figures 4.26a, i, j)

Common rafter run *a'd'* in plan on blade. Distance *a'f'* on tongue, lip cut bevel at (10). A practical method is shown in Figure 4.26(j) for finding the distance *af'*.

Note. To explain the principles of the steel roofing square, dimensions have been used exclusively, but some makers of roofing squares use degrees. For greater detail of application, consult the instruction booklets issued by the manufacturer of the particular square you are using.

4.11 TRUSSED ROOFS

Where suitable intermediate supports for purlins were not available, trusses were introduced into the roof structure to carry the purlins at intervals along their

length. A roof containing trusses was called a **tripple roof**, see Figures 4.3(f, g). The joint efficiency of these trusses was very low and consequently large sectioned timbers had to be used to transfer the loads from one member to the next across the joints, which were subject to compressive failure, if the bearing areas were insufficiently large (Figure 4.27).

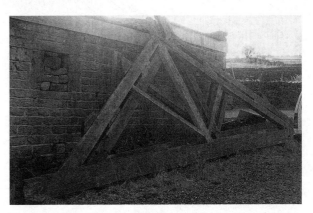

Figure 4.27 Nineteenth-century king post roof truss

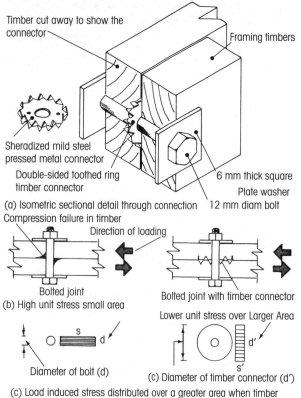

Timber cut away to show the connector

Framing timbers

Sheradized mild steel pressed metal connector

Double-sided toothed ring timber connector

6 mm thick square Plate washer
12 mm diam bolt

(a) Isometric sectional detail through connection

Compression failure in timber

Direction of loading

Bolted joint

(b) High unit stress small area

Bolted joint with timber connector

Lower unit stress over Larger Area

Diameter of bolt (d)

(c) Diameter of timber connector (d′)

(c) Load induced stress distributed over a greater area when timber connectors are used s × d = s′ × d′ = load

Figure 4.28 Increased efficiency of bolted joints by using timber connectors

Joints used were stub mortice and tenons and inclined bridle joints [BENCH AND SITE SKILLS, SECTION 7.5.1], and those under tension were held together with wrought iron straps. These trusses were spaced along the length of the roof at from 3 to 4 m intervals to provide intermediate support for the purlins. In some older buildings this type of truss can still be seen (Figure 4.27). During the twentieth century ways were found to increase the joint efficiency in large carpentry structures by the development and use of metal timber connectors (Figure 4.28a). The use of timber connectors had the effect of spreading the stresses over a wider area and thus reducing their intensity (Figures 4.28b, c). Because of increased joint efficiency, smaller sectioned timbers could be used, and lightweight trusses were developed. Figure 4.29 shows a TRADA domestic roof truss fabricated from 100 × 50 mm softwood. As the principal rafter of the truss was in line with the common rafters in the roof, they displaced every fourth pair of rafters.

4.12 TRUSSED RAFTER ROOFS

A further development in the late 1960s was the trussed rafter. Figure 4.30 shows that each pair of common rafters is replaced by a light weight truss manufactured from strength graded [BENCH AND SITE SKILLS, SECTION 1.8.4] wrot softwood with sections varying between 35 × 72 mm to 47 × 145 mm depending on the span and loads to be carried. All the members lie in a single plane and abut to each other, being jointed together by the use of:

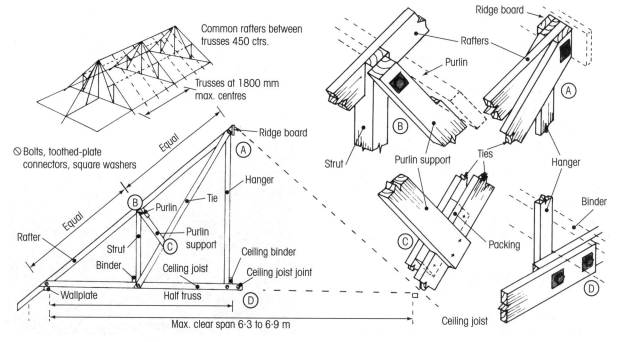

Common rafters between trusses 450 ctrs.

Trusses at 1800 mm max. centres

⊘ Bolts, toothed-plate connectors, square washers

Equal

Ridge board

Hanger

Tie

Purlin

Purlin support

Rafter

Strut

Binder

Ceiling joist

Wallplate

Half truss

Max. clear span 6·3 to 6·9 m

Ceiling binder

Ceiling joist joint

Ridge board

Rafters

Purlin

Strut

Purlin support

Ties

Hanger

Binder

Packing

Ceiling joist

Figure 4.29 TRADA domestic roof truss

(a) Nailed plywood gusset plates (Figure 4.30a).
(b) Nailed plates of galvanised or sheradised 18 or 20 gauge mild steel (Figure 4.30b).
(c) Punched plates of galvanised or sheradised 18 or 20 gauge mild steel (Figure 4.30c).

In corrosive conditions stainless steel plates can be used, but they are much more expensive. The plates are fixed to both faces of the joints in the trusses which form a triangulated frame of strutts (members in compression) and ties (members in Tension). On site it is only necessary to fix the factory fabricated trusses to the wall plates at from 450 to 600 mm centres, or as specified. Purlins and ridge board are not required, but the trusses do have to be tied and braced together.

4.12.1 Trussed Rafter Construction

The trussed rafters are manufactured in factories under conditions of strict quality control. A structural engineer is usually responsible for the design of the trusses. Wrot, strength graded timber members are automatically cut to the correct lengths and bevels, and assembled in large jigs. Connecting plates are then positioned, and in the case of punched metal plate fasteners the trusses are passed through a press to force the points of the plate fasteners into the timber. To meet design requirements

Figure 4.31 Trussed rafter manufacture pressing punched plates into truss members

the projections on the plates must be pressed home so that the total area of the plate surface is in close contact with the timber (Figure 4.31). Trusses are then transported from factory to building site by lorry on which they are stacked flat or vertically depending on their size and shape.

4.12.2 Site Storage of Trussed Rafters

Punched plate timber fasteners are very efficient in transferring axial compression and tensile forces across the joints in the trusses, but there is little resistance to lateral forces that cause the points to pull out of the timber (Figure 4.32). It is important on site that

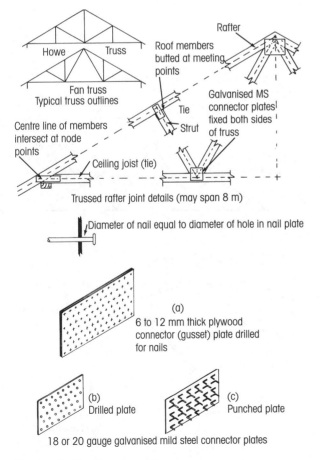

Figure 4.30 Trussed rafters

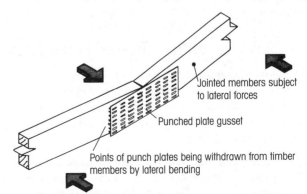

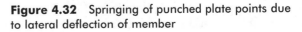

Figure 4.32 Springing of punched plate points due to lateral deflection of member

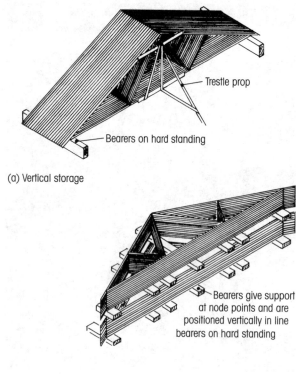

(a) Vertical storage

(b) Horizontal storage

Figure 4.33 On-site storage of trussed rafters

particular care is taken that the trusses are handled and stored in such away that this does not occur. The trusses should be supported vertically clear of the ground at the node points (points of attachment) between the rafter and the ceiling joist. To keep them in an upright position they should be supported by inclined, or trestle props (Figure 4.33a). As an alternative the trusses can be stacked flat, but they must be supported on a flat base clear of the ground with bearers positioned at the node points and at intermediate positions between nodes (Figure 4.33b). If stacked in the open they should be protected from the rain, but without restricting ventilation. It is good practice to keep the time the trusses have to be stored on site to an absolute minimum.

4.12.3 Erection of a Trussed Rafter Roof

The following is a general guide, but if fixing instructions are issued by the truss manufacturer they should be followed. Site bracing of the trusses should conform to BS5268:1985:Pt 3 for the bracing of domestic trussed rafters.

1. Bed the wall plates and mark out the positions of the trusses at 450 or 600 mm centres and fix truss clips to the wall plates. Figure 4.34(a) shows galvanised pressed metal truss fixing clips drilled for 3.35 mm diameter (10 gauge) round head plated nails.

2. Mechanically hoist (Figure 4.34c) or by correct physical handling (Figure 4.34b), lift in position a truss approximately half the span along the wall from the gable end and nail to the fixing clip. Plumb the first truss and temporally brace from the wall plate (Figure 4.34d).

3. Place in position and fix the next truss, working towards the gable wall (Figure 4.34e). Secure to the first truss with temporary ties to ensure that it is plumb and parallel to the first truss, and that the ceiling joist (tie) is not bowed in a lateral direction (Figure 4.34e).

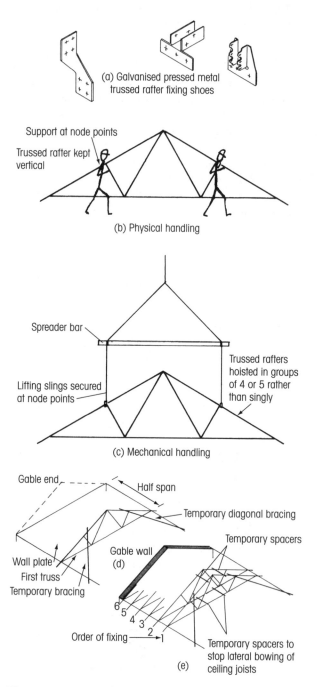

(a) Galvanised pressed metal trussed rafter fixing shoes

Support at node points

Trussed rafter kept vertical

(b) Physical handling

Spreader bar

Lifting slings secured at node points

Trussed rafters hoisted in groups of 4 or 5 rather than singly

(c) Mechanical handling

Gable end

Half span

Temporary diagonal bracing

Temporary spacers

Wall plate

Gable wall

(d)

First truss

Temporary bracing

Order of fixing

Temporary spacers to stop lateral bowing of ceiling joists

(e)

Figure 4.34 Handling and position of trussed rafters

4. Fix the remaining trusses along the roof as described in (3).
5. Fix permanent bracing to the underside of the rafters from wall plate to roof apex using 25 × 100 mm soft wood running diagonally at an angle of not less than 45°, nailed twice to each rafter with 65 × 3.5 mm diameter galvanised round head nails. If the diagonal bracing has to be in two or more pieces each brace should overlap sufficiently to be attached to two rafters (Figure 4.35).

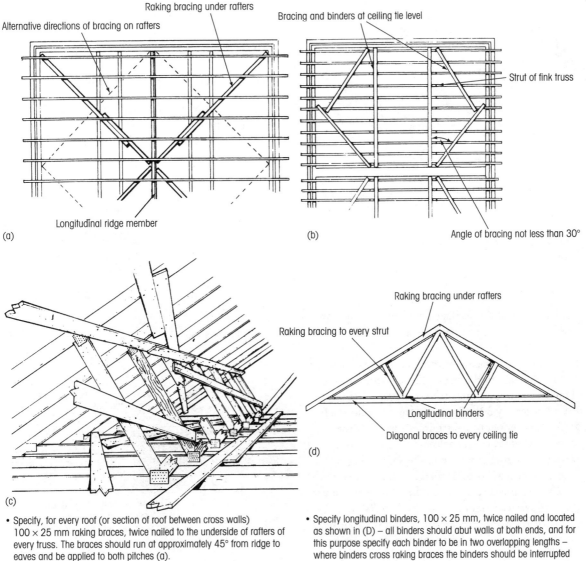

• Specify, for every roof (or section of roof between cross walls) 100 × 25 mm raking braces, twice nailed to the underside of rafters of every truss. The braces should run at approximately 45° from ridge to eaves and be applied to both pitches (a).

• Specify, when the distance between centres of cross walls is not more than 1·2 × trussed rafter span, at least two 100 × 25 mm diagonal braces, twice nailed to every ceiling tie in every roof (or section of roof betweeen cross walls) as shown in (b).

• Specify (unless trusses are less than 5 m span), for every roof (or section of roof between cross walls) 100 × 25 mm raking bracing twice nailed to every internal strut (c).

• Specify longitudinal binders, 100 × 25 mm, twice nailed and located as shown in (D) – all binders should abut walls at both ends, and for this purpose specify each binder to be in two overlapping lengths – where binders cross raking braces the binders should be interrupted and plated, as in (C).

• Specify that all lap joints in braces and binders are to be lapped, and nailed over at least two rafters.

• Specify all nailing to be 3.35 × 75 mm galvanised round wire nails.

• Specify that no bracing or binders shall penetrate a separating wall.

Note: roof pitches up to 30°, spans up to 11 m.

Figure 4.35   Bracing to domestic trussed rafters

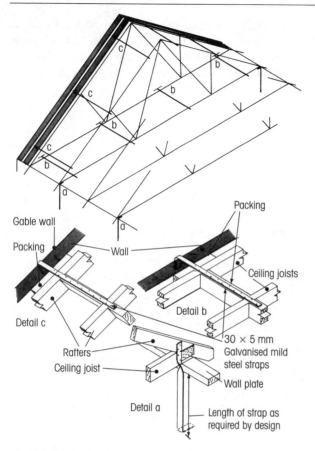

Figure 4.36 Anchoring a roof to the masonry structure

6. Fix longitudinal ties at the side of the intermediate node points on the ceiling joists and rafters (Figures 4.35b, c).
7. Secure diagonal bracing to the ceiling joists (ties) and chevron bracing to the struts. The chevron bracing should span at least three trusses (Figure 4.35).
8. Fix all necessary 30 × 5 mm galvanised mild steel vertical and horizontal restraining straps, at the specified centres, to tie the masonry gables to the roof, and the roof structure to the supporting walls. This is important as on low and mono pitched roofs negative air pressure occurs as the wind sweeps over the roof surface (Figure 4.36) (see also *Building Regulations Approved document "A"* and the roof truss manufacturer's fixing and anchoring specifications).

4.12.4 Projections Through Roof Surfaces

Openings may be required in the sloping surface of a roof for projections such as chimneys, skylights and dormer windows. If the projection passes in the space between the trusses it is only necessary to provide trimmers on the high and low side of the projection (Figure 4.37a). The space

between trusses rafters can be increased by 10% without over stressing the trusses. See the plan in Figure 4.37(a).

Note. If a chimney passes through the roof there must be a minimum gap of 40 mm between the chimney and the timber trimming to the opening [*Building Regulations Approved Document "J"*].

Openings up to 1200 mm wide on centres can be formed, if they are set out as shown in Figure 4.37(b) and provision is made to trim the ends of the truss members.

If a pair of trusses each side of an opening are nailed together with nails spaced at 600 mm along all abutting members, the opening can be increased to three times the normal spacing provided provision is again made for trimming the ends of the trussed members (Figures 4.37c, d).

4.12.5 Attic Trussed Rafters

Dormer windows in a roof indicate that a room is being formed in the roof space (attic) and in this case attic trussed rafters would be used to give the necessary space

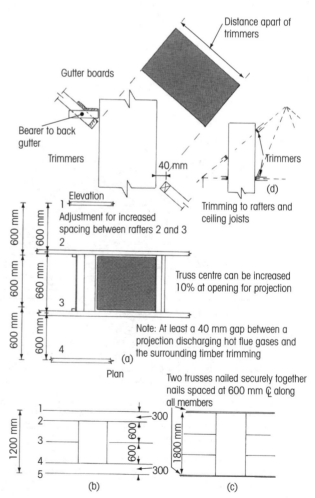

Figure 4.37 Trimming to roof openings

for a room in the roof void. Attic trussed rafters are of two types:

1. Trusses made up of large sectioned members where the tie forms the floor joist. These trusses are heavy and expensive and are only used when there is no alternative.

2. An upper suspended timber floor is first formed (Section 2.1), then lightweight attic trusses are fixed to the joists (Figure 4.38a).

Figure 4.38(b) shows how the framing for a dormer window is formed in attic trussed rafters.

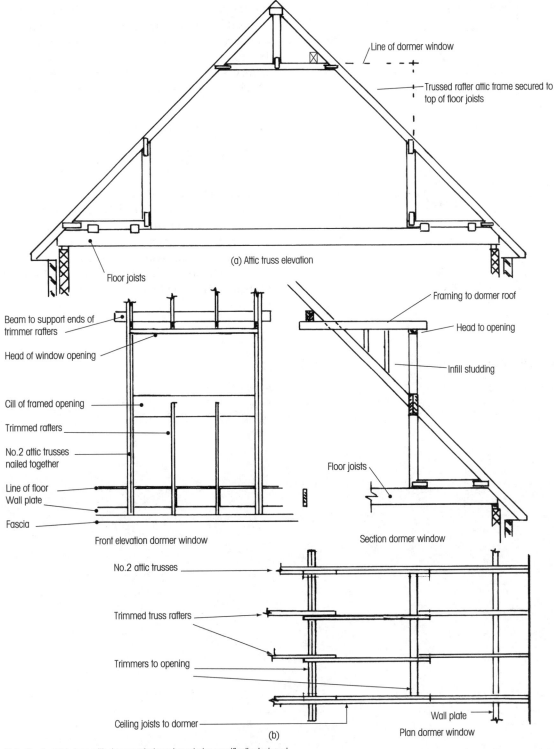

(a) Attic truss elevation

Line of dormer window

Trussed rafter attic frame secured to top of floor joists

Floor joists

Beam to support ends of trimmer rafters

Head of window opening

Cill of framed opening

Trimmed rafters

No.2 attic trusses nailed together

Line of floor

Wall plate

Fascia

Front elevation dormer window

Framing to dormer roof

Head to opening

Infill studding

Floor joists

Section dormer window

No.2 attic trusses

Trimmed truss rafters

Trimmers to opening

Ceiling joists to dormer

(b)

Wall plate

Plan dormer window

Note: Roofs of this type with dormer windows have to be specifically designed.

Figure 4.38 Forming a dormer window in attic trusses

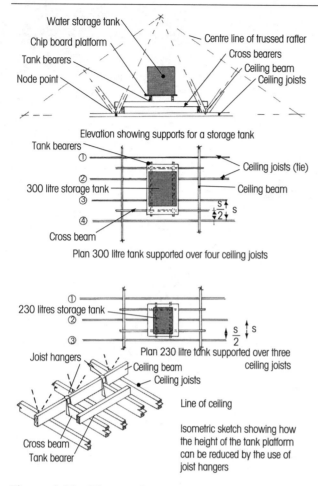

Elevation showing supports for a storage tank

Plan 300 litre tank supported over four ceiling joists

Plan 230 litre tank supported over three ceiling joists

Isometric sketch showing how the height of the tank platform can be reduced by the use of joist hangers

Figure 4.39 Water tank supports

4.12.6 Support for Water Tanks in a Roof Space

Water tanks of up to 230 litres (approximately 50 gallons) can be supported over three trusses. Tanks of 300 litres capacity (70 gallons) are supported over four trusses. It is normal practice for the tanks to be positioned in the centre of the roof between the inclined hangers, as this gives space for the plumber to work over the top of the tank. If the pitch of the roof is low, special attention must be given to arranging the framework that holds the tank so that sufficient working space is retained above the tank (Figure 4.39).

4.13 VENTILATION OF INSULATED ROOFS

All roofs must be thermally insulated and any cold spaces in the roof above the insulation must be ventilated if condensation is not to occur. Ventilation and heat loss insulation of roofs must comply with the requirements of the [*Building Regulations Approved documents "F" and "L"*]. Figures 4.13 and 4.14 show how ventilation and heat loss insulation is arranged in a roof space.

5
Temporary work

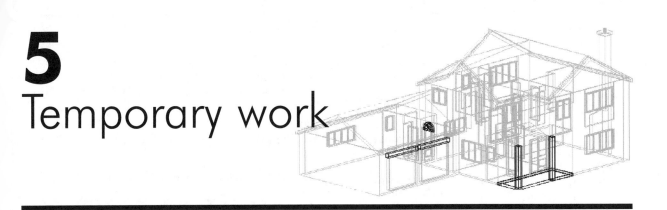

5.1 INTRODUCTION

As the name suggests this is carpentry work of a temporary nature used to support other structural materials under the following conditions:

- Moulds to support and profile concrete while it sets and gains strength – formwork (Figure 5.1a).
- Masonry gravity structures in the process of being built – centring (Figure.5.1b).
- Masonry gravity structures which are failing or will probably fail during modification work – shoring (Figure 5.1c).
- A framework of timber used to support the earth sides of trenches and embankments during excavations – timbering to trenches (Figure 5.1d).

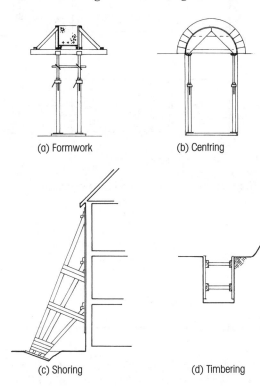

(a) Formwork   (b) Centring

(c) Shoring   (d) Timbering

Figure 5.1   Temporary work

5.2 FORMWORK

A series of mould-box forms which hold the liquid concrete to the profile required. These are held in place by a structural framework of struts, ties and braces, this gives horizontal and vertical support until the concrete has set, hardened and gained structural strength.

5.2.1 Assembly of Formwork

The formwork should be designed and constructed so that:

- The forms can be eased (pressure released between formwork and concrete faces) and struck (removed completely from the concrete) without the formwork or concrete being damaged. During assembly consideration must be given as to how the formwork is to be dismantled in a progressive and safe way [*BS5975:1996 CoP for falsework*]. It is easy to discover when dismantling the formwork that it has become trapped after the concrete has set (Figure 5.2).
- The forms and supporting structure must resist the live loads imposed by the workers and their equipment including spreading, tamping and vibrating the

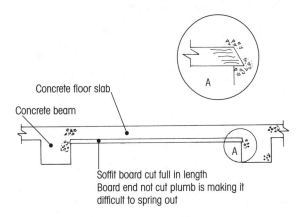

Concrete floor slab

Concrete beam

Soffit board cut full in length
Board end not cut plumb is making it
difficult to spring out

Figure 5.2   Trapped soffit forms

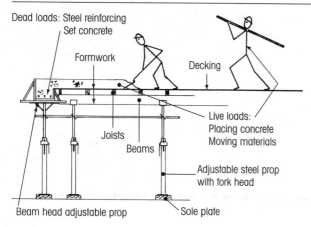

Figure 5.3 Dead and live loads

concrete into position, and the dead loads induced by formwork, steel reinforcement and the concrete (Figure 5.3).

• The formwork should have:
  1. Strength – it will not collapse when loaded.
  2. Stability – it will not rack or twist (Figure 5.4).
  3. Stiffness – it will not deflect under partial loading (Figure 5.5).

5.2.2 Safety

Recognition and implementation of the *Construction (Design and Management) Regulations 1995* ensure that a safe plan for the work is produced. The means that risks are not taken and the work is carried out in the safest way possible, following the *Health and Safety at Work Act* and all its supporting codes and regulations.

5.2.3 Characteristics of Concrete

Concrete is a mixture of graded aggregates (sharp sand and gravel) and hydraulic cement. When the mixture is combined with water it develops flow patterns very similar to that of a liquid, until a chemical reaction in the cement initiated by the presence of water causes it to solidify (set) and harden. This binds the aggregates together as a solid with high compressive (crushing), but low tensile (stretching) strength (Figure 5.6).

When in a plastic state, concrete like a liquid exerts a uniform pressure in all directions from any given point (Figure 5.7) and this pressure increases proportionally with depth (Figure 5.8). These pressure characteristics influence how the formwork is constructed and supported to give stability and stiffness (Figure 5.8b), the density of mass concrete is 2400 kg/m^3 and up to 2500 kg/m^3 when reinforced with steel bars.

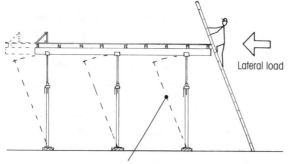

Figure 5.4 Possible effects of partial loading

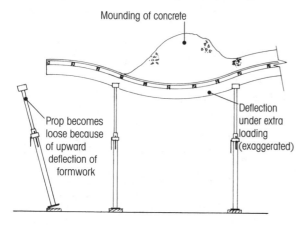

Figure 5.5 Deflection under partial loading

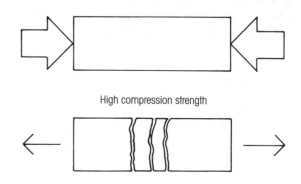

Figure 5.6 Strength characteristics of concrete

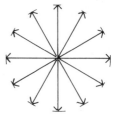

Concrete before setting exerts a uniform pressure in all directions from a given point

Figure 5.7 Pressure characteristics of unset concrete on the spacing of formwork restraints on wall and floor columns (*Continued opposite*)

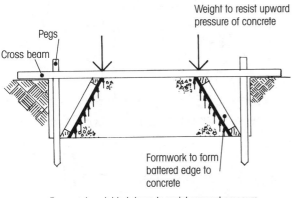

Pegs

Cross beam

Weight to resist upward pressure of concrete

Formwork to form battered edge to concrete

Formwork weighted down to resist upward pressure of concrete

Figure 5.7 Pressure characteristics of unset concrete on the spacing of formwork restraints on wall and floor columns

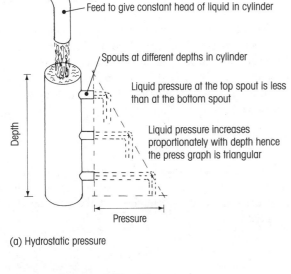

Feed to give constant head of liquid in cylinder

Spouts at different depths in cylinder

Liquid pressure at the top spout is less than at the bottom spout

Liquid pressure increases proportionately with depth hence the press graph is triangular

Depth

Pressure

(a) Hydrostatic pressure

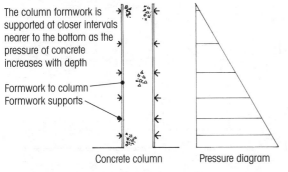

The column formwork is supported at closer intervals nearer to the bottom as the pressure of concrete increases with depth

Formwork to column

Formwork supports

Concrete column   Pressure diagram

(b) Influence on the way concrete moulds are supported

Figure 5.8 Pressure characteristics of unset concrete

Concrete can be:

(a) Pre-cast. That is cast in moulds in a casting yard or area set aside for this work and then when set, transported to its final position in a structure (Figure 5.9a).

(b) Cast in situ. The concrete is cast directly into its final position in the structure (Figure 5.9b).

Concrete passes through three stages in the curing process before it can be used as a structural material.

(a) The initial set of the cement takes from 20 to 60 min, and the concrete should be placed and consolidated in this time (Figure 5.10).

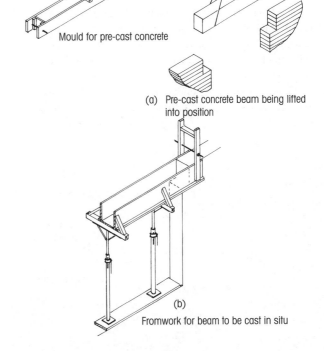

Mould for pre-cast concrete

(a) Pre-cast concrete beam being lifted into position

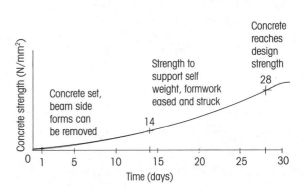

(b)

Fromwork for beam to be cast in situ

Figure 5.9 Pre-cast and cast in situ concrete

Concrete reaches design strength

Strength to support self weight, formwork eased and struck

Concrete set, beam side forms can be removed

28

14

Concrete strength (N/mm²)

Time (days)

Figure 5.10 Strength gain in concrete over time

(b) Setting of the concrete takes about 24 hours, after which time the side forms from beams can be removed.

(c) Development of maximum strength, this will occur at about 28 days after casting, but within 14 days the concrete will have gained sufficient strength to support its own weight and soffit forms can be eased then struck, provided some propping of soffits remain.

Note. The structural engineer will make the final decision when to ease and strike the formwork, as this will depend on the atmospheric conditions of temperature and humidity at the time of casting and curing the concrete.

5.2.4 Mild Steel Bar Reinforcement to Concrete

Concrete can resist compression, but not tensile forces (see Figure 5.11) and as structural elements have to resist both types of force it is necessary to reinforce with mild steel rods, any concrete subject to tensile forces (Figure 5.11).

The designer of the concrete structure will determine the size, number and the distribution of mild steel bars within the concrete, but the principle is that the

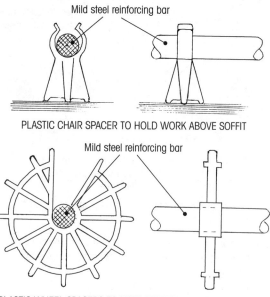

PLASTIC CHAIR SPACER TO HOLD WORK ABOVE SOFFIT

PLASTIC WHEEL SPACERS TO HOLD VERTICAL BARS AWAY FROM VERTICAL FORMWORK FACES

Figure 5.12 Plastic spacers for mild steel reinforcing bars

reinforcing steel is always where the concrete is subject to tensile forces (Figure 5.13).

Care must be taken when working on formwork after the steel bars have been placed not to displace them and to ensure they are surrounded by at least 25 mm of concrete. To make sure this happens plastic spacers (Figure 5.12) are clipped round the steel bars to give the necessary clearance (Figures 5.14a, b).

5.2.5 Release Agents

Their function is to stop the concrete from adhering to the formwork faces by forming a barrier between the concrete and the face of the forms. They are manufactured from:

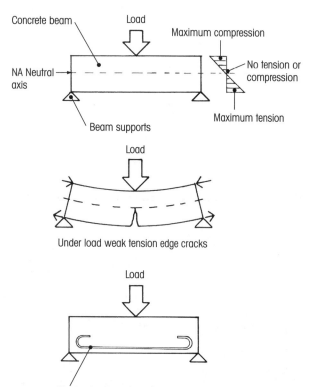

Figure 5.11 Reinforcement of concrete

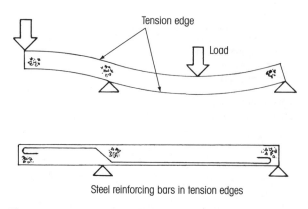

Figure 5.13 Steel reinforcing bars in tension edges of concrete

(a) Modified mineral oils. A wetting agent is used in the formulation to reduce the surface tension of the oil, this discourages the formation of oil beads on the formwork surface.

(b) Emulsions. A fine dispersion of oil in water. A good general-purpose release agent as the water evaporates leaving a coating of oil on the face of the forms.

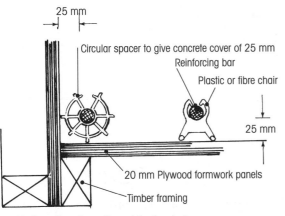

(a) Plastic spacers to position reinforcing steel

(b) Spalling of concrete face caused by the corrosion of mild steel reinforcing bars due to insufficient concrete cover

Figure 5.14 Need for sufficient concrete cover over reinforcing bars

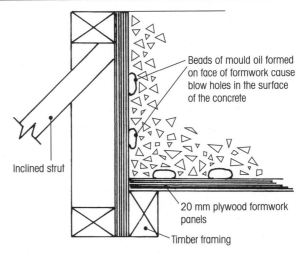

Figure 5.15 Blow holes on the face of concrete caused by the beading of mould oil

(c) Wax emulsions. A fine dispersion of wax in a solvent that coats the surface with wax as the solvent evaporates.

(d) Chemical release agents. A chemical suspension in a low viscous oil forms a soap film between concrete and formwork. If excessively applied it may increase the setting time of the concrete surface.

Note. Neat mineral oils are discouraged as they contribute to the formation of blow holes in the finished surface of the concrete (Figure 5.15).

5.2.6 Application of Release Agents

They can be applied by brush or spray to the formwork surfaces in contact with the concrete. Sufficient mould oil should be applied to avoid adhesion between the two surfaces, but application of excessive amounts of mould oil is wasteful and may cause blow holes, or retard the setting of the concrete surfaces. Steel reinforcement should not be contaminated with release agents (mould oil). The mould oil should be applied as close in time as possible to the placing of the concrete, to make sure it is not washed off by rain or contaminated by dirt.

5.3 TIMBER AND TIMBER-BASED MATERIALS USED FOR FORMWORK

These materials can be:

(a) Solid timber.

(b) Water-proof plywood with or without a resin-coated face.

(c) Oil-tempered hardboard

5.3.1 Timber Boarding

At one time softwood timber boards 25–38 mm thick and 150–200 mm wide were used almost exclusively, because of their cheapness and availability, their good strength to weight ratio and ease of working. Unfortunately timber is not the inexpensive material it once was and is also a hygroscopic material [BENCH AND SITE SKILLS, SECTION 1.7.4] changing dimension in cross-section depending upon the amount of moisture present (Figure 5.16). The boards would be held together by nailing them to cleats or battens (Figure 5.17).

Spacing and positioning of the cleats would depend on the arrangement of the forms and their supporting structure. Boards of different surface textures and thicknesses can be used to give a textured finish to the surface of the concrete, and is known as board finish.

5.3.2 Plywood

The plywood must be weather proof and usually 19 mm thick, although other thicknesses can be used down to 6 mm, especially if curved formwork is being constructed. The surface of the plywood in contact with the concrete should be solid [BENCH AND SITE SKILLS, TABLES 2.1–2.4].

Plywood can be worked with tongues and grooves on their abutting edges to be used as decking for soffits (Figure 5.18a) as square edged panels secured to a timber frames (Figure 5.18b), or as infill panels to steel or aluminium frames (Figure 5.18c). Face veneers of the plywood in contact with the concrete should be at least 2.5 mm thick unless resin or fibre-glass coated.

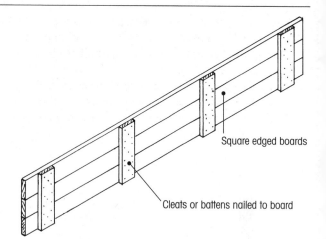

Figure 5.17   Formwork panel built up of boards

5.3.3 Fibre Glass – GRP

Normally used for concrete forms in pre-cast concrete requiring intricate mouldings such as concrete balustrade columns, or to form the indents in waffle floors (Figure 5.19). Making formwork moulds of this type is expensive and you would expect to produce a large number of castings from one mould to justify the expenditure.

5.3.4 Mild Steel

A good long-lasting material provided the surface of the panels are not indented with a sharp blow which stretches the metal making the indent difficult to hammer out. The sheet steel to [BS EN 10025] is welded to mild steel angle frames to form a panel, or pressed and bent to form a one-piece panel (Figure 5.20).

Table 5.1 gives a range of sizes in which formwork panels are made.

5.4 SUPPORT AND BRACING FOR THE FORMWORK

Timber was the traditional material used for the supporting structure and fine adjustments were made by the use of folding wedges (Figure 5.21).

Although timber is still used it is being replaced by adjustable metal props, beam and column clamps, and adjustable metal floor centres (Figure 5.22).

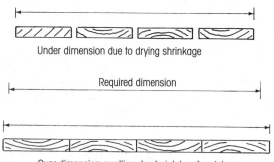

Under dimension due to drying shrinkage

Required dimension

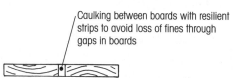

Over dimension swelling due to intake of moisture

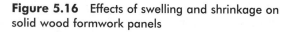

Caulking between boards with resilient strips to avoid loss of fines through gaps in boards

Figure 5.16   Effects of swelling and shrinkage on solid wood formwork panels

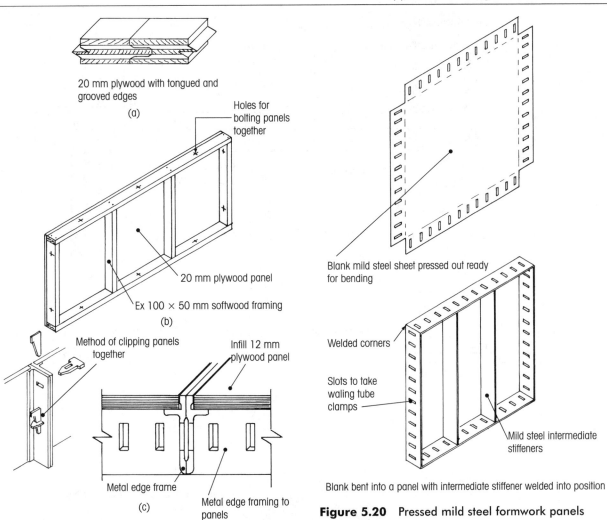

20 mm plywood with tongued and grooved edges

(a)

Holes for bolting panels together

20 mm plywood panel

Ex 100 × 50 mm softwood framing

(b)

Method of clipping panels together

Infill 12 mm plywood panel

Metal edge frame

Metal edge framing to panels

(c)

Figure 5.18 Plywood formwork panels

Blank mild steel sheet pressed out ready for bending

Welded corners

Slots to take waling tube clamps

Mild steel intermediate stiffeners

Blank bent into a panel with intermediate stiffener welded into position

Figure 5.20 Pressed mild steel formwork panels

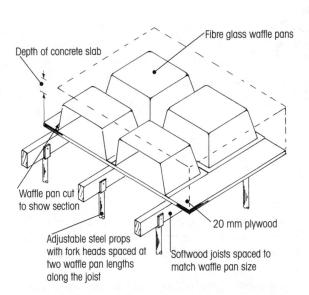

Depth of concrete slab

Fibre glass waffle pans

Waffle pan cut to show section

20 mm plywood

Adjustable steel props with fork heads spaced at two waffle pan lengths along the joist

Softwood joists spaced to match waffle pan size

Figure 5.19 Fibre glass GRP moulds used to form a waffle floor slab in reinforced concrete

Table 5.1 Standard sizes for metal formwork panels

Standard	Fractional widths	Height (mm)	Width (mm)	Width (mm)
2400	600	500	400	300
2100	600			
1800	600	500	400	300
1500	600	500	400	300
1200	600	500	400	300
900	600	500	400	300
600	600			

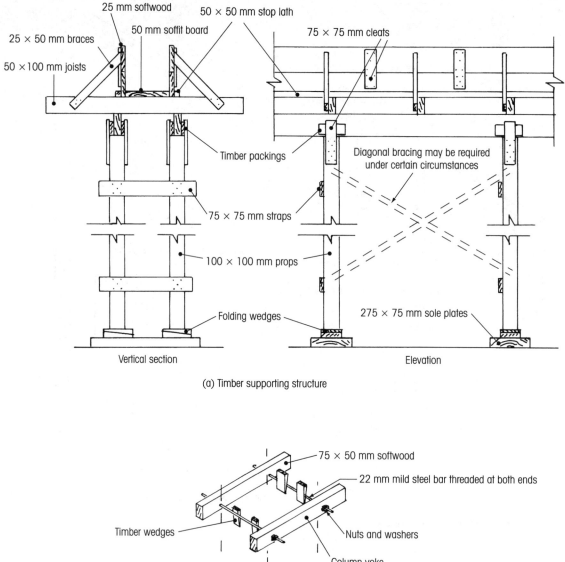

25 mm softwood
50 × 50 mm stop lath
50 mm soffit board
25 × 50 mm braces
75 × 75 mm cleats
50 ×100 mm joists
Timber packings
Diagonal bracing may be required under certain circumstances
75 × 75 mm straps
100 × 100 mm props
Folding wedges
275 × 75 mm sole plates

Vertical section
Elevation

(a) Timber supporting structure

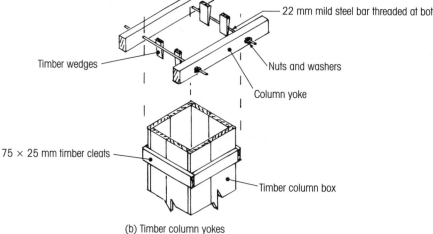

75 × 50 mm softwood
22 mm mild steel bar threaded at both ends
Timber wedges
Nuts and washers
Column yoke
75 × 25 mm timber cleats
Timber column box

(b) Timber column yokes

Figure 5.21 Timber supports for formwork

5.4.1 Formwork Spacers

These are used to keep the faces of formwork in beams a uniform distance apart. The simplest system was to use timber blocking pieces trapped between the two faces of the formwork, held together by passing a wire loop through the forms and tensioning the loop by twisting it with a large nail. The timber spacers would be removed as the concrete was poured and the twisted wire loops broken off as the beam side forms were removed (Figure 5.23).

A number of proprietary spacer and tie systems for use with formwork are available. Figure 5.24 illustrates a range of these.

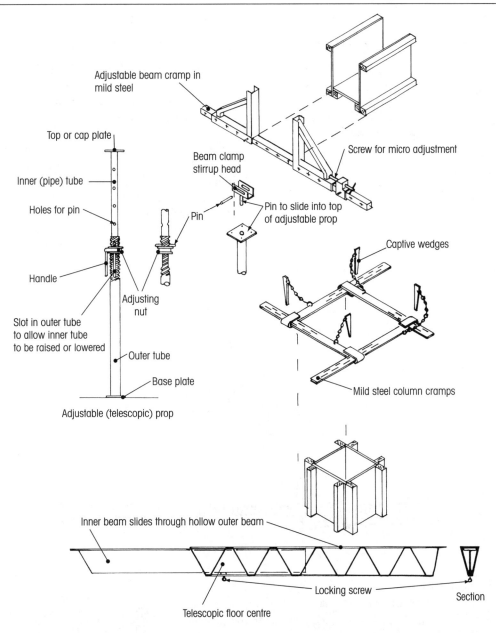

Figure 5.22 Metal supports for formwork

5.4.2 Plastic Spacers for Reinforcing Mild Steel Bars

These are used to keep the mild steel reinforcing bars at the correct distance from the edge of the concrete. A range of types are shown in Figure 5.12.

5.5 TYPES OF CONCRETE MEMBERS

Formwork used for pre-cast and cast in situ concrete. In both cases the following major components can be cast, using either method

(a) Beams. Those supported by the ground (ground beams) and those supported on columns.
(b) Columns.
(c) Walls.
(d) Floors.
(e) Staircases.

5.6 FORMS OR MOULD BOXES FOR PRE-CAST CONCRETE

The quality of pre-cast components will depend upon:

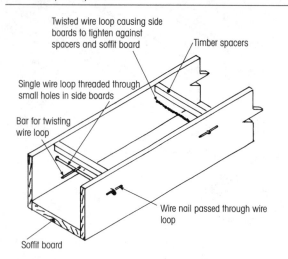

Figure 5.23 Use of simple twisted wire loops and timber spacers as ties

(a) The quality of the concrete and the way it is placed and vibrated in the moulds.

(b) The stiffness of the mould box, its dimensional accuracy and the finish of the surface in contact with the concrete. Open joints or poor detailing of the mould box will lead to poor surface finishes on the finished product.

Pre-cast concrete usually consists of a large number of identical items of consistent quality, which can be cast in multiple moulds (batteries). Think of the number of pre-cast concrete tunnel rings (segments) required for the Channel Tunnel.

When large numbers of pre-cast units are required special steel moulds would be designed and

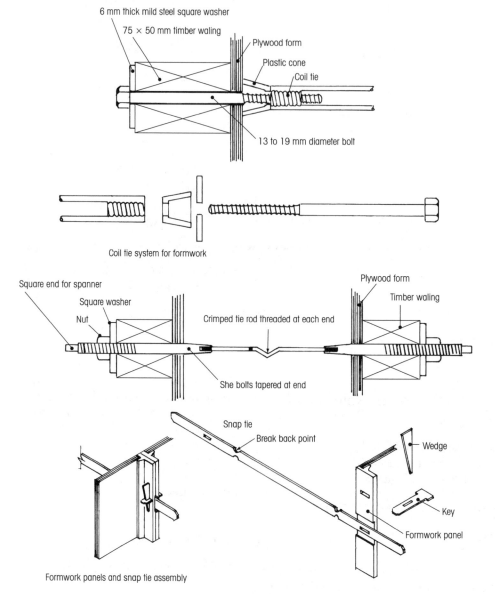

Figure 5.24 Metal ties for formwork

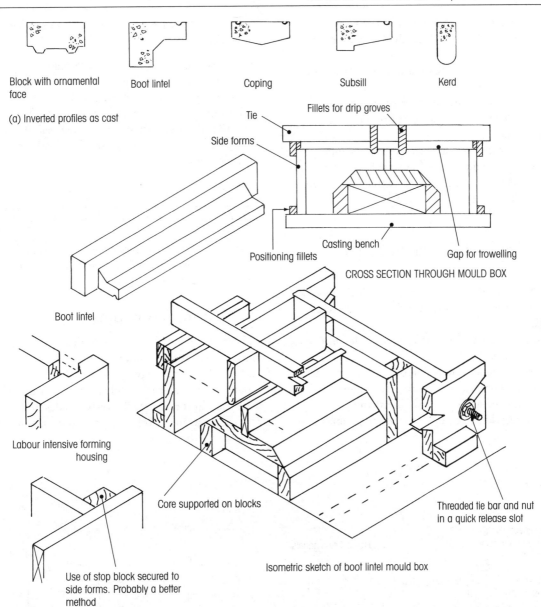

Block with ornamental face

Boot lintel

Coping

Subsill

Kerd

(a) Inverted profiles as cast

Tie

Fillets for drip groves

Side forms

Positioning fillets

Casting bench

Gap for trowelling

CROSS SECTION THROUGH MOULD BOX

Boot lintel

Labour intensive forming housing

Use of stop block secured to side forms. Probably a better method

Core supported on blocks

Threaded tie bar and nut in a quick release slot

Isometric sketch of boot lintel mould box

(b) Construction details of a mould box for casting a pair or boot lintels

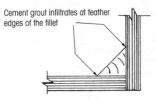

Cement grout infiltrates at feather edges of the fillet

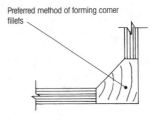

Preferred method of forming corner fillets

(c) Forming corner fillets in formwork

Figure 5.25 Mould box detailing

manufactured to withstand repeated assembly, filling, vibrating and striking without any deterioration in mould quality.

Many pre-cast concrete units are cast upside down for convenience (Figure 5.25a).

Figure 5.25(b) shows the construction of a mould box for the casting of a pair of boot lintels.

Note. If corner fillets are used to form chamfers, care must be taken to make sure (cement slurry) grout does not get behind the feather edges of triangular fillets (Figure 5.25c).

5.6.1 Pre-cast Concrete Stairs

These would normally be cast on edge, so that the mould forms could be removed much earlier and take up less room in the casting area than if they were cast flat.

Figure 5.26 shows the arrangement of formwork for a pre-cast staircase.

To provide good compaction of the concrete in the moulds, vibrating pokers are used (Figure 5.27a), or in the case of small pre-cast objects the formwork would be anchored to vibrating tables (Figure 5.27b).

Remember when constructing formwork for pre-cast concrete it must be:

(a) To the correct dimensions and shape.
(b) Sufficiently rigid so as not to deform when the concrete is poured and vibrated.
(c) Designed to be able to be easily and quickly stripped from the set concrete, and can be reused many times without deterioration of the formwork or clamping mechanisms.

5.7 CAST IN SITU CONCRETE

For cast in situ work the formwork is assembled on site and erected in the actual position where the concrete is to be placed in the structure.

5.7.1 Ground Beams

These are reinforced concrete beams where the ground acts as formwork for the underside of the beam. They are usually cast in trenches, with the area round the beams backed filled with earth after they are cast. In some cases the vertical sides of the trench act as formwork to contain the concrete (Figure 5.28).

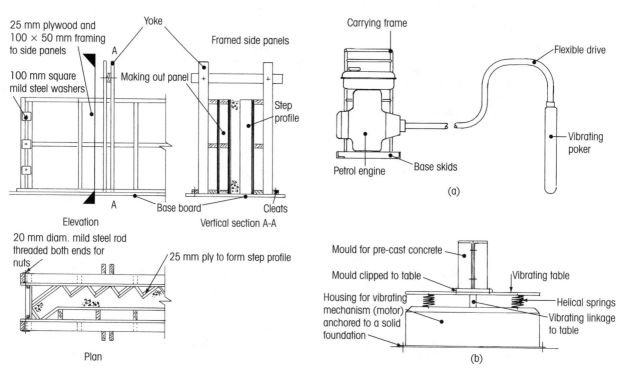

Figure 5.26 Formwork for pre-cast concrete stairs cast on edge

Figure 5.27 Mechanical compaction of concrete

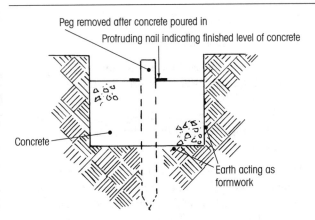

Peg removed after concrete poured in

Protruding nail indicating finished level of concrete

Concrete

Earth acting as formwork

Figure 5.28 Earth acting as formwork for concrete ground beam

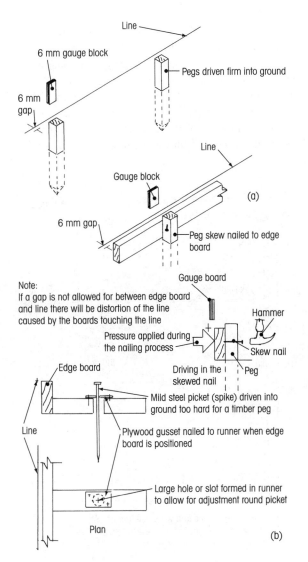

Line

6 mm gauge block

6 mm gap

Pegs driven firm into ground

Line

Gauge block

(a)

6 mm gap

Peg skew nailed to edge board

Gauge board

Note:
If a gap is not allowed for between edge board and line there will be distortion of the line caused by the boards touching the line

Hammer

Pressure applied during the nailing process

Skew nail

Edge board

Driving in the skewed nail

Peg

Mild steel picket (spike) driven into ground too hard for a timber peg

Line

Plywood gusset nailed to runner when edge board is positioned

Large hole or slot formed in runner to allow for adjustment round picket

Plan

(b)

Figure 5.29 Edge forms for concrete upstands

Narrow-edge forms up to about 300 mm deep can be supported by being nailed to stout pegs driven into firm ground if the beam projects above ground level. Care being taken to keep the face plumb, level and in line. If a builder's line is being used as a guide it should be set forward about 6 mm and a 6 mm thick piece of plywood used as a gauge, so that the edge of the formwork does not touch and distort the line (Figure 5.29a).

Should the ground be so hard that timber pegs are of no use, steel pins can be used, but difficulty may be experienced in aligning them. Figure 5.29(b) shows a method of achieving alignment.

If the edge form is at the side of a trench the pegs may have to be set back by using a piece of timber as a bridging member (Figure 5.30).

As a typical example Figure 5.31 shows the cross-section of formwork for a rectangular ground beam being cast in a trench, while Figure 5.32 shows how a kicker for the start of a concrete wall is formed as part of a strip foundation.

A pad foundation having anchor bolts for a steel stanchion cast in it is shown in Figure 5.33. In this type of formwork both the horizontal and vertical positioning of the formwork is important. To allow for any slight misalignment the anchor bolts cast into the pad have limited movement of about 25 mm radius within holes formed in the concrete which are filled with cement grout after the steel work is bolted to the concrete.

Note. Steel-work fabrications have a non-accumulative tolerance of 3 mm in 3 m.

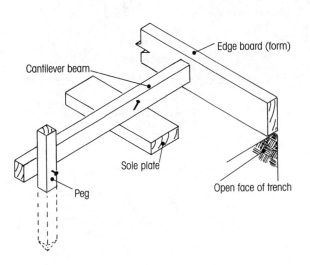

Edge board (form)

Cantilever beam

Sole plate

Peg

Open face of trench

Figure 5.30 Wall forms

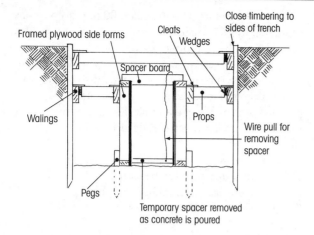

Figure 5.31 Formwork for a rectangular ground beam

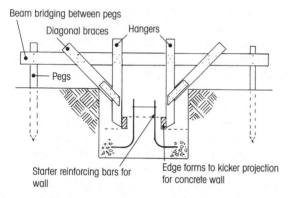

Figure 5.32 Formation of a kicker in a strip foundation for the start of a concrete wall

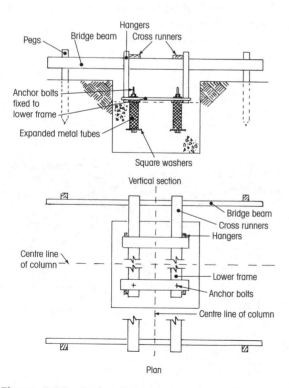

Figure 5.33 Anchor bolts in a pad foundation

5.7.2 Formwork For Walls

There are two types of wall.

(a) Retaining walls where the earth forms one supporting surface and wall formwork the other (Figure 5.34a).
(b) Dividing walls where both faces of the wall have to be supported until the concrete has set (Figure 5.34b).

Wall forms are made up into large rectangular panels which are then aligned plumbed and stiffened by the use of horizontal members walers and vertical members soldiers (strong backs). The wall form panels and supporting structure of walers and soldiers are held apart by through ties of various types (see Figure 5.23) and the whole is then held in the vertical position by adjustable props set at an inclined angle.

5.7.3 Fabrication of Plywood Wall Forms

The formwork panels are made up of a framework of 100 × 50 mm studding to which 19 mm formwork grade plywood is attached. Wall length and height and the methods available to move the formwork will determine the size of the panels. The base on which the wall forms are assembled must be flat and out of twist, and if a number of plywood sheets are used to form one panel then sheet end butt joints must be on the studs

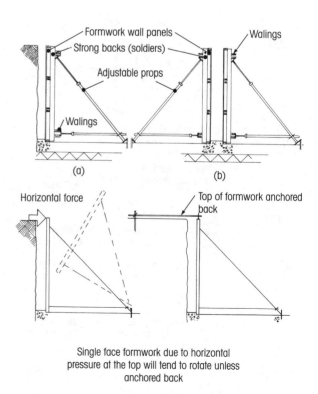

Single face formwork due to horizontal pressure at the top will tend to rotate unless anchored back

Figure 5.34 Single- and double-sided wall forms

and staggered between studs. If the wall forms are to be handled manually then their size is limited to a sheet size of 1.2 × 2.4 m which are then bolted together to form larger panels.

5.7.4 Assembly of the Wall Forms

To locate the wall forms in their correct position (Figure 5.35), they are placed against an upstand (kicker) previously cast in the concrete (Figure 5.32). The sequence of erection of the wall forms will depend upon.

(a) The wall reinforcing steel being in place, or having to be put in.
(b) The type of ties being used.
    1. Bolt ties with cones (Figure 5.37).
    2. She bolt ties (Figure 5.36a).
    3. Through ties (Snap off) (Figure 5.41).

In general for she bolts and through ties the two wall forms are simultaneously placed in position (Figure 5.36a).

For bolts and ties incorporating cones, one wall form can be erected before the other (Figure 5.37). This allows the steel worker to support the steel reinforcing rods off the formwork. Figure 5.38 shows wall forms in position and braced ready to receive the concrete.

If the wall forms have to be raised in a number of lifts to give the height of the wall required, Figure 5.39 shows how this is achieved by using the ties already set in the concrete.

5.7.5 Using Proprietary Wall Forms

A number of companies manufacture formwork systems based on metal panels (pans), or rectangular metal

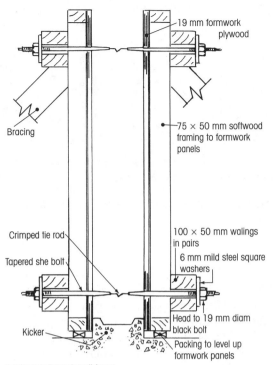

(a) Section through wall forms

(b) Wall forms on a construction site

Figure 5.36 Wall forms erected simultaneously

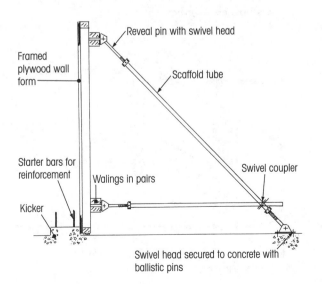

Figure 5.35 Location of a wall form against a kicker

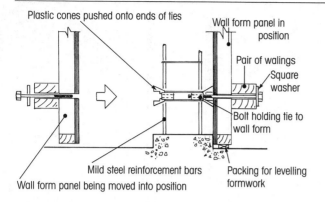

Figure 5.37 Wall formwork erected one side at a time

frames infilled with plywood panels (Figure 5.40) which can be clipped together using simple wedge type fittings, these can also hold the snap ties (Figure 5.24). For alignment and stiffening the panels can be attached to scaffold tubes forming walers and soldiers by the use of proprietary 'U' form clips (Figure 5.40). The general arrangement of wall forms of the metal panel type is shown in Figure 5.41.

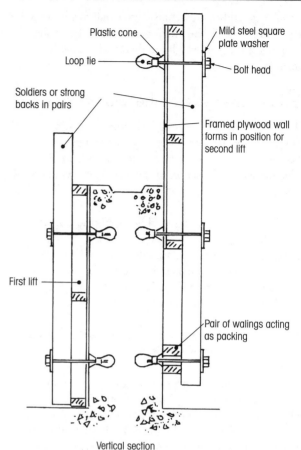

Vertical section

Figure 5.39 Formwork for a concrete wall built up in a number of lifts

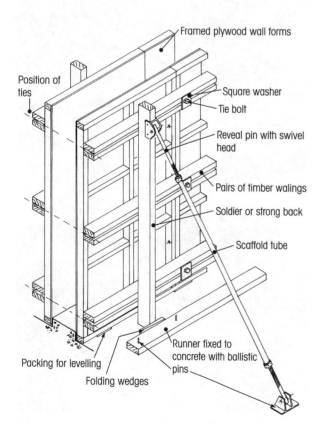

Figure 5.38 Wall forms tied and braced ready for concrete

5.7.6 General Sequence of Assembly and Dismantling of Wall Forms

(a) Assembly
   1. Apply mould oil, offer up and bolt or clip the wall panels together.

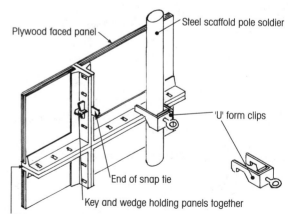

Figure 5.40 Proprietory panel formwork to walls

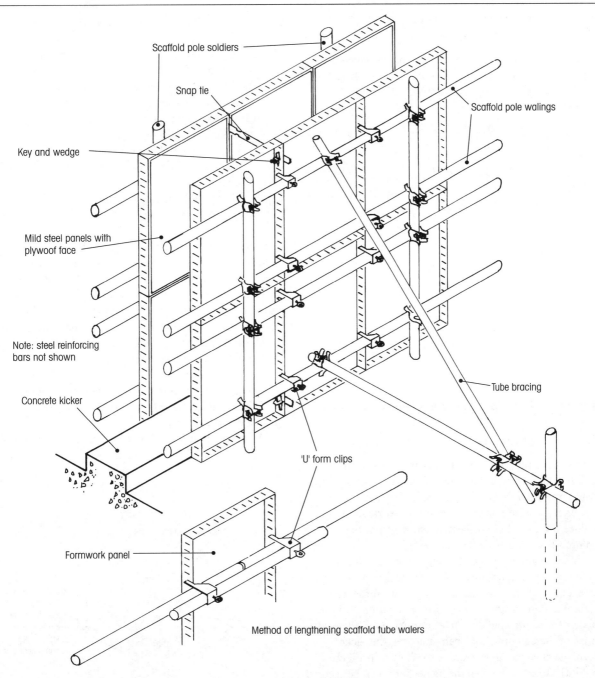

Scaffold pole soldiers

Snap tie

Key and wedge

Scaffold pole walings

Mild steel panels with plywoof face

Note: steel reinforcing bars not shown

Concrete kicker

Tube bracing

'U' form clips

Formwork panel

Method of lengthening scaffold tube walers

Figure 5.41 General arrangement of wall forms using metal panels

2. Level panels.
3. Temporarily strut while attaching walings and/ or soldiers (strong backs) to line up the panels.
4. Thread the wall ties.
5. Final adjustment and tightening of inclined props and ties to keep wall forms parallel and plumb.

(b) Dismantling
1. Slacken off wall ties.

2. Ease inclined props while keeping in place.
3. Remove a section at a time walers and soldiers and wall ties as necessary. If the wall panels are large make sure they are attached to the crane slings as the supporting walings, soldiers and props are removed.
4. Clean down the formwork and make any adjustments ready for the next use.

5.8 COLUMN FORMWORK

Columns can be cast in a variety of shapes depending on what is required by the designer. Figure 5.42 shows the outline of a variety of columns of differing shapes.

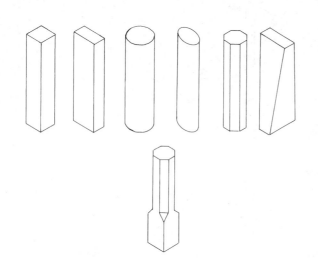

Figure 5.42 Various concrete column profiles

5.8.1 Kickers for Concrete Columns

As for walls they are concrete upstands approximately 100 mm high cast in the floor slab at the positions where the columns will be required. They act as location points for the base of the column formwork and help reduce the loss of grout (cement slurry) from the column base as the concrete is poured.

When making and positioning the kickers for the columns, it is important to make the kicker the correct size and shape and accurately position it on the floor slab. Accurate positioning is achieved by lining up the kicker forms on a builder's line running along the centre lines of the columns, and then attaching the kicker forms to the column starter bars as shown in Figure 5.43(a). As an alternative the kicker form can be secured in position by use of a ballistics tool to pin the form to the concrete floor [BENCH AND SITE SKILL, SECTION 8.19.4] (Figure 5.43b).

5.8.2 Column Formwork Making and Erecting the Forms

(a) The sides of the column forms are framed together from:
 1. Solid boards held together by nailing to battens (cleats) (Figure 5.44a), which also shows how a cleaning trap to allow the removal detritus from the base of the column is formed.

 2. Plywood nailed to a frame work of timber studs (Figure 5.44b).
 3. Steel or composite panels clipped together (Figure 5.44c).
(b) Once the column side panels are made and treated with mould oil they are placed against the kicker being tacked or clipped together and then firmly held in place by using
 1. Timber yokes (Figure 5.45a).
 2. Adjustable mild steel column clamps (Figure 5.45b).
 3. Mild steel banding or strapping (Figure 5.45c).

The spacing of the clamps along the height of the column will depend on the pressure exerted by the concrete and the deflection of the column forms under load. Usually the clamps are positioned closer together at the bottom of the column than at the top. This is because the pressure exerted by the liquid concrete is greater at the bottom than the top, see Figure 5.45(d).

Note. When using strapping to hold a column form in place the bearers should be so arranged that the strapping forms as near a circle as possible (Figure 5.45f), or horizontal stiffeners are used on framed plywood forms.

When erecting the column forms, care must be taken to make sure that they do not twist in their height and

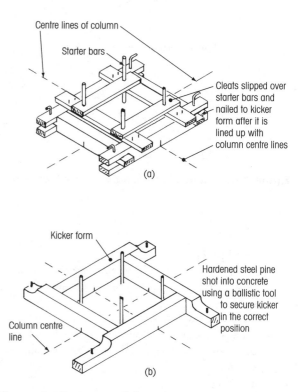

Figure 5.43 Formwork for column kickers

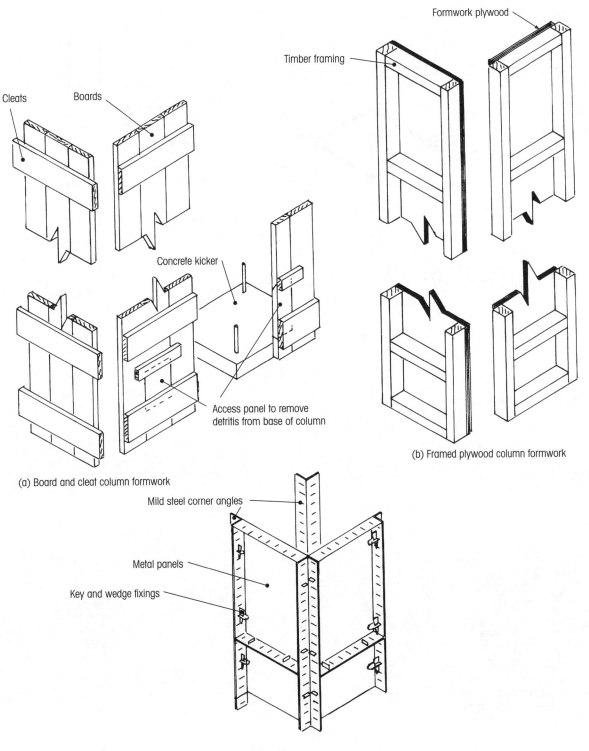

Figure 5.44 Types of column formwork

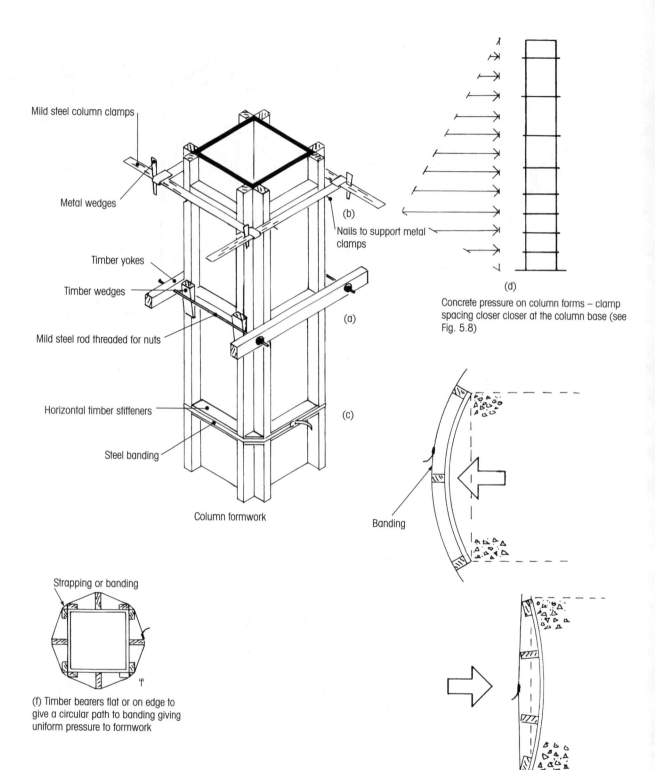

Mild steel column clamps

Metal wedges

(b)

Nails to support metal clamps

(a)

Timber yokes

Timber wedges

Mild steel rod threaded for nuts

(c)

Horizontal timber stiffeners

Steel banding

Column formwork

(d)

Concrete pressure on column forms – clamp spacing closer closer at the column base (see Fig. 5.8)

Banding

Strapping or banding

'f'

(f) Timber bearers flat or on edge to give a circular path to banding giving uniform pressure to formwork

(e) Banding will distort plywood formwork if there are no horizontal stiffeners

Figure 5.45 Securing column formwork

the cross-section is not distorted (Figure 5.46a). The columns forms are plumbed on two faces using a theodolite, or line and plumb bob (Figure 5.46b). The column formwork is then held in position by inclined props (Figure 5.46c).

If the column is at the edge or corner of a slab one or two of the inclined struts will have to resist both tension and compression (Figure 5.47).

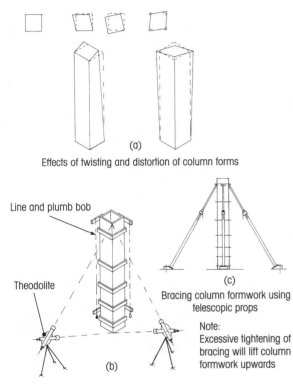

(a)

Effects of twisting and distortion of column forms

Line and plumb bob

Theodolite

(c)

Bracing column formwork using telescopic props

Note:
Excessive tightening of bracing will lift column formwork upwards

(b)

Plumbing column formwork using plumb bob or optical instrument (theodolite)

Figure 5.46 Plumbing and support of column formwork

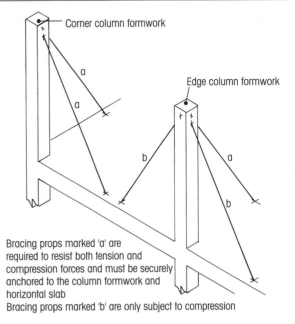

Corner column formwork

Edge column formwork

a
a
b
a
b

Bracing props marked 'a' are required to resist both tension and compression forces and must be securely anchored to the column formwork and horizontal slab
Bracing props marked 'b' are only subject to compression

Figure 5.47 Bracing props subject to tension and compression

5.9 PROPPED FORMWORK FOR SUSPENDED FLOORS

Formwork for suspended slabs is subject mainly to uniformly distributed vertical forces acting in a downward direction, although there are some horizontal pressures that have to be resisted at the edge of the slab. One of the big problems is high point loading as the concrete is discharged in a heap [*BS5975:1996 CoP for falsework*]. This limits the concrete discharge to double the depth of the finished concrete over an area of 1 m^2.

Joists used to support the decking should be as near square in cross-section as possible as there is a tendency for deep narrow joists to roll over. Joist widths should not normally be less than 45 mm, and their height should not be more than twice their width, unless thy are laterally restrained by nailing to the decking or using solid blocking or herring bone strutting (see Table 5.2 and Chapter 1).

5.9.1 Erecting Formwork for a Concrete Floor Slab

The spacing of props and the supporting structure of joists and beams for the decking will depend upon what deck forms are used, metal panels (pans) or 19 mm plywood sheets. If the concrete slab is 200 mm thick it will exert a force of 5 kN/m^2 and the spacing of any underframing directly supporting the 20 mm thick

Table 5.2   Lateral stiffening of joists supporting formwork

DEPTH TO BREADTH RATIO	REQUIREMENTS	ILLUSTRATIONS
Not greater than 2:1 Width not less than 45 mm	No lateral support or fixing	
3:1 Width not less than 45 mm	Ends of joists secured	
4:1 Width not less than 45 mm	Joists secured at ends and at intermediate supports	
5:1 Width not less than 45 mm	Positive fixing of joists at the ends and intermediate supports	
6:1 Width not less than 45 mm	Blocking or strutting of joists at the end and intermediate supports	

plywood decking must not be placed at centres greater than 600 mm (see Figure 5.48).

Note. Spacing of props and the size and spacing of support members would be determined by the structural engineer.

To save on the number of props used, a further underframe of secondary beams or runners approximately 75 × 225 mm in cross-section and spaced at about 1.2 m centres would be supported on adjustable props set at about 2–2.4 m centres (Figure 5.49).

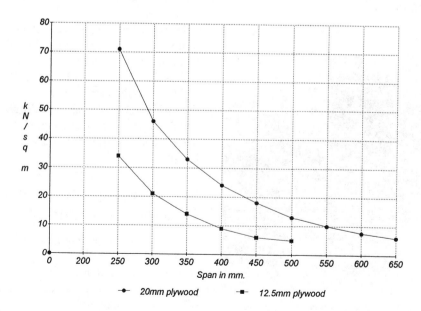

Figure 5.48   Joist spacing to support formwork plywood

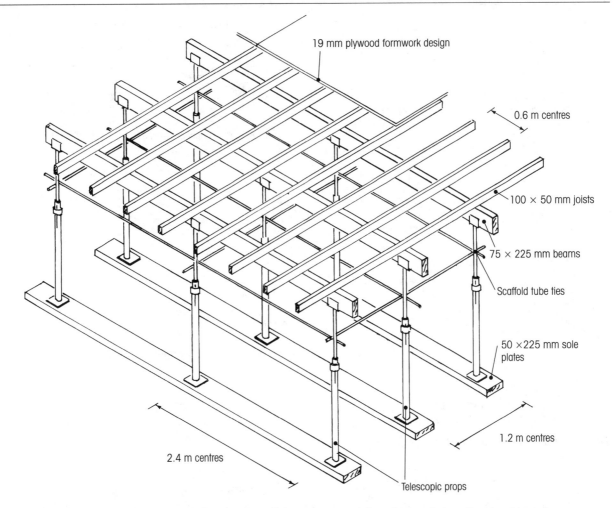

19 mm plywood formwork design

0.6 m centres

100 × 50 mm joists

75 × 225 mm beams

Scaffold tube ties

50 × 225 mm sole plates

1.2 m centres

2.4 m centres

Telescopic props

Figure 5.49 Assembly of formwork for concrete slabs using props (spacing and size of support members determined by the structural engineer)

5.9.2 Assembly of the Slab Formwork

(a) Timber sole plates are set out on the supporting substructure in line with the position of the props to give them a fixing and firm foundation thus avoiding some of the problems illustrated in Figure 5.50. The sole plates also distribute the force from the props over a wider area.

(b) A line of adjustable metal props at the appropriate spacing would be set up and a beam dropped into and secured to the fork head of the prop. This would be continued across the space enclosed by the walls, adjusting the props for height and lacing them with horizontal scaffold tubes as the props are set up (Figure 5.49). The first and last beam would be set close to the walls running parallel to the line of props and could be hung from adjustable wall hangers (Figure 5.51).

In the case of metal edged panels (pans) the same procedure would be adopted except that the joists carrying the panels would be spaced to support the edge of the panels.

(c) If the soffit of the slab overhangs or finishes flush with the outer face of the supporting wall the formwork for the edge of the slab would be constructed as shown in Figure 5.52.

(d) As an alternative to supporting the decking on a system of joists, beams and props, mild steel adjustable floor centres can be used, being hung directly from the supporting walls. As an alternative the ends of the adjustable floor centres can be supported by bearers and adjustable props (Figure 5.53).

(e) Stripping the soffit formwork in an enclosed space can be a dangerous operation unless care is taken.

Working a bay at a time, shorten the adjustable props to ease the panels from the soffit of the concrete, giving enough room to move just sufficient joists or bearers to allow the plywood formwork panel to be pulled clear.

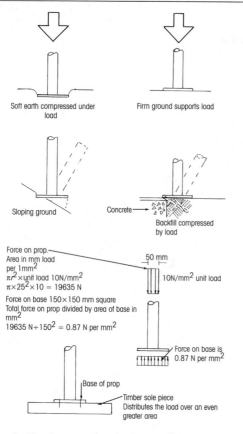

Soft earth compressed under load

Firm ground supports load

Sloping ground

Concrete → Backfill compressed by load

Force on prop.
Area in mm load
per 1 mm^2
$\pi r^2 \times$ unit load 10N/mm^2
$\pi \times 25^2 \times 10 = 19635$ N

Force on base 150×150 mm square
Total force on prop divided by area of base in mm^2
19635 N $\div 150^2 = 0.87$ N per mm^2

50 mm

10N/mm^2 unit load

Force on base is 0.87 N per mm^2

Base of prop

Timber sole piece
Distributes the load over an even greater area

Figure 5.50 Support for the base of props

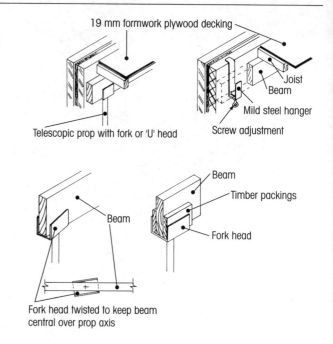

19 mm formwork plywood decking

Joist
Beam
Mild steel hanger
Screw adjustment

Telescopic prop with fork or 'U' head

Beam
Timber packings
Fork head

Beam

Beam

Fork head twisted to keep beam central over prop axis

Figure 5.51 Wall hangers and 'U' or fork head props

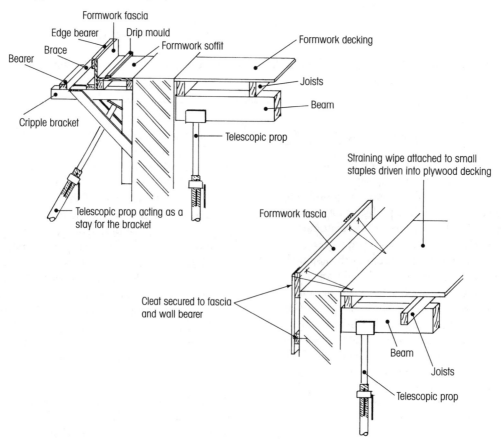

Formwork fascia
Edge bearer
Drip mould
Brace
Formwork soffit
Formwork decking
Bearer
Joists
Beam
Cripple bracket
Telescopic prop
Telescopic prop acting as a stay for the bracket

Straining wipe attached to small staples driven into plywood decking

Formwork fascia

Cleat secured to fascia and wall bearer

Beam
Joists
Telescopic prop

Figure 5.52 Flush finish and overhanging concrete roof slab

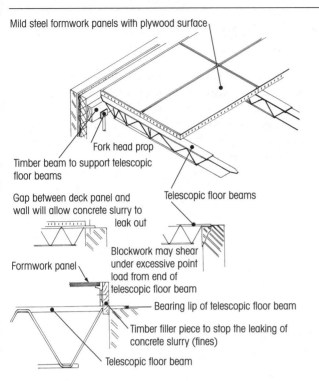

Mild steel formwork panels with plywood surface

Fork head prop

Timber beam to support telescopic floor beams

Telescopic floor beams

Gap between deck panel and wall will allow concrete slurry to leak out

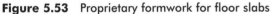

Blockwork may shear under excessive point load from end of telescopic floor beam

Formwork panel

Bearing lip of telescopic floor beam

Timber filler piece to stop the leaking of concrete slurry (fines)

Telescopic floor beam

Figure 5.53   Proprietary formwork for floor slabs

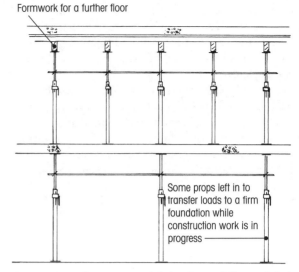

Formwork for a further floor

Some props left in to transfer loads to a firm foundation while construction work is in progress

Figure 5.54 Some props left in place while work is in progress above

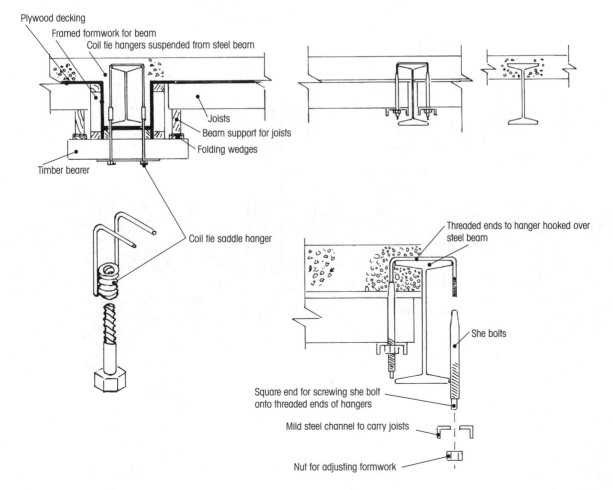

Plywood decking

Framed formwork for beam

Coil tie hangers suspended from steel beam

Joists

Beam support for joists

Folding wedges

Timber bearer

Coil tie saddle hanger

Threaded ends to hanger hooked over steel beam

She bolts

Square end for screwing she bolt onto threaded ends of hangers

Mild steel channel to carry joists

Nut for adjusting formwork

Figure 5.55   Suspended formwork

Remove the decking and joists working systematically across the room, and as the beams cease to carry the joists, they are lowered and removed from the fork heads while the base plates of the props are freed from the sole plates.

When stripping out the formwork it is usual to leave a grid of props in position until the concrete has gained its full design strength, and to make sure that any additional loading caused by a further lift of formwork is transferred through the structure to a foundation that will support the loads without failing (Figure 5.54).

5.9.3 Suspended Formwork For Floor Slabs

In steel framed buildings with in situ concrete floor slabs the formwork for the floors can be hung from the steel beams, which form part of the permanent floor structure. Figure 5.55 shows how the formwork would be suspended from the steel beams by the use of metal saddle hangers.

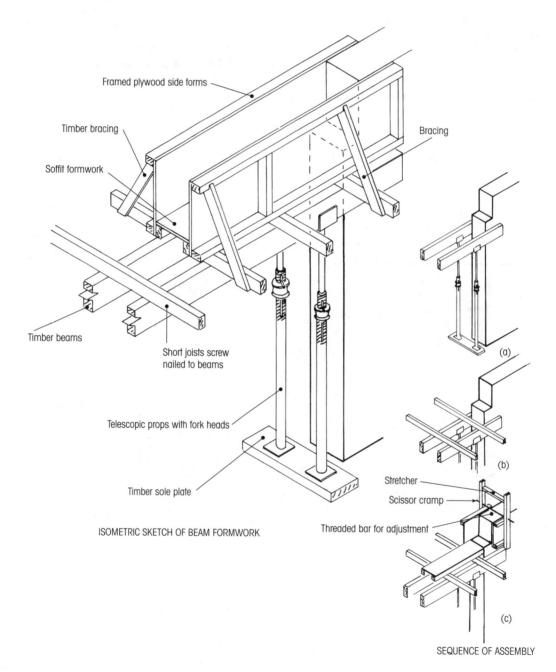

Framed plywood side forms

Timber bracing

Soffit formwork

Bracing

Timber beams

Short joists screw nailed to beams

Telescopic props with fork heads

Timber sole plate

ISOMETRIC SKETCH OF BEAM FORMWORK

(a)

(b)

Stretcher

Scissor cramp

Threaded bar for adjustment

(c)

SEQUENCE OF ASSEMBLY

Figure 5.56 Formwork for an isolated beam

5.10 FORMWORK FOR ISOLATED BEAMS

The sequence for the erection easing and striking of the formwork for isolated beams is as follows.

(a) Construct the formwork panels for the sides and bottom (soffit) of the beam.
(b) Check the foundation (surface on which the props are to rest) is level and of adequate and uniform strength to support the loads transferred from the props (Figure 5.50). Loads from the props can be distributed over a wider area and the bearing capacity of the foundation improved if timber sole plates are used (Figure 5.56).
(c) Place the props in position and tack to the timber sole plate, place the beams in the fork heads and level up, and lace the props with scaffold tubes if necessary (Figure 5.56a). Space out the bearers (joists) at right-angles along the beams at between 600–1200 mm centres depending on how the formwork panels are constructed (Figure 5.56b).
(d) The beam soffit and side forms are then placed in position, lined up, plumbed and braced (Figure 5.56c) and treated with mould oil. The reinforcing steel will be put in place and the concrete poured.
(e) After about a day when the concrete has set the beam side forms are taken down and cleaned while any green concrete adhering to the surface of the forms is easy to remove. Fourteen days later or when the concrete has gained sufficient strength to carry its own weight the soffit formwork and supporting structure is eased and if all is as it should be, the soffit form is removed and the supporting structure dismantled (joists first then beams and finally props and sole plates).

5.10.1 Combined Beam and Slab Formwork

In general the methods of constructing the formwork is as for floor slabs except that the beam is formed to project below the soffit line of the floor slab (Figure 5.57). The formwork is so assembled that the side forms of the beams can be removed before the soffit formwork. Figure 5.57(a) shows the formwork for a floor edge beam while Figure 5.57(b) illustrates the formwork for a beam with in the floor slab.

5.11 RANGE OF FORMWORK PROFILES

While an attempt has been made to cover the basic

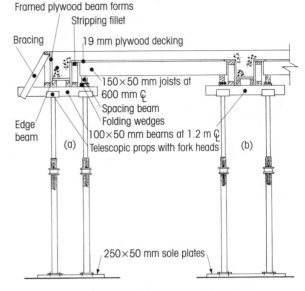

Figure 5.57   Integrated beams and floor slabs

types of formwork, the profile of concrete elements are almost infinitely variable and beyond the scope of a section in a book on carpentry and joinery. If an in-depth study of the subject is required there are numerous text books available which deal exclusively with formwork.

5.12 TIMBER CENTRES FOR ARCHES UP TO 1.5 m SPAN

Up to a span of 1.5 m timber is the most economical material to use for the construction of centres. In larger spans metal supporting structures would be used.

Centres for arches are temporary structures fabricated and set up to support stone, brick or concrete arches during their construction or repair. The profile of the centre is that of the soffit, or intados (underside profile of the arch) (Figure 5.58). Stone and brick arches are gravity structures, which are formed so that when constructed the force of gravity and the mass of the structure they support holds them in place, and the centres and their propping (temporary supports) used during construction can be eased and struck (taken down).

5.12.1 Common Arch Centre Profiles

(a) Flat or camber arches are supported by turning pieces and adjustable props. The rise is limited to about 0.01 (one hundredth of the span) (Figure 5.59). A level soffit would appear to sag.

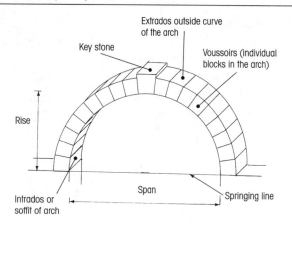

Extrados outside curve of the arch

Key stone

Voussoirs (individual blocks in the arch)

Rise

Intrados or soffit of arch

Span

Springing line

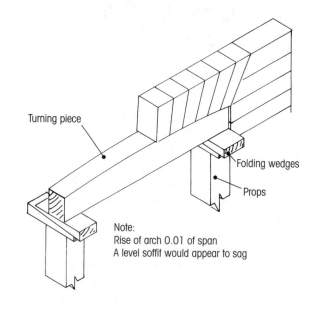

Turning piece

Folding wedges

Props

Note:
Rise of arch 0.01 of span
A level soffit would appear to sag

Figure 5.59   Turning piece for a flat or chamber arch

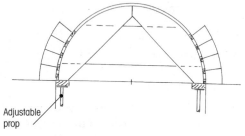

Adjustable prop

Figure 5.58   Arch centres

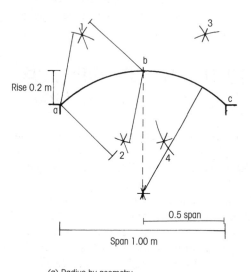

Rise 0.2 m

0.5 span

Span 1.00 m

(a) Radius by geometry
Strike arcs from a and b
Draw a line through the intersecting arcs 1 and 2
Strike arcs from b and c
Draw a line through the intersecting arcs 3 and 4
Where the two lines intersect is the centre for the curve

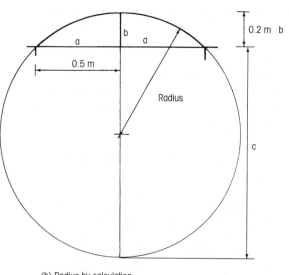

0.2 m b

a

0.5 m

Radius

c

(b) Radius by calculation
$a \times a = b \times c$

$\text{Radius} = \dfrac{b+c}{2}$

$c = \dfrac{a \times a}{b}$

$c = \dfrac{0.5 \times 0.5}{0.2} = \dfrac{0.25}{0.2} = 1.25 \text{ m}$

$\text{Diameter} = c + b = 1.25 + 0.2 = 1.45 \text{ m}$

$\text{Radius} = \dfrac{1.45}{2} = 0.725 \text{ m}$

Figure 5.60   Setting out the curve for a segmental centre

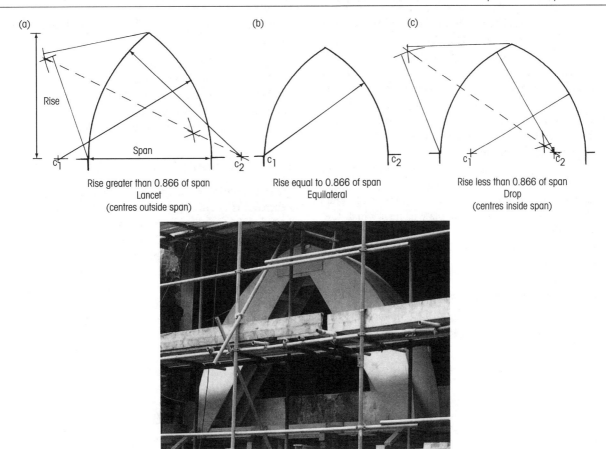

Figure 5.61 Arch outlines for pointed arches

(b) Segmental arches form the segment of a curve
[BENCH AND SITE SKILLS, SECTION 6.5] where the rise
of the arch is less than half the span (Figure 5.60).

(c) Semi-circular arches have a rise which is equal to
half the span.

(d) Pointed arches which can be subdivided into lancet
equivalaleral and drop arches (Figure 5.61).

(e) Elliptical (approximate) arches. The truly elliptical
soffit line for an arch is rarely used as it means that
each individual block in half the arch has to be cut
to a different shape and true elliptical outlines
cannot be drawn parallel. Therefore a curve approxi-
mating to an ellipse is drawn using segments of
circles. The simplest approximately elliptical curve
to draw is based on three centres (Figure 5.62). The
greater the number of centres used the nearer the
profile is that of an ellipse, but it is not usual to use
more than five centres.

5.12.2 Construction of Centres for Arches

The centre is built up of the following members
(Figure 5.63).

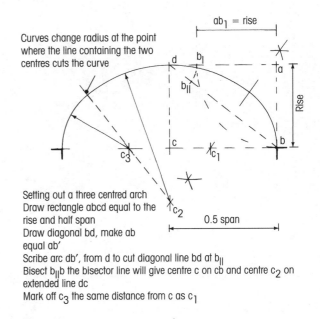

Curves change radius at the point
where the line containing the two
centres cuts the curve

Setting out a three centred arch
Draw rectangle abcd equal to the
rise and half span
Draw diagonal bd, make ab
equal ab'
Scribe arc db', from d to cut diagonal line bd at $b_{||}$
Bisect $b_{||}b$ the bisector line will give centre c on cb and centre c_2 on
extended line dc
Mark off c_3 the same distance from c as c_1

Figure 5.62 Setting out a three-centre
approximately elliptical arch centre

- *Ribs*. Made from 19 mm plywood or solid timber about 20 to 25 mm thick jointed together using gussets or by being built up of overlapping segments nailed together to form the required outline and made up in pairs.
- *Ties*. Usually horizontal members tying the feet of the ribs together to stop them spreading under load.
- *Struts*. Vertical or inclined members used to support the ribs and transfer compression forces to the supports for the centre.
- *Bracing*. Diagonal timbers running between the inner and outer ribs to hold them square.
- *Bearers*. Flat members, which tie the inner and outer ribs together and form a sole plate against which the top of the props can rest.
- *Laggings*. Can be made of sheet material such as hardboard and or plywood nailed to the pair of ribs to form the arch profile, or laths nailed across and at right angles to the pairs of ribs.

The length of laggings should always be 20 mm shorter than the arch thickness if bricks or stone is being used to form the arch. Batten (lath) laggings can be nailed:

(a) Close together to give a solid surface (Figure 5.63a).
(b) Spaced so that there is a lagging width space between adjacent laggings (Figure 5.63b).
(c) Spaced so that two laggings give support to each stone voussoir (Figure 5.63c).

5.12.3 Setting Out an Arch Centre

As the majority of centres are symmetrical about their centres it is usually only necessary to set out just over half the elevation on sheets of plywood or medium density fibreboard (MDF), although it is helpful to have a full outline when assembling the ribs.

From the outline of the arch, the shape and number of rib segments required and their width, which should be between 150 and 200 mm can be determined (Figure 5.64b). Templates of the ribs are then cut accurately to shape in hardboard or thin plywood. The rib shapes using

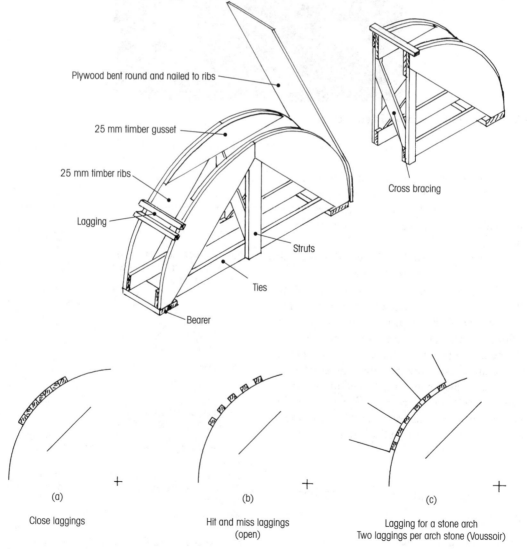

Plywood bent round and nailed to ribs

25 mm timber gusset

25 mm timber ribs

Lagging

Bearer

Ties

Struts

Cross bracing

(a)	(b)	(c)
Close laggings	Hit and miss laggings (open)	Lagging for a stone arch Two laggings per arch stone (Voussoir)

Figure 5.63   Parts of an arch centre

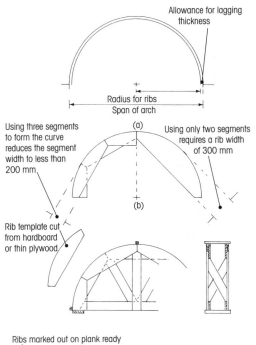

Allowance for lagging thickness

Radius for ribs
Span of arch

(a)

Using three segments to form the curve reduces the segment width to less than 200 mm

Using only two segments requires a rib width of 300 mm

(b)

Rib template cut from hardboard or thin plywood

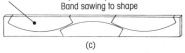

Ribs marked out on plank ready
Band sawing to shape

(c)

Figure 5.64   Setting out an arch centre

the template are than transferred to the prepared timber blanks ready for band-sawing to shape (Figure 5.64c).

Note. Always remember when setting out the ribs for the centre to allow for the thickness of the laggings (Figure 5.64a).

Prepare a cutting list of the size and length of the materials required

5.12.4 Assembly of the Arch Centre

The ribs are overlapped or gusseted and nailed together to form the outline of the arch. They are made as a pair to form the back and front profile of the arch. Check their profile against the setting out and attach the ties and braces by nailing.

Secure the bearers to the back and front of the centre and temporarily tack two or three laggings round the arch outline to keep them parallel. Cut and nail in any diagonal bracing. Space and nail the laggings in place, keeping them square with the face of the centre.

5.12.5 Erecting Easing and Striking of Arch Centres

When the wall has been built up to the height of the springing line of the arch (Figure 5.65).

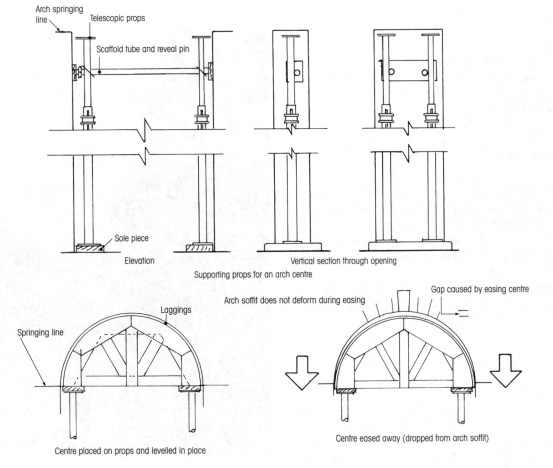

Arch springing line
Telescopic props
Scaffold tube and reveal pin

Sole piece

Elevation

Vertical section through opening

Supporting props for an arch centre

Springing line
Laggings

Centre placed on props and levelled in place

Arch soffit does not deform during easing

Gap caused by easing centre

Centre eased away (dropped from arch soffit)

Figure 5.65   Erecting and removing an arch centre

(a) Place the sole plates level and stable into the opening and against the reveals. Wall thickness of up to 400 mm will only require a prop at each side of the opening. For thicker walls pairs of props tied and braced together should be used.

(b) Lift the centre on to the top of the stayed props and secure by nailing through the holes in the prop head plate. Vertically adjust the props so that the centre sits level and plumb ready for the arch to be built.

(c) Once the arch is formed and the mortar set the centre is eased away from the arch soffit by lowering the props 3 or 4 mm. If the arch settles without distortion the centre is further lowered to allow it and its associated propping to be struck (taken away) without damaging the arch.

5.13 SHORING

When brick or masonry structures have to be modified, or if over time weaknesses occur the structure will require support while remedial work is carried out.

If walls have to be removed or new openings cut into them they will have to be supported while the work is undertaken. This temporary support work is called shoring.

Shoring consists of a framework of large, usually second-hand timbers which are cut and fitted in place with adjustments being made by the use of folding wedges, or adjustable props.

Shoring is divided into three types which can be used independently, or in conjunction depending on what support is required.

(a) *Raking shores*. Inclined props which are used to keep a wall in the vertical position during alteration or strengthening work (Figure 5.66a).

(b) *Flying shores*. Carry out the same function as raking shores but are framed to support a pair of facing walls (Figure 5.66b).

(c) *Dead shores*. Support the structure above as a new opening is cut in a wall and a supporting girder or beam is put in place (Figure 5.66c).

5.13.1 Raking Shores Cutting and Erecting

(a) The wall is inspected to see how many sets of raking shores are required, their spacing, usually 1.8–3.0 m apart and the number of rakers in a set. This will depend on the spacing and size of openings and the number of floors in the building plus the roof. Once

the spacing of the shores is set the position of the floors, and whether the joists are carried by the wall or run parallel to it will be marked on the outside of the wall. A header, or half a brick will be cut out of the wall in the course of bricks nearest to the

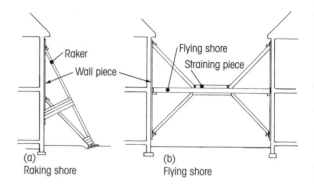

(a) Raking shore

(b) Flying shore

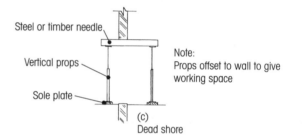

Note:
Props offset to wall to give working space

(c) Dead shore

(d) Raking shore

Figure 5.66   Types of shoring

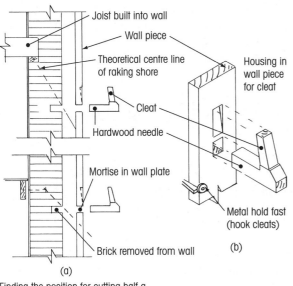

Joist built into wall
Wall piece
Theoretical centre line of raking shore
Housing in wall piece for cleat
Cleat
Hardwood needle
Mortise in wall plate
Metal hold fast (hook cleats)
Brick removed from wall

(a)
(b)

Finding the position for cutting half a brick out of the wall

Figure 5.67 Cutting out header bricks for the needles

position shown in Figure 5.67(a). A wallplace from 50–75 mm thick will have mortise holes cut in its length to match the holes cut in the walls. The wallpieces are offered up and hardwood needles are passed through the wall pieces and into the holes in the wall. To secure the wall pieces to the wall, hook cleats are driven into the bed joints of the brickwork and nailed to the wall pieces (Figure 5.67b).

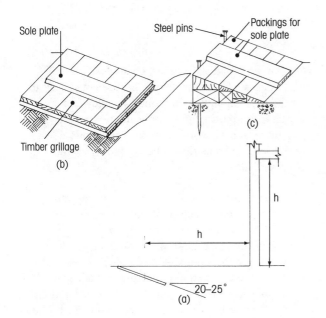

Sole plate
Steel pins
Packings for sole plate
Timber grillage
(b)
(c)

h
h
20–25°
(a)

Figure 5.68 Forming a foundation for the sole plates

(b) While the wall pieces are being fixed to the wall the sole plates can be positioned. If the ground can be dug out the sole plate is laid at an angle of approximately 25° to the horizontal and ideally about the height of one storey away from the wall requiring support (Figure 5.68a), but this will depend upon the space available. Should the ground have a low bearing capacity it may be necessary to form a grillage of two layers of short boards running at right-angles to each other under the sole plate (Figure 5.68b). If the sole plate is on a hard surface (concrete) a frame work of timber packings is arranged to hold the sole plate at the appropriate angle (Figure 5.68c).

(c) When the sole and wall pieces are in position the length and bevels for the rakers are found usually starting with the shortest raker, but this will depend on custom and practice. The person on the ladder holds the tape to the needle and a piece of plywood. When the tape is pulled taut the angle it makes is marked on the plywood, while the person at the sole plate does the same. The dimension on the tape when held in the correct position on the sole plate gives the length of the raker (Figure 5.69a). As an alternative vertical and horizontal distances can be measured and a scaled drawing produced to give the length and bevel of the rakers (Figure 5.69b). Rakers are then cut to length and bevel, and an open slot is cut in the top of the raker to fit round the needle to stop lateral movement of the rakers (Figure 5.70a). A notch is cut in the bottom of the raker to give purchase for a crowbar used to lever the raker into position (Figure 5.70b).

(d) Once the rakers are in position they are tightened to give support to the wall without pushing it over in the opposite direction (this can easily be done so take care). The feet of the rakers are then held in position on the sole plate by the use of spacing cleats (timber packing pieces) and metal dogs (Figure 5.71a). Rakers are stiffened in their length by bracing timbers (Figure 5.71b) and diagonal bracing between shores (Figure 5.71c).

5.13.2 Flying Shores Cutting and Erection

This system has much in common with raking shores, except that the rakers are supported off a flying shore, a horizontal beam set one or two storeys above the ground being supported on wall pieces and wedged between two adjacent walls (Figure 5.72).

(a) Form mortise holes in the brick work and cut the wallplaces and fix as described for raking shores.

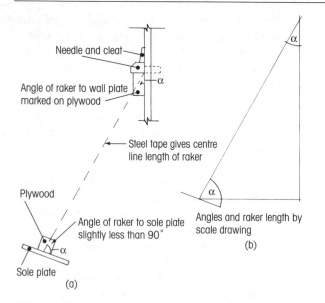

Figure 5.69   Raking shore lengths and bevels

(b) The flying shore is cut to length raised on to the appropriate needles manually, or with a crane. Folding wedges are then used to tighten the shore between the walls 5.72(a). A straining head and sill is then spiked to the top and underedge of the flying shore. This can be done before the shore is raised into position.

(c) Lower rakers are cut and fitted into position between the straining head and needles, using folding wedges to tighten them in place (Figure 5.73b).

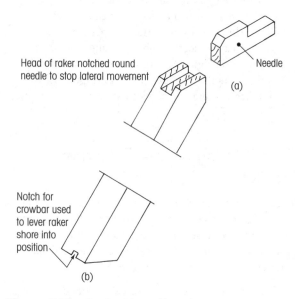

Figure 5.70   Cutting the raking shores

(d) The upper rakers are then cut and placed in position and tightened between the straining sill and needles using folding wedges. Metal dogs are then hammered into the various members to hold them in place.

5.13.3 Dead Shoring Cutting and Erection

Dead shoring is used to give support to walls above an opening while it is being cut out, and a beam or lintel being inserted to give permanent support to the structure. Needles, horizontal members of timber or steel, are passed through holes cut through the walls and supported at their ends by dead shores (vertical props). Dead shores are placed at about 1.5 m along the wall but can be placed closer together if the wall being supported is in a poor condition. Raking shores can also be used if the wall needs supporting vertically.

(a) Prop the ends of all floors being carried by the wall, using adjustable props, and make sure the props are carried down to a firm foundation (Figure 5.73a). Support window openings with wallplates and struts and give support with turning pieces to flat arches (Figure 5.73a).

(b) Cut the holes through the wall and pass through the needles of square sectioned timber or short steel girders (Figure 5.73b).

(c) Put down sole pieces on a firm foundation to support the ends of the dead shores, even if it means removing floor-boards of a suspended timber ground floor. If the dead shores are of timber they will rest on folding wedges which allows for height adjustment (Figure 5.73d). It is the more common practice to use adjustable steel props.

Note. The dead shores must be as close to the wall as possible, so that the needles are not subject to excessive bending, but space must be allowed for working and putting any supporting beams in position. It is usual to have working space only on one side of the wall, and this should be kept to a minimum (Figure 5.73b).

(d) After ensuring the shores are supporting the weight of the structure the new opening can be cut, lintels or beams inserted with new brickwork being built up and wedged to the existing wall.

(e) Once the new and old structure is stable and self-supporting, the dead shores can be eased and if all is well can be struck and the needles removed. The floor props are then taken out and the holes in the floors and walls made good, the window reveal supports removed and any raking shores used taken down.

General note. On all types of shoring continual checks must be carried out to make sure the structure being supported remains stable, and the shoring systems remain tight without distorting the structure they are supporting.

5.13.4 Alternative Materials Used for Shoring Systems

For large shoring systems of the three types which are to be in place for long periods of time, framed tubular steel scaffolding and steel beams are used as an alternative to timber. When using tubular steel scaffolding timber wall pieces are still used.

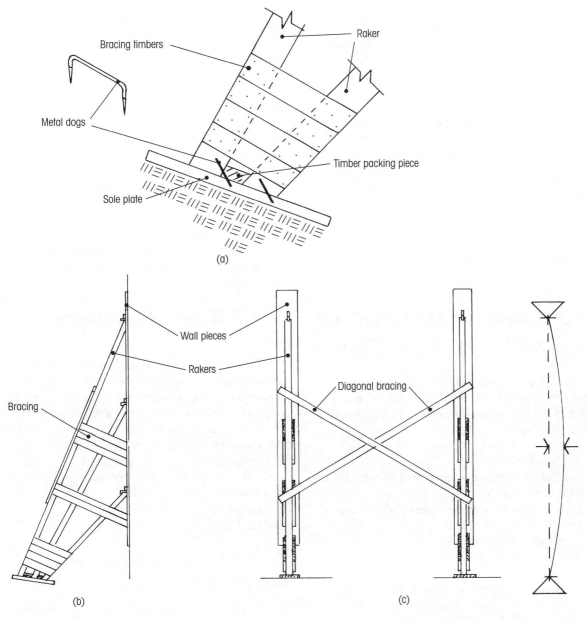

Note:
As the rakers are in compression bracing helps reduce lateral bending

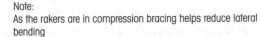

Figure 5.71 Bracing to raking shores

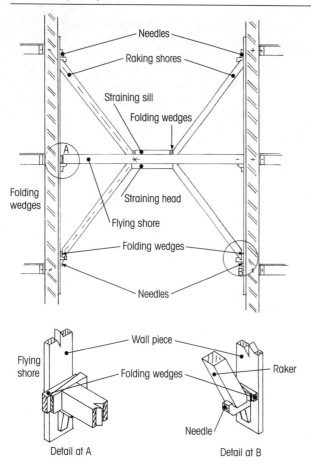

Figure 5.72 Flying shore

5.14 TIMBERING TO TRENCHES (NOT MORE THAN 2 m DEEP)

Excavated trenches will eventually collapse in on themselves unless they are supported in some way. This failure of the trench side will be caused by a number of factors such as:

(a) Traffic vibration.
(b) Additional pressure from excavated material at the edge of the trench.
(c) Loss of moisture from the exposed sides of the trench.
(d) Poor cohesion of the earth.
(e) Water logged ground.

A trench only 1 m deep can trap a person stooping over to work if the sides collapsed (Figure 5.74a). Fatalities occur each year because risks are taken and the side of trench excavations are not supported.

The majority of trenches are now mechanically excavated to their full depth (Figure 5.74b) rather than in depth stages of 1.5m as when being manually dug.

Earth can be divided into the following rough groupings

(a) *Rock*. Which may need supporting if the bedding plane lies at an inclined angle (Figure 5.74c).
(b) *Firm ground* – chalk and compact gravel.
(c) *Moderately firm ground* – stiff clay cohesive earth.
(d) *Poor ground* – loose non-cohesive earth and water logged ground.

A combination of the various types of ground can be found along the length and depth of an excavation and advice from a competent person should be taken when planning trench excavations and associated timbering.

5.14.1 Timbering in Firm Ground

As soon as a sufficient length of trench is excavated and people can work without danger from the moving arm of the excavator, pairs of poling (vertical boards) 38–45 mm thick are dropped down the sides of the trench and held tight to the sides by timber struts (Figure 5.75a), or trench props (Figure 5.75b) when expanded. These pairs of poling boards are placed along the length of the trench at 1.8 m intervals, with the operative always working forward from the timbered section of the trench.

5.14.2 Timbering in Moderate Ground

As the trench is excavated and there is safe working room a pair of poling boards and struts are placed in position, with the next pair being placed from 200 to 300 mm away from them and strutted (Figure 5.76a). Once a 2–3 m length of trench has been supported in this way, pairs of walers (horizontal members) are introduced and strutted against the poling boards, with the strut spacing being approximately 1.8 m. The temporary struts to the individual poling boards are then removed to give working space (Figure 5.76b).

5.14.3 Timbering in Loose, Soft and Water-logged Ground

In these conditions close timber boarding (sheeting) would be driven into the ground prior to the excavation (Figure 5.77a), as the excavation proceeds the runners would be progressively supported by walings and struts. The toe end of the runners should always be 300–400 mm below the bottom of the excavation, as this

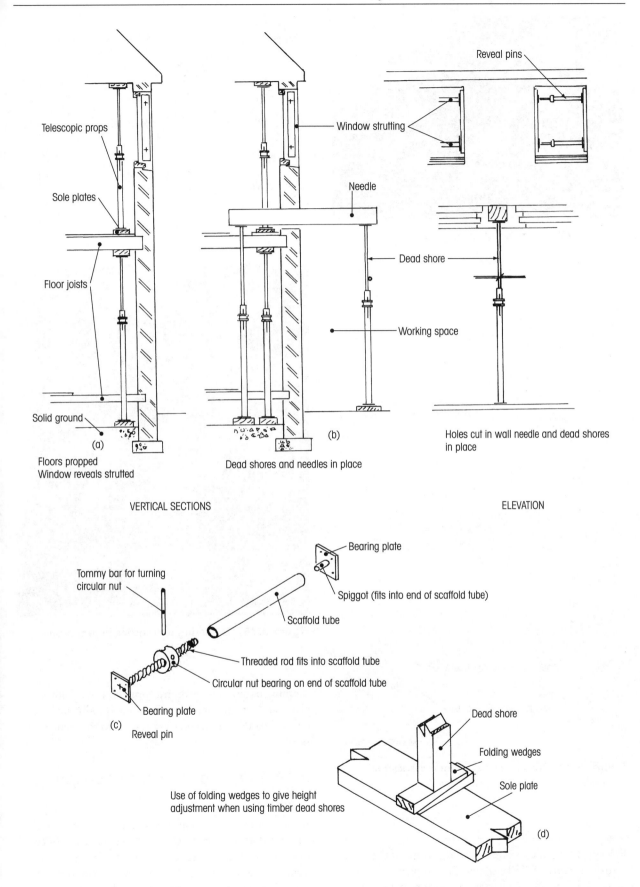

Telescopic props

Sole plates

Floor joists

Solid ground

(a)

Floors propped
Window reveals strutted

VERTICAL SECTIONS

Window strutting

Needle

Dead shore

Working space

(b)

Dead shores and needles in place

Reveal pins

Holes cut in wall needle and dead shores
in place

ELEVATION

Tommy bar for turning
circular nut

Scaffold tube

Threaded rod fits into scaffold tube

Circular nut bearing on end of scaffold tube

Bearing plate

(c)

Reveal pin

Bearing plate

Spiggot (fits into end of scaffold tube)

Dead shore

Folding wedges

Sole plate

Use of folding wedges to give height
adjustment when using timber dead shores

(d)

Figure 5.73   Dead shoring

Figure 5.73 Examples of raking shores

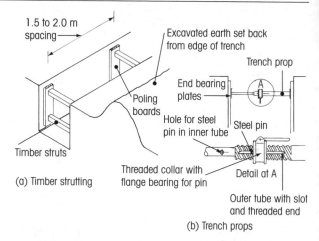

(a) Timber strutting

(b) Trench props

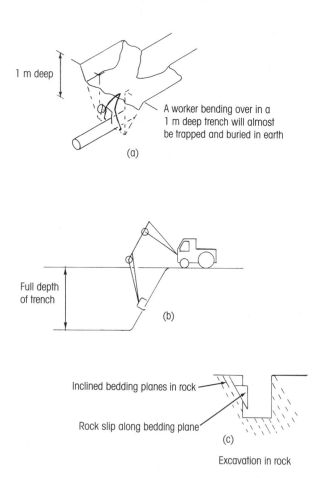

Figure 5.74 Trench excavation and associated dangers

Figure 5.75 Timbering to trenches in firm ground

slipping, lipping pieces on the struts to hold them in place, and wedges between walings and individual poling boards to keep then tight back to the trench sides.

5.14.4 Proprietary Trench Support Systems

A number of trench support systems have been developed using pairs of stiffened steel panels connected to hydraulic struts. After excavating a suitable length of trench, using a crane or the arm of the excavator, the pair of panels with the struts retracted are lowered into the trench. From outside the trench an operative using a

reduces the cantilever effect on the bottom of the close boarding. Figure 5.77(b) shows the complete excavation with the timbering in position including vertical puncheons to stop the walings from

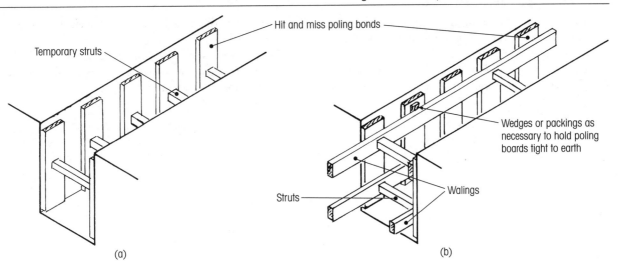

Temporary struts

Hit and miss poling bonds

Wedges or packings as necessary to hold poling boards tight to earth

Struts

Walings

(a)

(b)

Initial arrangement of timbering

Final arrangement of timbering

Figure 5.76 Trench timbering in moderate ground

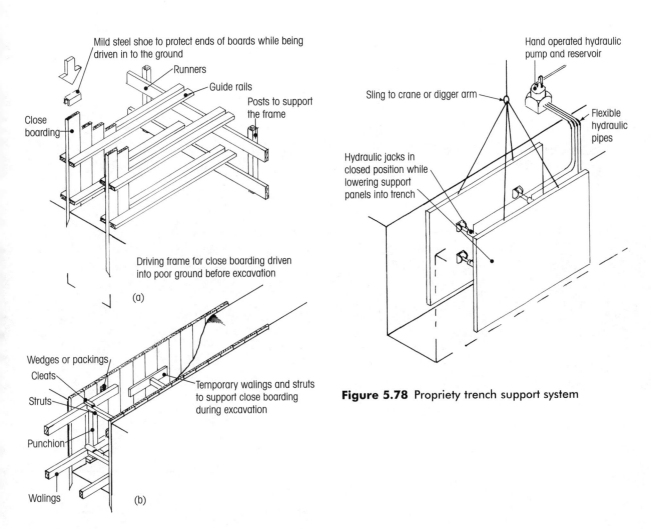

Mild steel shoe to protect ends of boards while being driven in to the ground

Runners

Guide rails

Posts to support the frame

Close boarding

Driving frame for close boarding driven into poor ground before excavation

(a)

Wedges or packings

Cleats

Struts

Temporary walings and struts to support close boarding during excavation

Punchion

Walings

(b)

Figure 5.77 Timbering to a trench in poor ground

Hand operated hydraulic pump and reservoir

Sling to crane or digger arm

Flexible hydraulic pipes

Hydraulic jacks in closed position while lowering support panels into trench

Figure 5.78 Propriety trench support system

manually operated hydraulic pump expands the struts which in turn press the panels outwards against the side of the trench (Figure 5.78).

Note. Whichever system is used for supporting the sides of trenches there must be a daily inspection to see that the timbering is tight and secure against the sides of the trench.

Never take chances – always timber the sides of open trenches. Saving a few pounds by taking a risk and not timbering may cost you a limb or even a life.

6
Wooden doors

In this chapter we will be looking at different types and style of door, their hanging, and support framework, together with any necessary hardware to enable them to operate.

6.1 GROUPING DOORS

Doors can be divided into four groups according to the appearance or method of construction, for example:

1. Unframed matchboarded door:
   • Ledged and braced battened (matchboarded) door.
2. Framed doors:
   • Framed ledged and battened (matchboarded) door.
   • Framed ledged and braced battened (matchboarded) door.
   • Panelled (solid and/or glazed) door.
   • Louvered door.
3. Moulded panel door.
4. Flush doors:
   • Plain door.
   • Embossed door.
   • Glazed (vision panel) door.
   • Blank door.

Figure 6.1 shows how some of the above doors may appear.

They can be further subdivided according to the location, function, and operation. For example:

1. Location:
   • Exterior.
   • Interior.
2. Function:
   • Fire resistance.
   • Security.
   • Permanent ventilation (louvered).
   • Sound insulation.
   • Mobility.
3. Operation:
   • Side hung single leaf door (inward or outward opening).
   • Side hung double leaf door (inward or outward opening).
   • Sliding single leaf door.
   • Sliding double leaf door.
   • Swing doors.
   • Up-and-over door (garage door).
   • Sliding garage and industrial doors.
   • Folding garage and industrial doors.

Exterior doors – must be designed to prevent the elements (rain, wind and snow) from entering the building, and be strong enough to provide security – they will therefore be thicker heavier and more substantially built than interior doors.

Interior doors – the primary function is to close openings in room divisions and to offer privacy from within. They are not normally built for strength, but there are, however, a few exceptions. For example, the outer doors to flats or apartments, although technically may well be sited within the building they will require to be secure, therefore standard exterior doors would be used.

Fire doors (also see Section 6.1.8) – these doors are specifically designed to be fire resistant. In the event of the fire breaking out within a compartment (room) they would be expected to contain the fire for a prescribed period of time, thereby giving occupants of the building time to escape.

Louvered doors (also see Section 6.1.7) – permanently slatted opening within the door provide a means of allowing ventilation into to a building, or compartment.

Mobility doors – these are special doors which close openings wide enough for a wheelchair to pass.

Double doors – provide a wide means of access to adjoining rooms, a patio, or garage.

Depending on type, doors may be side hung, sliding, or as with garage doors supplied with an 'up-and-over' mechanism.

Door sizes – a range of sizes are shown in Tables 6.1–6.4, some of which are available in both Imperial and metric sizes.

6.1.1 Ledged and Braced Battened (Matchboarded) Doors

This type of door is usually associated with outbuildings, sheds and garages. Provided the door is well made it should withstand harsh treatment, and if regularly maintained remain serviceable for many years [*BS459:1988*].

Construction

Figure 6.2 shows in authographic projection a typical arrangement for this type of door. This very simply constructed door is made up of the series of 'V' or beaded tongued and grooved matchboard (Figure 6.2a) – known as a 'battens'. These battens are nailed to horizontal members called ledges (if the door opens outwards the top edges of the ledge should be bevelled

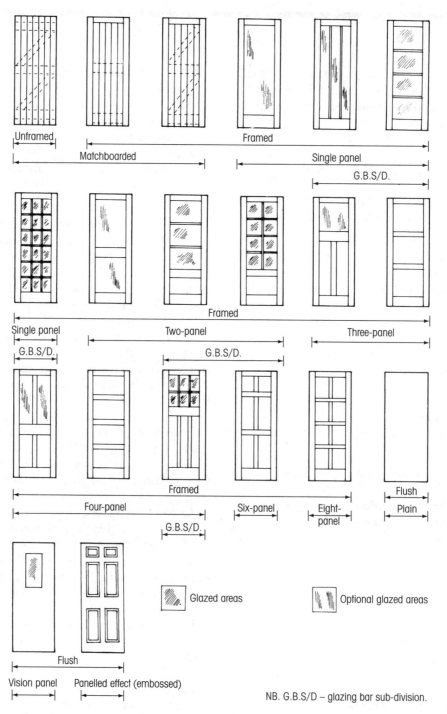

NB. G.B.S/D – glazing bar sub-division.

Figure 6.1   Examples of single leaf door patterns

Table 6.1 Exterior door sizes

Height	Width	Thickness
Exterior metric single leaf doors		
2000 mm	807 mm	44 mm
2040 mm	726 mm	44 mm
2040 mm	826 mm	44 mm
Exterior metric double leaf doors		
2000 mm	1106 mm	44 mm
Exterior imperial single leaf doors		
6'6" (1981)	2'0" (610)	1¾" (44)
6'6" (1981)	2'3" (686)	1¾" (44)
6'6" (1981)	2'6" (762)	1¾" (44)
6'6" (1981)	2'9" (838)	1¾" (44)
6'8" (2032)	2'8" (813)	1¾" (44)
Exterior imperial double leaf doors		
6'6" (1981)	3'10" (1168)	1¾" (44)

Table 6.2 Internal door sizes

Height	Width	Thickness
Interior metric single leaf doors		
2040 mm	626 mm	40 mm*
2040 mm	726 mm	40 mm*
2040 mm	826 mm	40 mm*
Interior imperial single leaf doors		
6'6" (1981)	2'0" (610)	1⅜" (35)
6'6" (1981)	2'3" (686)	1⅜" (35)
6'6" (1981)	2'6" (762)	1⅜" (35)
6'6" (1981)	2'9" (838)	1⅜" (35)
Interior imperial double leaf doors		
6'6" (1981)	3'10" (1168)	1⅜" (35)
6'6" (1981)	4'6" (1372)	1⅜" (35)

*Fire doors are 44 mm thick.

Table 6.3 Mobility door sizes

Height	Width	Thickness
Mobility single leaf doors		
2040	926	40*

*Fire doors are 44 mm thick.

Table 6.4 Garage doors (nominal door opening size)

Height	Width	Thickness
Double leaf side hung wood **garage** doors		
6'6" (1981)	7'0" (2134)	1¾" (44)
7'0" (2134)	7'0" (2134)	1¾" (44)
Single leaf timber clad up-and-over **garage** doors		
6'6" (1981)	7'0" (2134)	
7'0" (2134)	7'0" (2134)	
7'0" (2134)	7'6" (2286)	

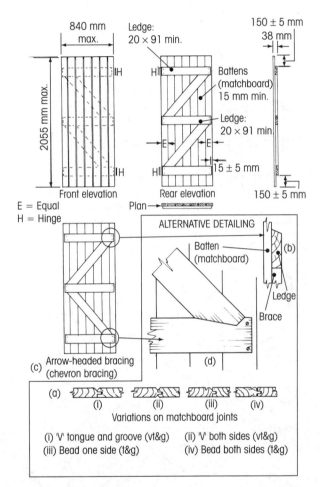

Figure 6.2 Ledged and braced battened (matchboarded) door

(Figure 6.2b) to help disperse any rainwater). These ledges are held square by similarly fixing diagonal members called braces.

Bracing – as can be seen from Figure 6.3 if the door as a whole is to remain square in-service, it will require bracing, and just as importantly the braced must be in the right position to support it – always pointing upwards away from hanging side (hinged side) of the door. Otherwise, as shown in the illustration, bracing the door in the opposite direction will more than likely result in the door sagging.

This would mean that this type of door should be handed (designed to hang on either its left- or right-hand side). To avoid this, manufacturers either supply the braces loose to be fixed in situ (on-site) to suit the handing, or fix them in a chevron pattern (Figure 6.2c) – in this way only one of the braces is fully effective. To further increase the effectiveness of the brace, particularly with heavier doors, ends of the braces can be cut into the ledges using a bird's mouth joint (Figure 6.2d).

Assembly

Figure 6.4 illustrates the various stages of making this type of door by hand, for example:

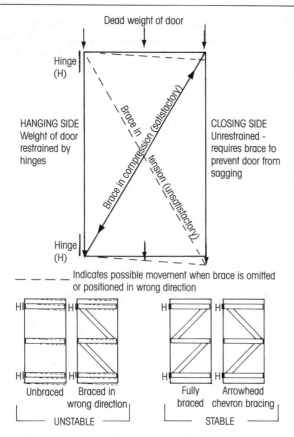

Figure 6.3   Principle of using a brace

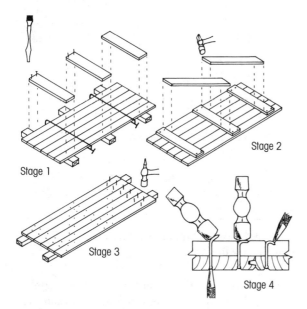

Figure 6.4   Sequence of assembling a ledged and braced battened (matchboarded) door

Stage 1

(a) Cut the battens and ledges to length.

    Note. Before cutting the ledges determine whether the door is to open inwards or outwards as shown in Figure 6.6, if it is to open inwards then the ledges will be full width of the door. Whereas if it opens

outwards ledges will be setback 15 ± 5 mm from the edge to clear the rebated door check.

(b) Check that the ledges are not twisted (plane them out of twist if necessary).

(c) Paint all the tongues and grooves and the ledge faces that will come in contact with the battens with a proprietary preservative.

(d) Rest the battens face down across bench bearers, then lightly cramp the battens together.

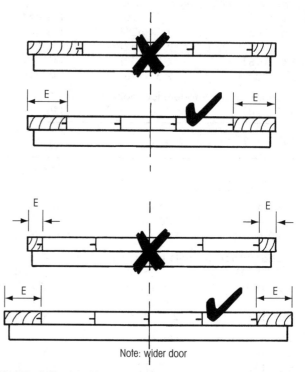

NB. Edge battens must be equal (E) and of sufficient width to allow for fixing - achieved by placing either a batten or a joint to the centre of each ledge

Figure 6.5   Cutting edge battens

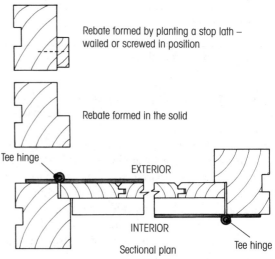

NB. Different width of rebate

Figure 6.6   Ledges in relation to door opening

Note. As shown in Figure 6.5 both edge battens must be equal and not less than half the width of a matchboard.

(e) Double screw each ledge to the cramped edge battens.

Stage 2
(a) Cut the braces to fit between the ledges.
(b) Edge nail the braces to the ledges (ensure that the door is kept square during this operation – if necessary, use ends stops).

Stage 3
(a) Turn the door over on to bench bearers, which have been placed lengthways.
(b) Double nail each battened to ledges and braces, taking care to avoid the bearers, as after the nails have been punched below the surface they will protrude 10 mm through the door.

Stage 4
(a) Turn the door back onto its face, were the protruding point of each nail will be visible.
(b) Clench each nail, by bending its over in the direction of the grain, and then punch the clench below the surface of the wood.

Hardware (ironmongery) for matchboarded doors (see Sections 6.9, 6.10.3, Figures 6.75, 6.84, 6.86).

6.1.2 Framed Ledged and Braced Battened (Matchboarded) Doors

The main difference between this type of door and the ledged braced and battened door is that the battens are constrained within the outer framework of the door, this makes for a much stronger door, which is less inclined to distort. An external door can be made from hardwood, but is more likely to made from softwood, particularly if used in outbuildings.

Construction (Figure 6.7)

(a) *Frame* – the framed part to the door is made up to stiles and a top rail of the same width and thickness. The bottom and middle rails are thinner by as much as the thickness of the battens plus 2 mm by which the battens are recessed to below the framework.

Joints are through mortise and tenons (barefaced tenons will have to be used on the middle and bottom rails with proportions as illustrated – glued, wedged, and pinned with hardwood or metal (star) dowels).

(b) *Bracing* – these must be 'handed' as with ledged and braced battened doors and closely fitted into the corners between stiles and rails. In some cases braces can be omitted.

(c) *Battens (matchboard)* – should be tongued-and-grooved with 'V' or bead joints and tongued or rebated into styles and top rail, then nailed or stapled (with a pneumatic nailing gun) to all framing members.

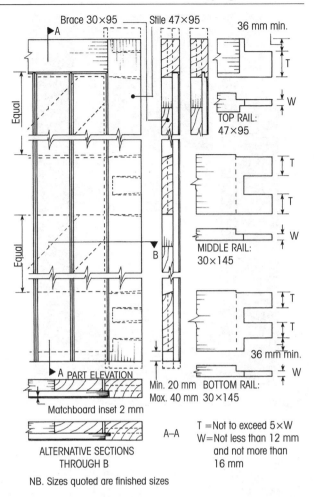

Figure 6.7 Framed ledged and braced battened (matchboarded) door

All tongues grooves and rebates should be pre-treated with a proprietary preservative before assembly.

6.1.3 Stable Door (Type)

Originally intended to contain horse or pony within its stable while still permitting open access at about shoulder level. Two door leaves in one, the top portion opens independently the bottom, which means that the bottom portion can remain closed while the top half is fully open. As shown in Figure 6.8, this type of door has become very popular as a domestic back door leading from the utility room to a garden area.

As can be seen in Figure 6.9 its construction is similar to that of the framed ledged and battened door with the exception that the middle rail is divided by rebates across the full width of the door.

Some manufacturers have replaced the battens (matchboard) with sheet MDF, which has been grooved on its faced to imitate the effect of matchboard. Glazed openings may also be incorporated into the design like the one shown in Figure 6.8.

Figure 6.8 Example of a stable door type

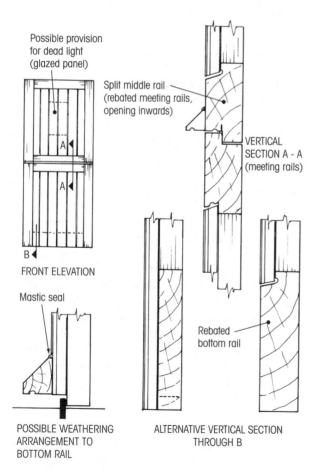

Possible provision
for dead light
(glazed panel)

Split middle rail
(rebated meeting rails,
opening inwards)

VERTICAL
SECTION A - A
(meeting rails)

B◄

FRONT ELEVATION

Mastic seal

Rebated
bottom rail

POSSIBLE WEATHERING
ARRANGEMENT TO
BOTTOM RAIL

ALTERNATIVE VERTICAL SECTION
THROUGH B

Figure 6.9 Stable door type with alternative detailing

6.1.4 Panelled Doors

Framed doors of hardwood or softwood which can
house one or more panels (Figure 6.1) in a variety of
styles. The panels may be of solid wood, manufactured
board, glass, or their combination.

Figure 6.10 identifies different members used in the
construction of a four-panelled door together with four
different panels' styles.

Door construction

Framework – two vertical stiles, with two or more
horizontal rails (top rail, bottom rail, one or more intermediate rails) in some cases intermediate vertical
members called 'muntins'. Joints between members are
mortised and tenoned or dowel jointed.

Horizontal and/or vertical glazing bars can be used to
subdivide panelled openings even further.

Joints (Figure 6.11)

(a) *Mortise and tenon joints* – top, bottom, and with the
exception of single panel doors, at least one rail
should be though jointed – stub tenons into stopped
mortise holes can be used to join any other rails, or
muntins to rails, also glazing bars to stiles and or
rails. As shown in Figure 6.12, foxed wedging
methods may be used as an alternative to just
gluing.

Tenon thickness should in nearly all cases be as
near to one-third of the thickness of the members
being joined as is practicable (depending on the
mortise chisel width). The tenon should not exceed
five times its thickness. Haunches should be as deep
as the tenon is thick and never less than 10 mm.
Tenons should be clear of the top and bottom
of the door by a distance of not less than 36 mm
(Figure 6.7).

All joints to be glued and wedged (two per tenon)
– adhesive type will depend on whether the door is
classified interior or exterior use [BENCH AND SITE
SKILLS, SECTION 10.1].

Where moulding has been worked (stuck) onto
the inside edge or edges of the door framework, rail
and bar ends can be through scribed (Figure 6.11a)
which is standard machine scribe practice, or if
cut by hand a stopped scribe can be used
(Figure 6.11b).

(b) *Dowelled through scribed joints* (Figure 6.11c) –
dowels should be of hardwood, or of the same
species as the members being joined – not less than
16 mm diameter, 125 mm long, and at centres not
exceeding 55 mm, with no less than the number of
dowels per joint as shown in Table 6.5.

Dowels to the bottom rail should be up from the
bottom of the door by not less than 45 mm.

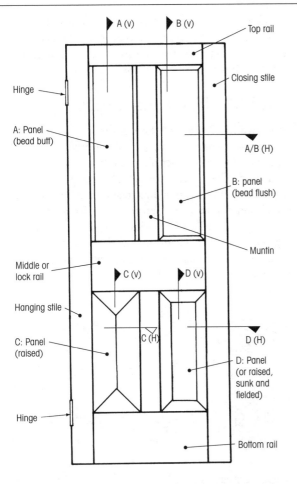

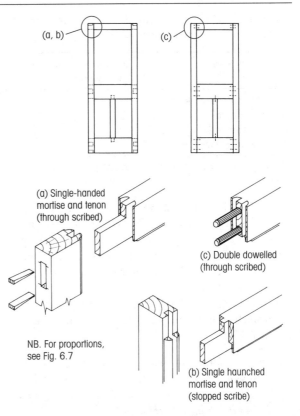

Figure 6.11   Corner joint alternatives

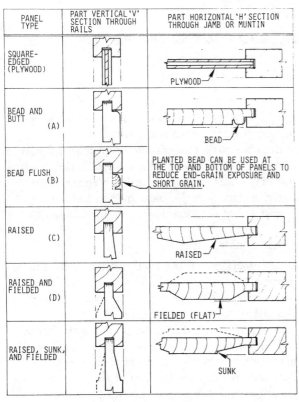

Figure 6.10   Traditional four-panelled door with alternative panel designs

All dowels must be fluted to allow air and surplus adhesive to escape from (prevents adhesive from becoming compressed) the bottom of the dowel hole.

Joints to moulded sections should be through scribed with a tongue to help to stiffen the joint and resist twist.

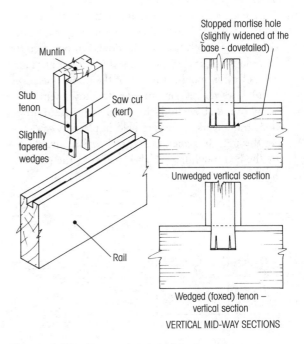

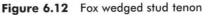

Figure 6.12   Fox wedged stud tenon

Table 6.5 Minimum number of dowels required between interconnecting members

Joints between members	Minimum number of dowels
Stiles to middle/lock rail and bottom rail	3
Stiles to top rail	2
Intermediate members	1*

*Provision must be made to prevent turning.

Panels – Figure 6.10 shows how different panel types fit into the door framework. Notice the clearance left between the panel edge and the bottom of the plough groove – a minimum of 2 mm for plywood and 3 mm for solid wood should be left all round to accommodate any moisture movement.

Framed edges may be left square, but it is more usual to finish them using one of the methods shown in Figure 6.13, for example:

(a) *Stuck moulding* – in this case a moulding (ovolo) has been worked on to both sides of the groove which houses a plywood panel – *panels must be built into the door as it is being assembled.*

(b) *Planted mouldings* – the groove in this case has been formed by nailing a planted mould (bead) onto both sides of the plywood or glass panel – *these panels are fixed after the door frame is assembled.*

(c) *Stuck and planted mouldings* – the rebate is formed with a stuck moulded ovolo, onto which the plywood or glass panel is checked and then held in place with a planted ovolo moulded bead – *again the panels are fixed after the door frame is assembled.*

(d) *Bolection moulding* – panels are built into the door framework as it is assembled – mouldings surround the panel on both faces of the door. On one face the bed moulding is slightly recessed, on the other face the bolection mould is proud of the door. Bolection moulds are rebated on one edge to allow them to stand proud above both the panel and surrounding door framework – they are held in place by screws set through the panel, the screw heads are masked by the bed moulding which is nailed to the panel framework.

Safety glass can be used as a panel not just in exterior doors, but also interior doors to enable light to be transmitted to those areas not so naturally well lit (using borrowed light) (see also Section 6.2.3).

6.1.5 Moulded Panel Doors

As shown in Figure 6.14 these doors are made up of a timber framework enclosing a cellular core which is clad on both sides with either plastics (PVC-u) or hardboard. These facings are moulded to give the appearance of a traditional panelled door. The design can also incorporate an aperture to receive glazing.

Moulded doors with plastics facings are for use as exterior doors, Hardboard moulded doors are for interior use, they are usually pre-treated with a primed textured wood grain finish suitable to receive a paint or wood stain.

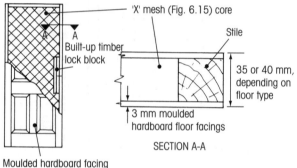

Figure 6.14 Moulded panel effect door

6.1.6 Flush Doors

As the name implies the outer surfaces of the door should, with the exception of protruding glazing beads or embossed features be perfectly flat.

Both interior and exterior qualities are available, their construction, and type of adhesive used in the manufacture of the facings being the main difference. Other factors include door thickness, rigidity, and the provision for fixing door hardware (ironmongery).

Door construction

Framework (Figure 6.15) – with the exception of those doors with a laminated solid wood core, some form of outer framework will be required to retain the core material, or to provide a fixing for rails.

All the rails, including those at the top and bottom will, in the case of hollow doors require a small venting notch or hole to allow movement of air through the door and help disperse any trapped air whilst the facings are being applied during manufacture.

Door framework is assembled in a very simple fashion. Corner and end joints may be mortised and

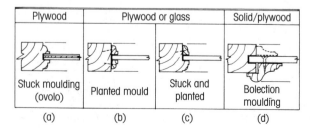

Figure 6.13 Detailing between frame edge and panel

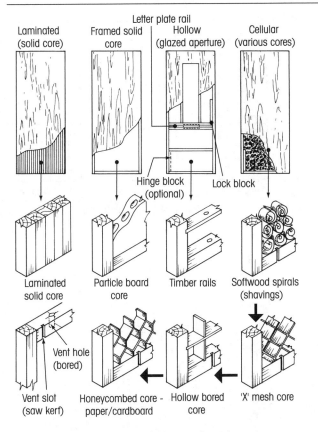

Figure 6.15 Flush door framework and cores

tenoned, but are more likely to be tongued and grooved or butted and stapled together. With cellular doors the main strength is derived from the stressed skin facings (plywood or hardboard) used to clad both sides of the framework and core.

Cores (Figures 6.14, 6.15) – solid cores include solid laminated strips of timber, particle boards (chipboard or flaxboard) [BENCH AND SITE SKILLS, SECTION 2.3]. Hollow cores include rails, which are either tenoned or tongued into mortised or grooved stiles. Cellular cores can be made from a variety of materials such as softwood shavings in the form of spirals formed by shaving short blocks of wood, hardboard strips in a box or X pattern, or even cardboard in a honeycombed pattern.

Facings (Figures 6.14–6.16) – generally made from plywood or hardboard, often with a decorative face

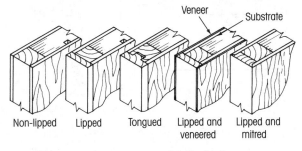

Figure 6.16 Edge treatment for flush doors

veneer of wood, or plastics film. Hardboard can also be lightly embossed (pressed) to give the appearance of a panelled door.

Exterior door facings should be of weather-resistant materials such as exterior grade plywood or oil-tempered hardboard. Facings are glued under pressure to the door framework and core.

Lippings (Figure 6.16) – these are thin laths glued to both edges of the door to cover the edges of the facings. Wood species [BENCH AND SITE SKILLS, SECTION 1.3.1] similar to that used as a facing veneer will usually be used if the door is to be polished. Several methods of applying edge lippings is shown in Figure 6.16.

Glazed flush doors – glazed openings not only allow light to be transmitted from one area to another, but they also provide a vision panel, and in some cases a means of preventing collision when opening the door.

Aperture size and shape can vary – square, rectangular, circular, oval or their combination are all possible with standard doors, *restrictions are however put on the*

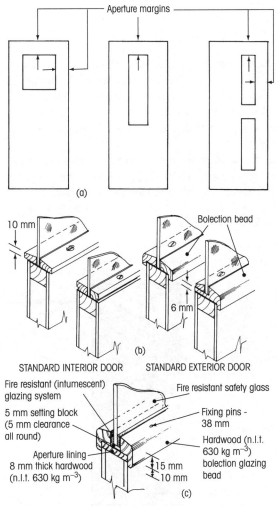

NB. Fire resistant flush doors will require minimum aperture margins of not less than (n.l.t.) 127 mm

Figure 6.17 Flush doors with vision panels (glazed apertures)

size and shape of apertures in fire-resisting doors (see Section 6.1.8). For the door to retain its rigidity, edge margins (Figure 6.17a) at the top and sides are usually kept to a minimum (excluding the thickness of the lipping) of 127 mm, and 200 mm at the bottom and separation of apertures. If the door is non-solid, then as shown in Figure 6.15 the blocking rail around the aperture will have to be provided to stiffen the door and provide a means of fixing the glazing beads (Figure 6.17b). Before exterior doors are beaded all the edges of the aperture should sealed with waterproof tape.

All glazing must be in accordance with current *Building Regulation* and *BS6206* (see Section 8.13, Figure 8.42) – with regard to the use of safety glass and its application. As shown in Figure 6.17(c) glazing to fire-resisting doors will requires special attention to the type of fire-resistant glass, intumescent glazing system, and type, size, and means of fixing the glazing bead (see Section 6.1.8). All must be applied in accordance with the manufacturer's instructions, and comply fully with the appropriate British Standard (Euro Code).

6.1.7 Louvered Doors

In most cases the purpose of a louver set into a door would be to provide permanent ventilation through it.

Figure 6.18 shows how this can be achieved without the loss of privacy and in the case of an exterior door (depending on blade pitch and projection) weather protection.

Figure 6.19 shows how a louver within an exterior door may be arranged and constructed, whereas Figure 6.20 shows how a louvered arrangement can be incorporated into an interior door to both functional and decorative effect.

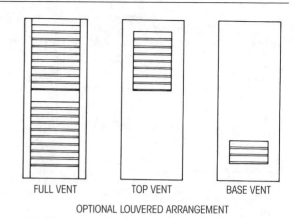

OPTIONAL LOUVERED ARRANGEMENT

FULL VENT   TOP VENT   BASE VENT

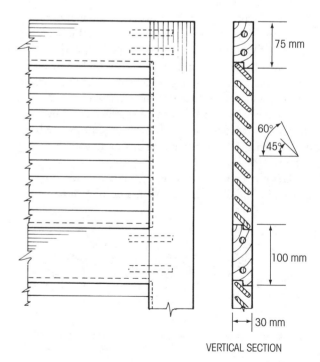

Figure 6.19   Louvered exterior doors

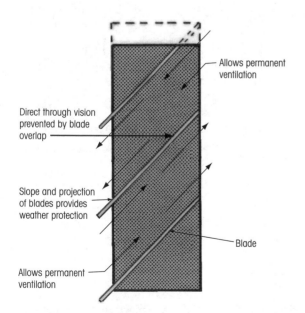

Figure 6.18   Principle of using a louvered opening

Figure 6.20   Interior louvered screen door or panel

6.1.8 Fire Doors [BS8214:1990]

Fire doors allow the normal movement of persons and objects between compartments. When closed, they must together with their associated hardware (ironmongery), provide a prescribed period of fire resistance to satisfy current regulations.

Should a fire break out a fire door should be capable of performing two main functions:

1. Contain the fire within the affected area while allowing persons to safely travel either horizontally, vertically, or both to vacate the building.
2. Help prevent the fire spreading and limit the movement of smoke into designated escape routes.

Fire doors must therefore be self-closing at all times.

Each type of door after testing is classified according to its fire resistance – its integrity is quoted in minutes, which could be as low as 20 or as high as 90. Chosen ratings will depend on their intended use [*BS5588*].

To achieve an approved fire-resistance certification, the door assembly which would include the sample door, fitted into its frame together with all relevant hardware (ironmongery), such as hinges, latches, and/or a self-closing device. Must undergo stringent fire tests by an approved testing organisation in accordance with the procedures set down in the relevant legislative documentation [*BS476: pt 8 or 22*] to ensure that their prescribed duration of fire resistance is met.

It is important that a correctly rated fire door is fitted in the specified location [*PD 6512 pt1*]. Therefore a means of site identification is needed. Identification is provided by a means of either a colour-coded label or plug let into the hanging stile of the door as shown in Figure 6.21 [*BS8214:1990*]. Table 6.6 shows a few common examples of how different rated fire doors (FD) are colour coded.

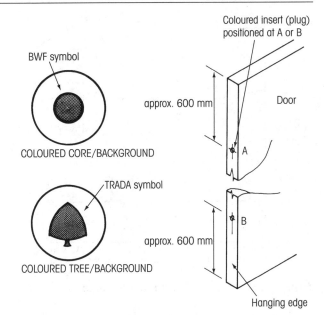

Figure 6.21   Fire door identification markings

Note. Remember that when a **red** core is used it means that an intumescent seal must be fitted at the time of installation.

Intumescent strips (seals)

An intumescent material is one which expands when subjected to high temperatures – these materials are available as paint, paste, mastic, or as in the case of door to frame edges in strip form which is usually encased in a plastics sleeve.

The object of using an intumescent strip between the edge of the door leaf and frame rebate is to seal the clearance gap around the door in the event of a fire. It also locks the door into the frame when the closing mechanism fails.

Table 6.6   Examples of some of the colour codes used as fire door performance identification

Plug core colour	Plug background colour	Fire resistance rating: Fire door integrity* (mm)	Requirements
Red	White	FD20	Intumescent seals **must** be site fitted
	Yellow	FD30	
	Blue	FD60	
Green	White	FD20	**No** additional intumescent seals required
	Yellow	FD30	
	Blue	FD60	
Blue†	White	FD20	**Without** intumescent seals fitted
	White	FD30	**With** intumescent seals fitted into door edge or frame.

FD – fire door.

   *Integrity – period of time (min) during which a door resists the passage of flame, or hot gasses, or when flaming appears on the face of the door away from the fire.

   †Only doors constructed to satisfy both FD20 (without intumescent seals fitted) and FD30 (with intumescent seals fitted) performance criteria are marked with this coloured plug.

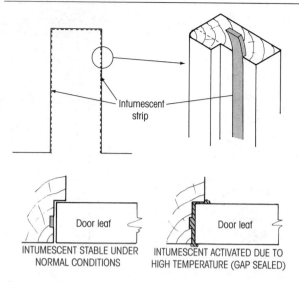

Figure 6.22   Fire door to frame treatment

These strips are as shown in Figure 6.22 usually fixed into grooves cut centrally into the reveal of both frame jambs and head. In some cases they may be fixed into the door leaf.

Most of these strips are capable of sealing gaps of up to 5 mm wide. Depending on product type they activate at high temperatures between 140 and 300°C, this causes them to foam and expand up to four or five times their volume. It is therefore important that fitting clearance gaps between jambs and head do not at any point exceed 4 mm. But as you will see in Table 6.7 we are looking at clearance gaps of around 2–3 mm.

Note. Intumescent seals are an integral part of the fire-resistance rating given to the door leaf in conjunction with its approved frame as with a door set (see Section 6.5).

Glazed Apertures [BS6262]

Apertures must not be cut into a door leaf for the purpose of glazing unless a test certificate has been issued for such a door of similar size, type, and construction. Some manufacturers allow apertures to be

Table 6.7   Door-leaf edge clearances

	Door-leaf edge clearance	
Frame/lining member	Internal doors	External doors
Jamb leg	2	2.5
Head	2	2.5
Transom	2	2.5
Threshold/Sill	3	3.5

Note. The permitted tolerance on clearances between the door-leaf and frame/lining members is +1.0 mm to –0.5 mm. For example, an accepted clearance of 2 mm (thickness of a 2p coin) may be increased to 3 mm or decreased to 1.5 mm.

cut providing all their restrictions are fully observed. These may include:

- Permitted opening size.
- Opening shape.
- Position of opening in relation to door leaf edges, or adjacent openings.
- Modification to core material.
- Special glazing beads and a fire-resistant glazing system.

Glass – Georgian wired, or special fire-resistant safety glass, may be specified. Figure 6.17(c) shows a typical arrangement for fixing the glass when using glazing beads and an intumescent glazing system.

Note. Glazing systems, as a whole should only be used if test evidence warrants its use – door manufacturers usually offer this assurance.

Door Hardware (Ironmongery)

All items such as hinges, latches, closers, and door furniture (see Section 6.9) should have the equivalent specification as described on the test certification.

Hinges (butt hinges) – they must be constructed of non-combustible material such as steel, or stainless steel. Some regulations will quote that the material must have a melting point greater than 800°C. The number of hinges per leaf will depend on the weight of the door, but not less than three (1½ pairs).

As shown in Figure 6.23 the hinges used with fire doors FD30 can interrupt the intumescent strip but, with doors rated as FD60 the intumescent seal needs to be continuous with an adequate amount by-passing the hinge (this will also apply to other items of ironmongery).

Mortise lock and latches – only those types approved under a fire test certification must be used. The latch body should be as slim as possible – thus reducing the amount of metal which would conduct heat in the event

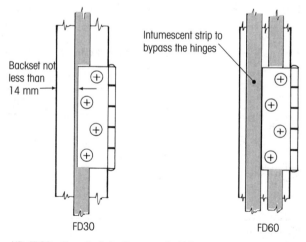

NB. FD60 will require twice the amount of intumescent strip than FD30

Figure 6.23   Butt hinges in relation to intumescent strip

of a fire. Inaccuracies (holes) should be made good with intumescent paste, mastic or other approved fire stopping material.

Knobsets (see Figure 6.67) should not be used unless satisfactory test evidence is provided because they require a large hole to be cut through the door leaf to house the lock/latch mechanism.

Closing devices (closers) – all fire doors will require an approved self-closing device (as previously mentioned rising butt hinges are not permitted for this purpose).

If the door is to be latched then the device must be strong enough to overcome the striking plate, if unlatched then able to retain the door leaf in the closed position during a fire.

Face fixed overhead closers (Figure 6.68) are the most common types of fire door closers – only those supported by a fire door approved certification should be used.

Spring link and chain types which are housed within a barrel cut into the hanging stile of the door may be used to latch a single door leaf, but as with overhead closers they must be supported by evidence of suitability for use in fire doors in the situation specified.

Barrel bolts – may be required to restrain one leaf of a double door set. In which case it should be face fixed with brass or steel screws of not less than 30 mm long.

Letter plates (letterbox) – any letter plate should be an approved type.

Under simulated fire conditions the upper portion of the doorset would be subjected to a positive pressure whilst the lower potion would be under negative pressure. For this reason as shown in Figure 6.24, the letter plate because it means cutting a hole through the door should be fixed within the neutral pressure zone of the door. The neutral zone can be taken to be between 800 and 1000 mm up from the threshold. The letter plate hole should not exceed 250 mm long by 38 mm wide.

Special fire-resistant letter plates are available and only these should be fitted (Figure 6.76).

Smoke Control Doors

Most casualties associated with fires are as a result inhalation of toxic fumes and suffocating smoke. It seems sensible therefore that smoke should be prevented from entering those areas designated as protected escape routes.

So those doorsets requiring smoke seals will also be fire-resisting doors. These doors can be identified on a specification by the suffix 'S' – so a fire door requiring smoke seals could for example read as 'FD30S'.

There are two main types of seal available:

1. *Compression seals* (Figure 6.25a) – usually fitted into the door stop as a compressible bead.
2. *Wiping seals* – in the form of either a blade or brush fitted to either the door leaf edge or within the frame reveal.

A combined wiping seal and intumescent strip is as shown in Figure 6.25(b) is available.

Door frames suitable for fire doors

Standard door frames are dealt with in the next section. These frames differ in that they are generally pre-grooved to receive intumescent strip (Figure 6.23) – one groove for FD30 doors, and two grooves (twice the amount of intumescent) for FD60 doors, and special attention paid to the type and condition of the timber.

FD20 and FD30 doorframes should be made from straight grained softwood or hardwood with a minimum density of 420 kg/m^3 [BENCH AND SITE SKILLS, SECTION 1.10.4b] free from twist, spring, bow, cup, splits, shakes, checks, and with no dead knots. Moisture content should be within the region of 12% ± 2%. Minimum section size (excluding door stop) 32 mm by 85 mm with a door stop depth of 12 mm.

FD60 doors – in this case we are looking at using timber with a density of not less than 650 kg/m^3 which puts it into the hardwood range – again the condition of the timber and its moisture content should be as FD30 doors. Section size will be increased to 41 mm by 85 mm.

Frame and lining jambs should be fixed back to their surrounding support material with steel screws at centres not greater than 500 mm.

Junctions between door framework and surrounding structure must be adequately sealed – ideally, no gaps should be present, but this is highly unlikely and often

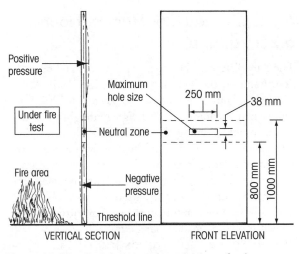

Figure 6.24 Determining the position of a letter plate within a fire-resisting door

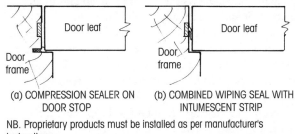

NB. Proprietary products must be installed as per manufacturer's instructions

Figure 6.25 Smoke control door seals

unpractical. Method and procedure for sealing these gaps will depend on the size of the gap and required fire resistance. For example, a FD30 doorset within a non-load-bearing wall with gaps smaller than 10 m wide all round. Can be protected by using 15 mm thick architrave, providing it is well fitted and overlaps the wall by at least 15 mm. If the gap exceeds 10 mm then the void will require packing with mineral or glass wool to a depth of at least 10 mm, or by applying a proprietary intumescent mastic sealant. With load-bearing walls and FD60 doors, direct reference should be made to BS8214:1990, Tables 2 and 3.

6.2 DOOR FRAMES

Door frames should be strong enough to carry a hung door or doors (door leaf(s) and any accompanying hardware) without relying solely on a wall for their support, although they will require fixing to it. They are usually associated with external doors, but as can be seen from Figure 6.26 this is not always the case.

Door frames can consist of as many as five components, for example:

- two jambs – vertical side members
- one head – top horizontal member.
- one sill (cill or threshold) – bottom horizontal member.
- one transom – intermediate horizontal member.

With the exception of large openings, door frames usually arrive on site ready assembled as a separate item, or as part of a door set (see Section 6.5).

6.2.1 Open Door Frames (Figure 6.26a, b)

The omission of a sill or threshold allows for clear uninterrupted movement through the doorway at floor level. This arrangement will be required for entrances to public buildings, shops, garages, etc., also where wheel chair mobility is required.

Assembly – head and jambs are joined together by a mortice and tenon. Figure 6.27(a) shows the assembly where the horns are used as a means of securing the frame head to the wall – cutting the horn on the splay (Figure 6.27b) allows the face wall to lap and totally enclose it and make ready for the lintel above. Alternatively, if the frame is be fixed into a pre-formed opening (Figure 6.28) then the horn will need to be cut back. Therefore (as shown in Figure 6.27c) it would be advisable to use a haunched mortice and tenon joint [BENCH AND SITE SKILLS, SECTION 7.5.7].

Assuming that the assembly has been pre-treated with a wood preservative [BENCH AND SITE SKILLS, SECTION 4.3.1], then the joints should be glued with a suitable adhesive [BENCH AND SITE SKILLS, SECTION

10.0], cramped, wedged, and doweled. If cramping is not practicable (as shown in Figure 6.27d), the joint could be 'draw bored' by driving a hardwood dowel through off-centred holes. In this way the shoulders of the tenon are pulled up tight. Because the door frame is missing a sill, two temporary distance pieces will be required. One at the foot of the jamb and the other midway of the jambs height. This will prevent them bulging as the wall is being built up. Top corners will require diagonal bracing to keep the frame square.

Fixing – if the frame is to be built into the structure then temporary propping will be required (Figure 6.27e). When accurately positioned – both level and plumb (vertical) – the permanent securing process can begin. The foot of each jamb is held in place by a galvanised steel dowel (peg). Which is either set into a pre-formed hole to be later grouted (filled with a mix of sand and cement) into position, or they may be left until the sand and cement floor screed is laid over the concrete floor slab (oversite concrete) – this task is usually undertaken by the plasterer. Figures 6.27(f–h) show different anchoring methods – the later one (concrete upstand) could be used in situations where the floor area is constantly being swilled, thereby reducing the risk of moisture uptake into end grain of the foot of the jamb.

As the walls are being built up, wall clamps or other fixing devices are attached to the back of the jambs and built into the wall at centres of between 500 and 600 mm depending on the type of door set (door leaf, frame and hardware).

If the frame is to fixed into a pre-formed opening produced by the use of a temporary door frame profile, like the one shown in Figure 6.28, other fixing devices will be used. For example, wood or plastics plugs (frame fixings) [BENCH AND SITE SKILLS, SECTIONS 11.6, 11.7], or wood pallets (Figure 6.33 – wood slips) inserted between mortar joints by the bricklayer as the wall is built-up.

6.2.2 Closed Door Frames (Figure 6.26c, d, g, h)

The only difference between these frames and open door frames is the inclusion of a sill (threshold). As shown in Figures 6.29 (section C–D) and 6.30(d), these frames may include in their design a fanlight (glazed and/or vented opening above the door). Or, side lights (glazed areas) to the sides of the door opening as with a French casement (French window), or vestibule frame.

Figure 6.29 shows how different frame members are arranged to suit either an inward or outward opening door. If the framework is constructed of softwood then the sill should be made from a durable hardwood [BENCH AND SITE SKILLS, SECTION 1.10.4e]. Corner joints may be mortice and tenoned or combed jointed (see Section 8.9.4 and Figure 8.22). Joints between transom

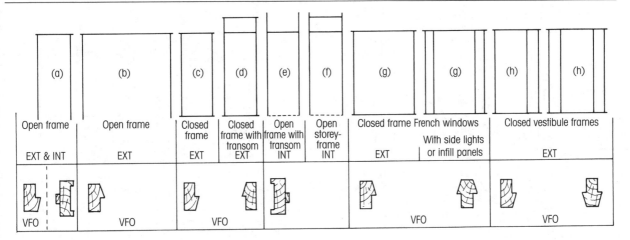

(a)	(b)	(c)	(d)	(e)	(f)	(g)	(g)	(h)	(h)
Open frame	Open frame	Closed frame	Closed frame with transom	Open frame with transom	Open storey-frame	Closed frame French windows	With side lights or infill panels	Closed vestibule frames	
EXT & INT	EXT	EXT	EXT	INT	INT	EXT		EXT	
VFO	VFO	VFO				VFO		VFO	

EXT = Exterior   INT = Interior   VFO = Viewed From Outside

Figure 6.26   Door frames

rails and jambs should be through mortice and tenoned. Adhesives should be at least – type 'boil resistance' (BR) [BENCH AND SITE SKILLS, TABLE 10.1].

Fixing options are the same as the open door frame, the only difference being that the sill horn will need cutting back similar to the head horns but the front potion of the sub-sill will have to protrude beyond the face wall and sideways past the reveals – the same

procedure is carried out with window sills (see Section 8.7.1, Figures 8.12a–c).

6.2.3 Storey frames (Figures 6.26d–f)

In this case the jambs are at least as long as the room is high – hence the name storey frame. The frame may incorporate a fanlight, which can be used to provide

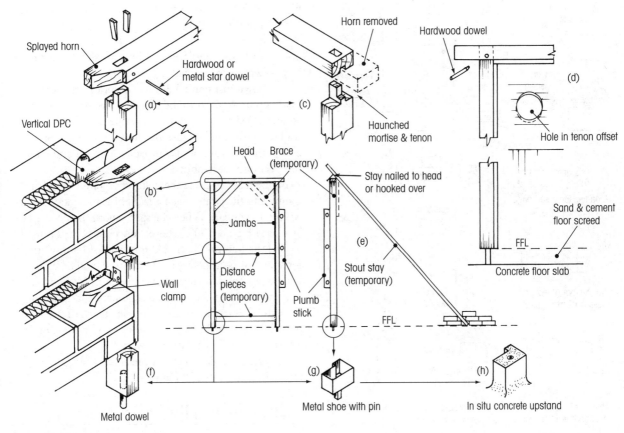

FFL = Finish Floor Level

Figure 6.27   Open door frames assembly and fixing

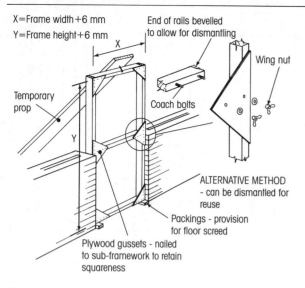

X=Frame width+6 mm
Y=Frame height+6 mm

End of rails bevelled to allow for dismantling

Wing nut

Temporary prop

Coach bolts

ALTERNATIVE METHOD - can be dismantled for reuse

Packings - provision for floor screed

Plywood gussets - nailed to sub-framework to retain squareness

Figure 6.28 Exterior temporary door frame profile

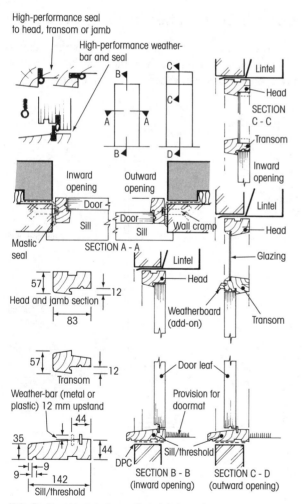

High-performance seal to head, transom or jamb

High-performance weather-bar and seal

Inward opening

Outward opening

Door

Sill

Door

Sill

Wall cramp

Mastic seal

SECTION A - A

57

12

Head and jamb section

83

57

12

Transom

Weather-bar (metal or plastic) 12 mm upstand

35

44

44

9   9

9

142

DPC

Sill/threshold

Lintel

Head

SECTION C - C

Transom

Inward opening

Lintel

Head

Glazing

Lintel

Head

Weatherboard (add-on)

Transom

Door leaf

Provision for doormat

Sill/threshold

SECTION B - B (inward opening)

SECTION C - D (outward opening)

Note: For the sake of clarity, mastic seals between framework and masonry have been omitted from the vertical sections

Figure 6.29 Sectional details through exterior door frames

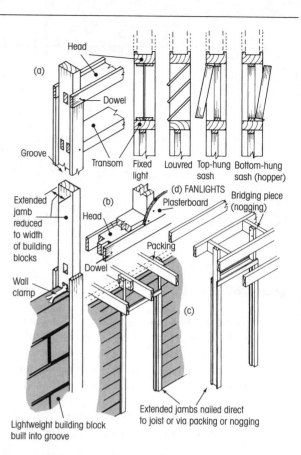

(a)

Head

Dowel

Groove

Transom

Fixed light

Louvred

Top-hung sash

Bottom-hung sash (hopper)

(d) FANLIGHTS

Extended jamb reduced to width of building blocks

(b)

Head

Plasterboard

Bridging piece (nogging)

Packing

Dowel

(c)

Wall clamp

Lightweight building block built into groove

Extended jambs nailed direct to joist or via packing or nogging

Figure 6.30 Internal storey frames within non-load-bearing partition walls

borrowed light, ventilation, or both to an adjacent room or passageway, or infilled.

As shown in Figure 6.30, framework in this case is made up of grooved timber. The groove provides lateral (sideways) support both to the frame and the partition wall set within it. Because the storey frame is erected before the partition it also provides a useful guide for the partitions assembly.

Figure 6.30(a) shows the method of joining head and transom to the jambs when a fanlight is to be incorporated into the design. When the space above the door is to be closed Figure 6.30(b) shows how plasterboard is used to clad the framework, should this space need insulating for sound or heat loss them the cavity should be infilled with an appropriate medium (see Sections 1.35 and 2.3.2).

Figure 6.30(c) shows an arrangement for attaching the extended jambs to the ceiling joists when the partition crosses the joists at either a right angle or, in the same direction as the doorway.

Variations on dealing with the space above the door (fanlight) are shown in Figure 6.30(d).

6.3 DOOR LININGS

Unlike the majority of door frames door linings cover the whole width of the doorway reveal. They are made-

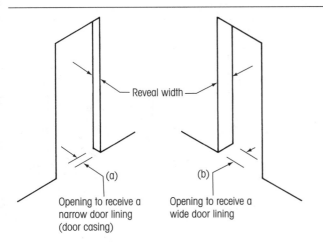

Figure 6.31 Doorway openings in relation to linings

up from thinner sectioned timber and their outer edges are in line with the face wall plasterwork.

We can place them into two categories of requirement:

1. Narrow linings (Figure 6.31a).
2. Wide linings (Figure 6.31b).

Narrow linings are as wide as a single leaf partition wall of brickwork, blockwork, timber or metal studwork together with any covering plasterwork, or other lining material.

Wide linings are built across wider reveals of pre-formed doorways.

The openings which are to receive door linings can be formed with an opening template (template profile) like the ones shown in Figure 6.43 (see Section 6.5, Framing openings), or left in the wall by the bricklayer as the wall is being constructed.

6.3.1 Narrow Door Linings (Often Referred to as Door Casings)

Figure 6.32 shows how a solid (with cut rebate – Figure 6.32a) and plain (planted rebate, formed by attaching a loose stop lath – Figure 6.32b) linings are sited in relation to the walls – also the different methods of making the corner joints (Figure 6.32c–f) between the head and jambs. The joints are assembled using a suitable adhesive [BENCH AND SITE SKILLS, SECTION 10.1] and dovetail nailing [BENCH AND SITE SKILLS, SECTION 11.1.3].

Figure 6.33 show a procedure for fixing an assembled door lining into a pre-formed opening (see also Section 6.5.1, Framing openings). Depending on the wall material fixings may be made by:

1. Nailing into wood pallets (thin timber slips positioned lengthways across the wall) implanted by the bricklayer as the wall is built-up.
2. Nailing into twisted wood plugs (rarely used these days) driven into seams of either brick or blockwork [BENCH AND SITE SKILLS, SECTION 11.6.1].

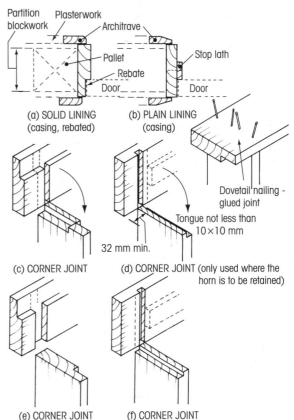

Figure 6.32 Narrow door linings (casings)

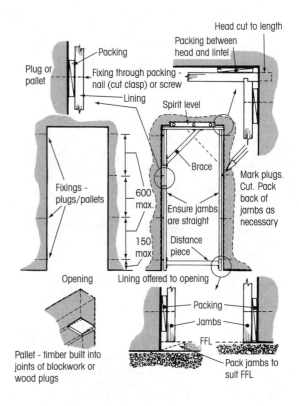

NB. Plastic plugs may not to be used with fire-resisting doors

Figure 6.33 Fixing a narrow door lining

3. Driving cut clasp nails [BENCH AND SITE SKILLS, TABLE 11.1] directly into lightweight building blocks as directed by the block manufacturer.

4. Screws and proprietary framing plugs [BENCH AND SITE SKILLS, SECTION 11.6] made from plastics or soft metal.

Note. The use of plastics plugs may not be permitted with fire-resisting doors.

Jamb linings must be in contact with the wall at all the fixing points. Fixing centres will be between 500 and 600 mm apart depending on the situation (fire-resisting door frames will unless otherwise specified be fixed near the foot and head as shown but with mid-fixings not to exceed 500 mm centres) (see also Section 6.5, Door sets).

6.3.2 Wide Linings

In many cases it is impracticable to use solid timber as a wide lining. Framed panels (Figure 6.34a) or a skeleton framework clad with board material (Figure 6.34b) can be used. In both cases they will require a sub-frame of timber grounds (Figure 6.34c). These are pre-made in the workshop to suit the reveal opening – any type of framing joints can be used [BENCH AND SITE SKILLS, SECTION 7.5], three examples are shown in Figure 6.34(d). Alternatively the outer framework could be clad in MDF (medium density fibreboard).

Much care is required when fixing timber grounds, because of the relatively small sectioned framework constant checks are needed to ensure the frames are not twisted or out of square with the face wall. Some useful checks are shown in Figure 6.35 – the same methods can be used with narrow linings.

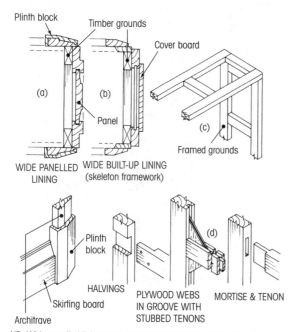

NB. Wide panelled linings and wide built-up linings require framed grounds

Figure 6.34 Wide door linings

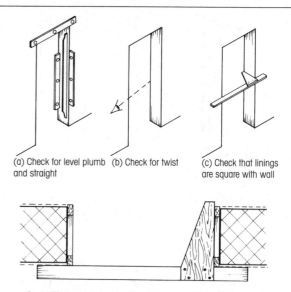

(a) Check for level plumb and straight   (b) Check for twist   (c) Check that linings are square with wall

Reversible template for checking linings are square with face walls

Figure 6.35 Checking linings for straightness level, plumb, twist and square

6.4 DOOR HANGING

We must first ensure that we have the right door pattern to suit the opening. This is found by cross-referencing the door schedule (see the door and hardware schedule in Table 6.8 which relates to the loose drawing in Appendix 1) [BENCH AND SITE SKILLS, TABLE 12.9] with the designers drawings, which will tell you which way round the door is to be hung.

Door hanging is the term used to describe the method of attaching the door to its surrounding structure (door frame or door lining) in such a way that it can be freely opened or closed. It is this opening and closing rotation we call 'door handing'. Figure 6.36 illustrates a method of door handing – it is however more commonly associated with choosing the correct hand of lock or door closer (see Sections 6.9.2 and 6.9.4).

Different type, size, and weight of door will require different hanging techniques. But for the purpose of this chapter we shall consider hanging a side hung standard domestic door into an opening of either a door frame or narrow lining.

6.4.1 Fitting the Door

Before fitting the door into the surround you should first establish whether the doorway is stepped – in other words is there a threshold (open or closed surround framework). Internal door linings may require a threshold strip (carpet step). This is a hardwood board about 12 mm thick set between the jambs and fixed down to the floor decking (nailed onto floor boarding or wood based decking) – fixing to a concrete floor will mean either plugging or the use of a cartridge operated tool [BENCH

Table 6.8 Door and door hardware (ironmongery) schedule example – refer to Appendix 1

| Description | Location | | | | | | | | | | | | | | | | Totals |
	D01	D02	D03	D04	D05	D06	D07	D08	D09	D10	D11	D12	D13	D14	D15	D16	
Doors																	
Number 2 hardwood framed and panelled up and over garage doors 2134 by 2286 × 44 mm	**2**																2
Garage door 2000 × 807 × 44 mm flush exterior grade, fire door to BS476. Cat No. DF30		1															1
Softwood framed ledged and braced door 2000 × 807 × 44 mm to store. Cat No. FL20			1														1
Hardwood exterior six panel front entrance door 2040 × 826 × 44 mm. Cat No. XD05				1													1
Hardwood exterior two panel door 2000 × 807 × 44 mm complete with beads for glazing upper panel. Cat. No. EGXP44					1												1
Internal flush doors faced in Red Oak. 2040 × 726 × 40 mm lipped on the long edges. Cat No. RO1FD897						1	1	1	1	1	1	1	1	1	1		10
Hardwood double glazed sliding patio doors. Complete with hardwood frame 2072 × 2387 mm. Locks and hardware. Cat No. HSPD 69																1	1
Hardware																	
Up and over mechanism locking handle and keys	2																2
100 mm brass steel washered butts no. 3 (1½ pair) per door		1		1	1												4.5 pair
100 mm steel butts no. 3 (1½ pair) per door			3														1.5 pair
75 mm brass butts no. 2 (1 pair) per door						1	1	1	1	1	1	1	1	1	1		10 pairs
Brass letter plate. Cat No. BLP533				1													1
5 lever motice locks. Cat No. ML5017		1		1	1												3
3 lever mortice locks. Cat No. ML3003			1														1
Brass lever handle furniture. Cat No. BLF037				1													1 set
Brass knob furniture. Cat No. BKN0639		1			1												2 sets
Mortice latches with brass face plates. Cat No. MIB853						1	1	1	1	1			1	1	1		8
Mortice latch locking bathroom sets complete with brass plated knob furniture. BBS008											1	1					2
Brass plated knob sets. Cat No. BPK031						1	1	1	1	1			1	1	1		8

Note: As shown in this example reference has been made to a catalogue (cat) code/number – in which case a supplier/manufacturer's name should be stated.

AND SITE SKILLS, SECTION 8.19]. Figure 6.37 shows an arrangement suitable as a carpet step/threshold.

The stages, which follow, should be read in conjunction with Figure 6.38(a–i).

(a) *Select and mark the door* (Figure 6.38a) – check that you have the correct door against the door schedule (Table 6.8 – see also the drawing of the house in Appendix 1), then from the drawings determine its face side and hanging edge. If it is a panel door mark a 'TH' on the hanging side face to indicate the 'top hinge' position. Before similarly marking a flush door determine the position of the lock block (Figures 6.14 and 6.15) If the door is a fire-resisting door also check the rating and position of the colour coded plug. Should the door have a pre-finished

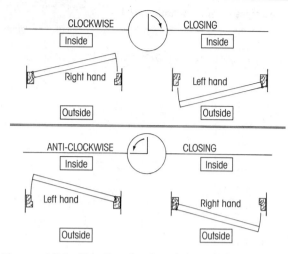

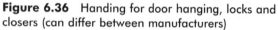

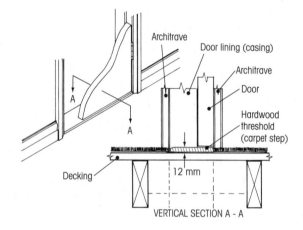

Figure 6.36 Handing for door hanging, locks and closers (can differ between manufacturers)

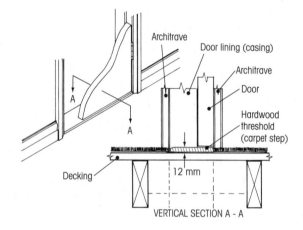

Figure 6.37 Door lining threshold (carpet step)

surface you could us masking tape as a maker (avoid using parcel tape as this can damage the door surface). Then establish clearly in your mind the door clearances you will be striving to achieve. Table 6.7 lists recommended door clearances – it should serve as a useful guide.

(b) *Corner protection* (Figure 6.38b) – remove all corner protection. This may involve carefully cutting off the horns left on both ends of the stiles, or simply untacking wood or plastics corner protectors.

(c) *False rebate* (Figure 6.38c) – before starting to fit the door between non-rebated linings (plain linings) two temporary door stops (checks) will be needed – one either side mid height to prevent the door falling through the opening as it is being fitted. Temporary stops consist of short end of timber about 12 mm thick set back by as much as the door thickness from the face edge of the lining. **It is important for your safety and that of any passer-by that the nails used to hold the stops in place are not left to protrude.**

(d) *Fitting the hanging edge* (Figure 6.38d) – start by positioning the door into the opening and assessing how much (if any) fitting will be required – it may just squeeze into the opening, or on the other hand be over size. Either way start by checking that the hanging edge is parallel with the side lining/frame it is to abut. Adjust as necessary by planing as shown in the illustration.

(e) *Fitting the closing edge* (Figure 6.38e) – reposition the door into the opening, and with the help of an assistant or handhold of an open panel door, and whilst held up against the hanging side scribe the full length of the closing door edge by the width of half a pencil (about 4 mm). When planed down to this line it should make provision for a parallel gap of 2 mm (the thickness of a 2p coin) on either side of the door. *Don't forget however that an exterior door will require a clearance 2.5 mm on either side* (Table 6.7).

(f) *Door bottom* (Figure 6.38f) – the next step will depend on whether the door frame/lining is open or closed. For example open interior doors will usually require clearances of about 10–12 mm for carpet clearances. Where as closed types, depending on the situation 3–3.5 mm clearance. Pre-rebating the bottom of an exterior door to accommodate the water (weather) bar could be carried out in the workshop by using a portable powered router [BENCH AND SITE SKILLS, SECTION 8.15]. Alternatively. First us a marking gauge to deep mark edges of the rebate across the door. Then with a tenon saw cut across as much of the stile end grain as possible before cutting away those portions of the rebate. Then remove the remaining portion with a rebate plane [BENCH AND SITE SKILLS, SECTION 5.4.2b], or make a double cut with a plough plane [BENCH AND SITE SKILLS, SECTION 5.4.2c].

(g) *Top and bottom clearance* (Figure 6.38g) – again these clearances can be achieved by using the width of a pencil as described at (e). Or if rear access or an assistant are not available, by transferring a 4 mm margin line from the head lining to the edge of the door to be joined up and then planed down to the line.

(h) *Providing a 'lead-in'* (Figure 6.38h) – now that all round clearance has been achieved all that remains is to provide a 'lead-in' to the closing edge, and the removal of all the sharp edges (arrises). The term lead-in refers to the removal of that portion of the wood, which would otherwise prevent the door from closing. In other words, because the hinging point (rotation point) of the hinge is offset to one edge of the door [BENCH AND SITE SKILLS, SECTION 7.6.1]. When the door swings (rotates) about this point a small portion projects outside the arc it would otherwise scribe. It is this portion we must remove (plane at an angle) otherwise the door, even though

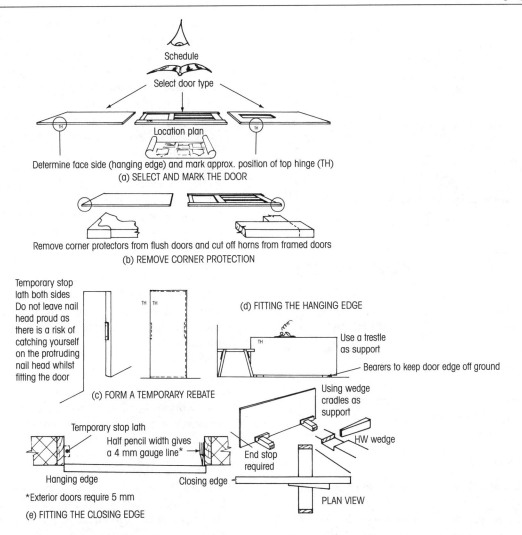

there was clearance when it was in line with its opening would be prevented from closing.

(i) *Removing the 'arris'* (Figure 6.38i) – the arris refers to the sharp external angle formed at the corners of a wood object or a piece of timber. This is achieves either by planing – a block plane [BENCH AND SITE SKILLS, SECTION 5.4.2a] is a useful tool for this operation, or abrasive paper on a sanding block. Apart from the obvious reasons for cutting back sharp edges, i.e. reduce the risk of splinters. By flattening, or preferably rounding this edge the adhesion of any the surface treatment such as paint is improved.

6.4.2 Hanging a Door

When satisfied that the door has been fully fitted and that clearances are within the set tolerances listed in Table 6.7 the door can be prepared to receive the hinges.

In the example that follows the leaf is to be hung into a frame or narrow lining using 1½ pairs of standard cranked steel butt hinges (Figure 6.48). The procedure should be read in conjunction with Figure 6.39.

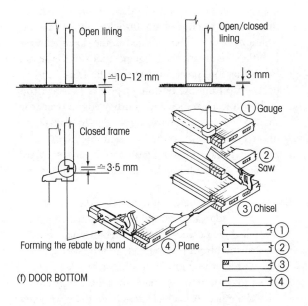

Figure 6.38(a–f) Fitting a door into its opening

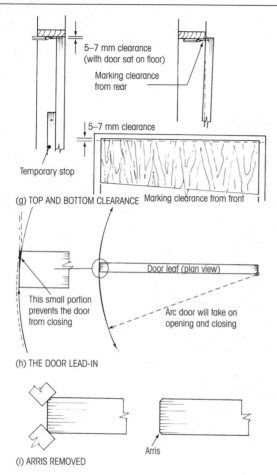

5–7 mm clearance
(with door sat on floor)

Marking clearance
from rear

5–7 mm clearance

Temporary stop

(g) TOP AND BOTTOM CLEARANCE   Marking clearance from front

Door leaf (plan view)

This small portion
prevents the door
from closing

Arc door will take on
opening and closing

(h) THE DOOR LEAD-IN

Arris

(i) ARRIS REMOVED

Figure 6.38(g–i)   Fitting a door into its opening

1. Decide on the position of the hinges, Figure 6.47 gives a recommendation when using 2040 mm high metric-sized doors. However, traditionally top hinges have been set down about 150 mm from the top of the door, and 225–250 mm up from the bottom, with the centre mid-way between them – hinge positions must always be clear of any through tenons. Actual positions may well be specified in the documentation. In either case mark the hinge positions onto the door.

2. With the door positioned within the frame and packed up from the bottom to give the correct top and bottom clearance and gently wedged over from the closing edge. Use a straight edge to transfer the hinge positions on to the frame/lining edge.

3. Set a marking gauge or combination try square to the width of the hinge leaf (see also the butt gauge Figure 6.40).

4. (a)   Gauge or rule the hinge-leaf width along the door edge and frame/lining (planted rebates) at all three positions. Where rebates have been formed from the solid, stage 4(b) will apply.

   (b)   Where the frame/lining has a struck rebate, provision must be made for rebate back clearance. A back clearance template (or gauge) made from a piece of hardboard or plywood

cut to width 'X' (door thickness – minus gauged hinge leaf width plus a back clearance of 1–1.5 mm – depending on the surface finish of the door) will enable the back line of the housing to be marked with a pencil or scored with a scriber or chisel.

5. Set the marking gauge to the hinge leaf thickness.

6. At each hinge position, gauge the door face and the frame/lining edge over the length of each hinge.

7. Holding the chisel vertically with a mallet accurately chop the ends of each hinge leaf position to the depth of the gauge line. Before recessing the hinge some joiners prefer to make a series of saw cuts along and across the housing with a dovetail saw.

8. Using a mallet and firmer chisel gently feather each hinge housing to the recess depth. Use the chisel with its under face uppermost

9. With the door firmly supported, pare (don't forget *both* behind the cutting edge of the chisel at *all* times) and clean the bottom of each hinge recess to the pre-scribed depth, but with a very slight slope (under cut) towards the back.

10. Fully screw each hinge to the door (use the hinge leaf with the least number of knuckles), after first using a bradawl or drill to form a pilot hole for the screw. If slot headed screws are to be used the slot should end up vertical (in line with the door's height) – in this way, paint and polish build-up is less likely, which could otherwise contribute towards the hinge binding, also hinge appearance is enhanced.

11. Offer the hinged door to the frame or lining recess. Pack-up as necessary and, using one screw per hinge, secure in position.

12. Check the door leaf for clearance and fit. If satisfactory, fix all remaining screws.

Butt Gauges

When fitting butt hinges into non-rebated or rebated door frames or linings a butt gauge like the 'Stanley' butt gauge shown in Figure 6.40 can provide an alternative to using a marking gauge at stages 3 and 5 in Figure 6.39 above.

For example, with a **non-rebated** door frame or lining the following will apply:

(i)   Using the gauge as a square, transfer the face positions of the hinge leaf to the door leaf edge and to the edge and the inside face of the lining (not shown) (Figure 6.40a).

(ii)   Set the leading spur of the double-ended scribing bar to the width of the hinge leaf. Scribe the edge of the door from the door-leaf face and the face of the door lining from its edge (Figure 6.40b).

(iii)   Set the single-ended scribing bar to the hinge leaf thickness (cranked butt hinge), then scribe the face of the door leaf and the edge of the door lining (Figure 6.40c).

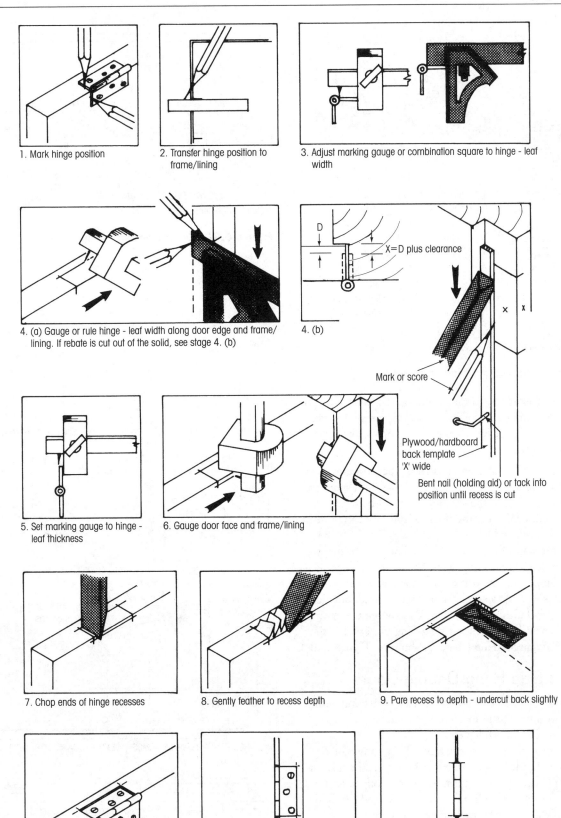

1. Mark hinge position

2. Transfer hinge position to frame/lining

3. Adjust marking gauge or combination square to hinge - leaf width

4. (a) Gauge or rule hinge - leaf width along door edge and frame/ lining. If rebate is cut out of the solid, see stage 4. (b)

4. (b)

X=D plus clearance

Mark or score

Plywood/hardboard back template 'X' wide

Bent nail (holding aid) or tack into position until recess is cut

5. Set marking gauge to hinge - leaf thickness

6. Gauge door face and frame/lining

7. Chop ends of hinge recesses

8. Gently feather to recess depth

9. Pare recess to depth - undercut back slightly

10. Screw hinge to door

11. Offer and screw door hinge to frame/ lining (one screw per hinge)

12. Check clearance and fit - if satisfactory, fix remaining screws

Figure 6.39 Hanging a door into a solid lining using standard (cranked) steel butt hinges

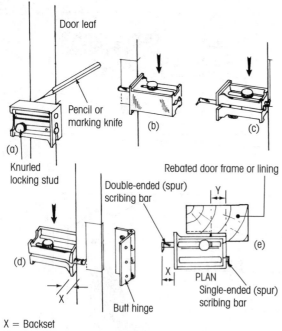

X = Backset
Y = Backset (X) plus door clearance (set automatically)

Figure 6.40   Using a hinge butt gauge

With a **rebated** door frame or lining, stages (i) and (iii) remain the same but, instead of setting the spur to the width of the hinge leaf as at stage (ii), make the following modifications:

- Remove the double ended scribing bar from its stock and reposition it upside down, so that both spurs are exposed. (Part of the scribing spur will be visible at one end, while only the point of the spur can be seen at the other end.)
- Set the spur of the protruding scribing bar to the distance (X) remaining across the door-leaf edge, as if the hinge were in position (as shown in Figure 6.40d). In this way, the spur of the rear scriber is automatically set to include door clearance (Y) between the door face and rebate check when scribing from within the door frame or lining rebate (as shown in Figure 6.40e).

6.4.3 Side Hung Double Doors

As the name implies – two door leaves are hung within one door frame (exterior door – French window, etc.), lining set (room divider, etc.).

Meeting stiles are usually rebated. As shown in Figure 6.41 the rebate can either be formed by fixing a bead (planted) to both meeting stiles on opposing sides. Or by forming them in the solid (stuck bead). In both cases the closing edges will require a slight leading-in bevel (Figure 6.38h).

6.4.4 External Door Weatherboards

Inward opening external doors will (as shown in Figure 6.42a) require a weatherboard unless they are protected

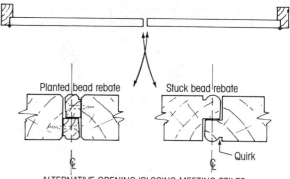

ALTERNATIVE OPENING/CLOSING MEETING STILES
Note the slight lead-in (see Fig. 6.38h)

Figure 6.41   Double door meeting stiles

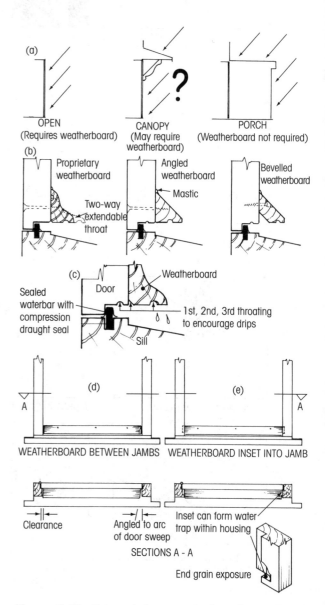

Figure 6.42   External door weatherboards

from inclement weather by an enclosed porch – protection offered by a canopy will depend on the amount of overhang.

A weatherboard redirects surface water running down the door external face and encourages it to drip (via a 'throating' (groove)) well beyond the water/weather bar. Several alternative methods are used three of which are shown in Figure 6.42(b). If (as shown in Figure 6.42c) water from driving rain is driven beyond the first throating there should always be provision for a second throating as a second line of defence.

Weatherboards can be positioned either between the jambs or let into them (as shown in Figure 6.42d, e). The problem with letting the ends into the jambs is that unless great care is given to detailing, the housing which will have some end grain exposure will become a water trap – this is a situation which should be discouraged at all times. Also inset ends can if the gap is small induce and encourage capillary attraction, which will eventually have an accumulative effect on increased moisture content of both the ends of the weatherboard and foot of the frame.

In any case the weatherboard is secured to the door with screws, it should first however be bedded with a suitable mastic compound behind the fixings then a bead of mastic run across the meeting edge with the door.

6.5 DOORSETS

A doorset can be regarded as an external or internal door leaf which has been matched to a frame or lining. Fully assembled, pre-hung under factory conditions with any glazing (glass and beads), hardware (ironmongery) (lock, latch, and or fixings), seals (intumescent, smoke control, etc.) in place, or provision for them (protruding hardware such as handles could interfere with packaging).

Doorsets can be delivered to site either fully assembled, or packaged in such a way that they can be easily assembled as and when required.

Because the doorset may have been surface finished (painted or polished) in the factory the use of detachable hinges (Figures 6.49 and 6.50) enables the door leaf to be detached after installation and thereby reduce the risk of on-site damage.

6.5.1 Framing Openings

Template (profiles) 10 mm wider and taller than the door set are built into the masonry to leave a slightly oversize opening on their removal. Templates should be detachable to facilitate their removal and reuse.

Figure 6.43 shows two types of template. The system shown in the left-hand side consists of a detachable framework lined with 12–18 mm plywood of a width equal to the thickness of the wall plus any plasterwork. The plywood liner is fixed to the blockwork, etc., as the

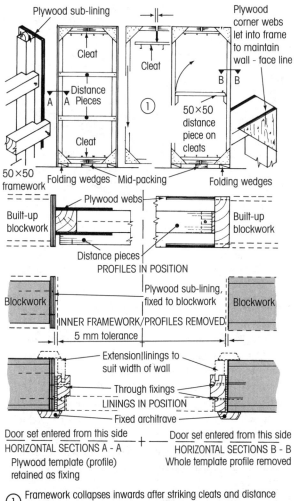

Figure 6.43   Template (profile) for doorset opening

wall is being built, and acts as a profile for the plasterer. After the liner framework is removed (when the wall is built and set), the plywood is retained as a fixing for the door set.

The system on the right-hand side of Figure 6.43 involves removing the whole (detachable) template to leave an opening of fair-faced blockwork, etc. In this case, the door frame or lining will require fixing to the wall as described in Section 6.3.1 or as described below.

Fixing an internal doorset into the opening

If necessary, assemble the doorset framework/lining including one set of architraves.

1. Drill the door frame/lining to receive fixing screws as shown in Figure 6.44.
2. Offer the door frame/lining to the opening – the fixed architrave acts as a restraint (check).
3. Level the head or transom rail – adjust the foot or framework as necessary.
4. Check for straightness, plumb and alignments.

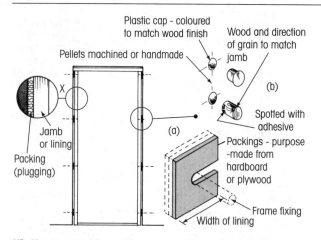

NB. 'X' gaps around fire-resisting door sets (5 mm each side, 10 mm at head) to be plugged with mineral wool, intumescent mastic, or other non-combustible filler.
Distance between jamb fixings for fire-resisting doors - not to exceed 500 cm.

Figure 6.44 Fixing a doorset frame

5. Using the holes in the jambs as a guide, drill into the base material (e.g. Blockwork). Insert framing plugs and screws [BENCH AND SITE SKILLS, SECTION 11.7.2].
6. Position any packing behind all fixing points between the back of the jamb and the base material. Tighten the screws. Check with a distance piece or the door leaf (detachable hinges) for correct clearance.
7. Remove door leaf (if supplied). As shown in Figure 6.44 plug all fixing points with wood pellets or plastics caps.
8. Fix loose architrave on the opposite side (Figure 6.43) and fanlight if fitted.

Note. Fire doorsets will require any gaps between the door frame/lining plugging with an approved non-combustible filler as shown in Figure 6.44.

With the exception of wall plugs the same procedure can be adopted for fixing into timber-studded partitions.

6.6 DOMESTIC INTERNAL SLIDING DOORS – (ALSO SEE SECTION 12.11.3)

Sliding a door across an opening provides a useful alternative to hinging, were floor space is restricted. There is however, the disadvantage that, unless the door slides into a wall, wall surface space is lost.

There are several proprietary systems on the market. In the main these consist of an overhead track from which the door is suspended and slid via a series of plastics wheels or metal rollers. The whole of the top sliding gear is covered with a pelmet, which should be made detachable to enable maintenance to be carried out without disturbing any surface decoration.

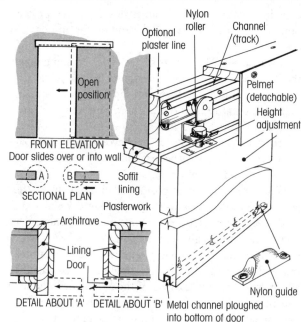

Figure 6.45 Domestic sliding door

The bottom edge of the door leaf may have to be grooved to receive a metal channel which as the door is slid open or shut, runs over a plastics guide block which is screwed to the floor.

Figure 6.45 shows a typical arrangement for lightweight sliding-door gear.

The same principles can be applied to a pair of doors where, if they are allowed to pass over the face wall they can abut one another on the same track. If they were set within a reveal, then a dual track would be used (Figure 12.35) allowing one door to pass over the other.

6.7 SWING DOORS

Swing doors use their hanging mechanism as a self-closing device. Single action devices (Figures 6.46a, b) move the door through 90°, where as double action devices (Figures 6.46c, d) move it around 180°. Both single and double doors can be used with these devices.

Figure 6.46(e) shows the arrangement for fitting a single action floor spring, the door is hung on offset or centrally positioned pivots (pivot positions can vary with type of hanging). The leading edges of the meeting stile(s) require to be bevelled. The amount of lead-in will depend on door width (this will affect the radius of the arc – see Figure 6.38) and thickness, and the projection of the pivot.

Figure 6.46(f) shows an arrangement for fitting a double action floor spring. In this case the pivots are central to the door and both the hanging and meeting stile(s) will need to be radiuses to suit the door width.

Figure 6.46(g) shows a double action helical sprung hinge, the position, number, type (sprung loaded or not)

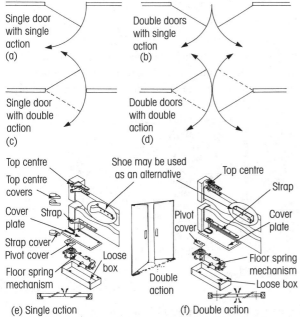

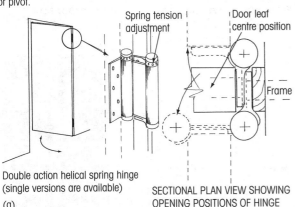

NB. For single action doors hung on offset pivots the leading edge of meeting stiles require to be bevelled back. The amount depends on width and thinkness of door and projection or pivot.

NB. For double action doors or centrally pivoting single action doors the heel and meeting stile should be radiused accordingly.

(e) Single action

(f) Double action

Double action helical spring hinge (single versions are available)

(g)

SECTIONAL PLAN VIEW SHOWING OPENING POSITIONS OF HINGE

Figure 6.46 Swing doors

and tensioning techniques will be supplied by the manufacturer. As shown in the illustration double action hinges operate by folding one single leaf over the other – single action hinges are also available.

6.8 ON-SITE STORAGE AND PROTECTION

Doors, their frames, and doorsets should not be delivered to site until they are required. If, on delivery, it is found that they have not been factory sealed or primed both faces and all four edges should be immediately treated with an appropriate sealer or primer.

All doors awaiting use should be close 'sticked' (not less than three bearers laid across each door) and stacked flat (out of twist) off the floor and under cover in well-ventilated building.

Door stacks should have a dust cover, but this should not restrict through ventilation. Doors, which have been shrunk-wrapped, should be left with the wrapping intact until required for conditioning.

Conditioning – doors should be gradually introduced into the climatic condition in which they will remain [BENCH AND SITE SKILLS, SECTION 2.12].

6.9 DOOR HARDWARE

Door hardware (often referred to as 'ironmongery') can be put into sections according to their use, for example:

Hardware	*Use*
Hinges	Door hanging
Locks and Latches	Door restraint and security
Knobsets	Combined lock/latch and knobs
Closers	Self-closing mechanism
Door furniture	Door and jamb attachments – handles, etc.

You should also refer to Section 6.1.8 (Fire doors). A guide to positioning hardware is given in Figure 6.47.

6.9.1 Hinges

The size and number of hinges per door leaf will depend on the size, and weight of the door. They can be made from mild steel, stainless steel, cast iron, or brass.

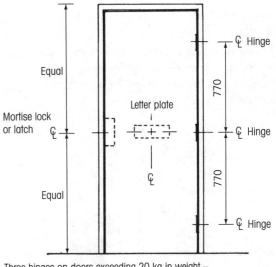

Three hinges on doors exceeding 20 kg in weight – recommended on all doors to help counteract any tendency for stile to warp

Figure 6.47 Guide to positioning door hardware

The type used will to a greater extent depend on the type of door and usage, for example, brass hinges will be used on self-finish hardwood doors just like painted doors could use steel or cast iron hinges.

Hinge construction will also effect its end use, for example its leaf may be cranked, flat, broad, narrow, or extended, or even sprung. Hinge knuckles can be washered or non-washered, and their pins may be fixed or detachable.

Extra security on the hinged edge can be achieved by using hinge bolts (Figure 6.55).

Examples of a few hinges and a hinge bolts are given below.

Standard cranked steel butt hinge (Figure 6.48)

The knuckle of the hinge must protrude beyond the face of the door leaf and the edge of the door jamb. Use 100 mm hinges on exterior doors and 75 mm on lighter interior doors.

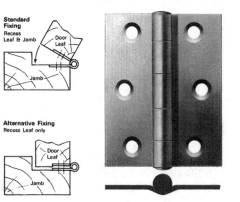

Figure 6.48   Cranked steel butt hinge

Loose cup head pin – butt hinge (Figure 6.49)

Once the door leaf is hung, it can be detached from the door frame/lining by removing the pins from the hinge knuckle – ideal for light-duty door sets.

Figure 6.49   Loose cup-head pin butt hinge

Lift-off butt hinge (Figure 6.50)

Handed steel cranked hinges – providing that the hanging edge of the door leaf projects beyond any projections (wall trims) when fully open, the door can be lifted off its hinges. Each pair of handed hinges consist of one short and one long pin to help assembly.

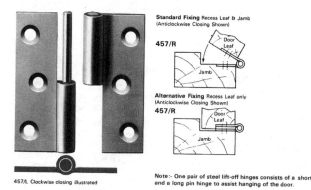

Figure 6.50   Lift-off butt hinge (handed)

Brass butt hinges (Figure 6.51)

Solid drawn brass uncranked hinge with either brass or steel pins, hinges may have steel or phosphor bronze washers between the knuckles.

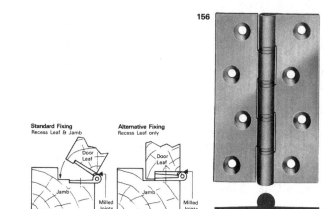

Figure 6.51   Brass butt hinge

Cast iron butt hinges (Figure 6.52)

A heavy duty uncranked hinge with fixed flush pins. Care should be taken when fitting these hinges – although they are very strong they are also quite brittle and can break if struck with a hammer.

Rising or skew butt hinges (Figure 6.53)

A lift-off hinge with a helical knuckle, which causes the door to rise as it is opened. This movement enables the bottom of the door to clear carpets, etc. and to fall shut under the doors own weight, thereby providing a simple self-closing device.

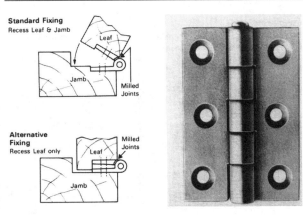

Figure 6.52   Cast iron butt hinge

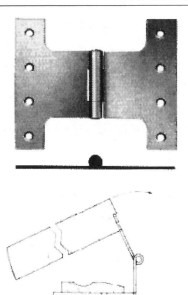

Figure 6.53   Rising or skew hinge

Figure 6.54   Parliament pattern butt hinge

Because the door starts to rise as soon as it is opened, the top inside corner (hinged side) of the door will need to be slightly splayed, as shown in Figure 6.53(a), to avoid catching the door head or transom rail.

Parliament pattern butt hinges (Figure 6.54)

A steel hinge with fixed pin with uncranked knuckle. They allow the door to open fully back against a wall, which would otherwise bind against wall trims such as architraves skirting boards or plinth blocks.

Hinge bolts (Figure 6.55)

Hinge bolts help in preventing the door being forced of its hinges. They are used to greatest advantage on outward opening doors where the hinges are most vulnerable to having their knuckle pin removed.

The bolt may take the form of a steel rod fixed to the male leaf of the hinge as shown in Figure 6.55(a) which when the door is closed is housed within the opposing leaf which is attached to the doorjamb. Figure 6.55(b) shows two other hinge bolt devices which are independent of the hinges.

(a)

(b)

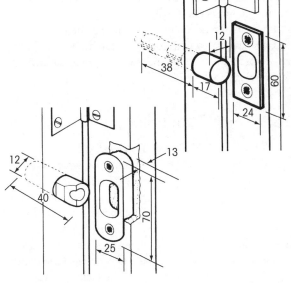

NB. All measurements are in millimetres

Figure 6.55   Hinge bolts: (a) bolt integral to the hinge and (b) two examples of 'Chubb' hinge bolts

6.9.2 Lock and Latches

Locks and latches are either mortised into the closing edge of a door, or fit onto its inside face edge (rim). The bolting mechanism is usually operated by a pair of handles or knobs and/or a key. The variations on type are many, but for the sake of this section they have been subdivided into the following sections.

1. Mortise locks (Figure 6.60)
   (a) Upright mortice lock (sash lock).
   (b) Upright mortice lock (sash lock) with roller bolt.
   (c) Cylinder mortice lock (sash lock).
   (d) Horizontal mortice lock.
   (e) Mortice dead lock.
   (f) Sliding door hook bolt mortice lock.
   (g) Sliding door claw bolt mortice lock.
2. Mortice locks for rebated doors (Figure 6.61)
   (a) Rebated upright mortice lock.
   (b) Rebated mortice lock set.
   (c) Fitting a rebate lock set.
3. Mortise latches (Figure 6.62)
   (a) Mortise latch.
   (b) Tubular latch.
   (c) Roller latch.
4. Rim locks and latches (Figure 6.63)
   (a) Rim lock.
   (b) Rim deadlock.
   (c) Rim latch.
   (d) Cylinder rim lock (cylinder rim night-latch).

There is one other we should mention at this stage, that is the 'Bored lock or latch set' more commonly referred to as a 'Knobset' because the door knobs are an integral part of the lock/latch mechanism. They are dealt with separately in Section 6.9.3.

Probably the best way of identifying the many features associated with locks and latches is to refer to an annotated list in alphabetical order of lock/latch terminology.

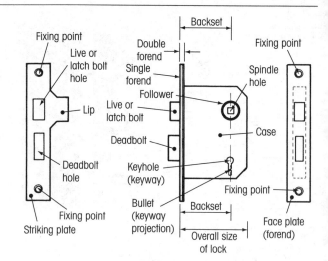

Figure 6.56 Mortice lock components and terminology

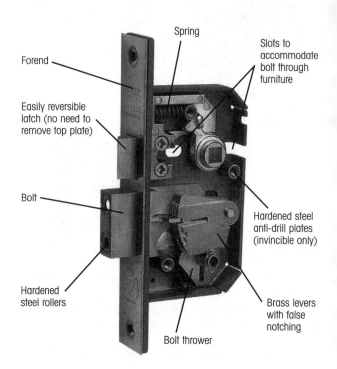

Figure 6.57 Mortice lock mechanism (ERA mortice locks)

Lock/Latch Terminology

Term	Figure	Explanation
1. Backset	6.56	Distance by which the vertical centre line of the keyhole and/or spindle hole is set back from the front of the outer forend.
2. Bitting	6.59	The shape or notching of the key blade to accommodate the locking mechanism.
3. Blank		An un-notched key (see key blank)
4. Bolt	6.56 6.58	That part of the lock or latch that protrudes from the case and enters a striking-plate or keep to lock or latch the door.
5. Bored latch set	6.67	A combined assembly of a tubular latch and knob furniture.
6. Bored lock set	6.67	A combined assembly of a tubular lock and knob furniture.

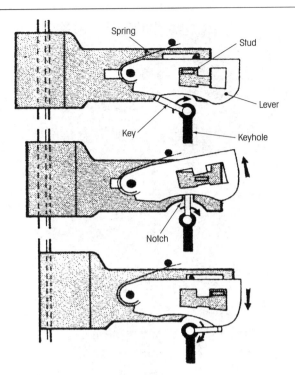

Figure 6.58 Moving a dead bolt via a levered mechanism

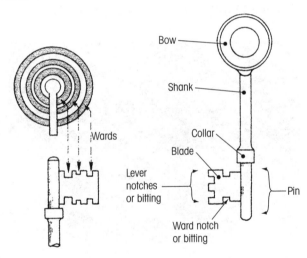

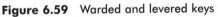

Figure 6.59 Warded and levered keys

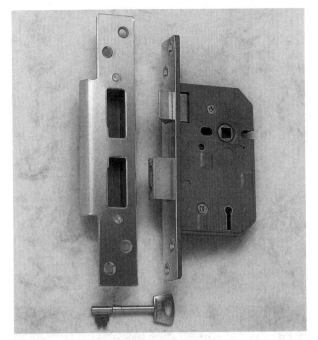

Figure 6.60(a) Upright mortice lock (ERA mortice locks)

Figure 6.60(b) Upright mortice lock with roller bolt (ERA mortice locks)

7.	Boxed striking-plate		A strong metal box formed as part of the striking-plate and mortised into the door jamb – provides greater security.
8.	Case	6.56	That which contains the lock and latch mechanism.
9.	Claw bolt	6.60(g)	Sections of this bolt move sideways and dovetail fashion as the bolt is shot (forced out of its case by a key). It is used on sliding doors. Also see 'Hook bolt'

10.	Cylinder	6.64	This contains the pin or disc tumbler and their spring mechanism. It may have a round or keyhole face.
11.	Cylinder lock or latch	6.63(d)	A lock or latch that uses a cylinder key mechanism.
12.	Cylinder rose	6.64	Metal disc supporting the outer face of a cylinder.

Figure 6.60(c) 'Union' cylinder mortice lock (sash lock)

Figure 6.60(d) 'Union' horizontal mortice lock

Figure 6.60(e) 'Chubb' 5 lever mortice dead lock

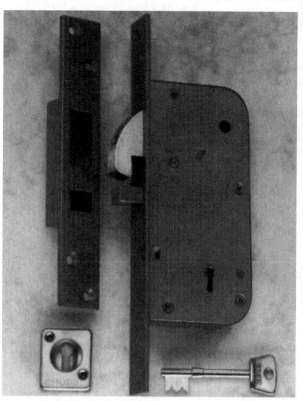

Figure 6.60(f) Sliding door hook bolt mortice lock (Chubb)

Figure 6.60(g) Sliding door claw bolt mortice lock (Union)

13.	Deadbolt	6.56	A square-ended bolt moved by a key, not a spring.
14.	Deadlatch	6.63(d)	A springbolt capable of being locked with a key or catch.
15.	Deadlock	6.60(e)	A lock with a square-ended bolt operated by a key or thumb turn.
16.	Disc tumbler lock		A cylinder lock with discs instead of pin tumblers – not included in this chapter.

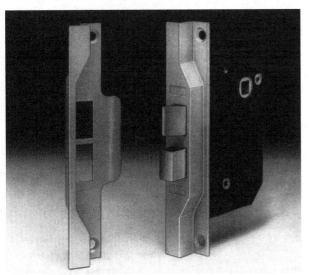

Figure 6.61(a) 'Union' light duty rebated upright mortice lock

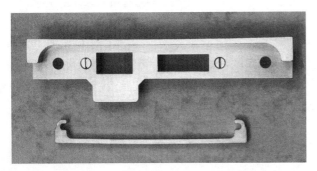

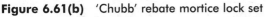

Figure 6.61(b) 'Chubb' rebate mortice lock set

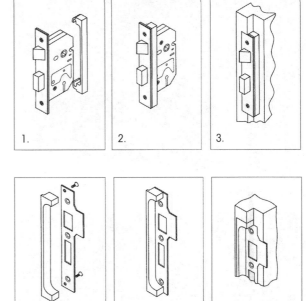

1. The forend rim can be fitted on either side of the lock, depending on hand
2. Forend strip in position, behind the forend. Left hand shown
3. Lock and forend strip applied to a rebated door, the forend strip is secured by the lock fixing screws.
4. The striking plate rim can be fitted either side of the striking plate (and secured by screws shown) depending upon hand.
5. The striking plate rim secured to the striking plate. Left hand shown.
6. The striking plate applied to a rebated door or frame.

Figure 6.61(c) Fitting a 'Union' reversible rebated forend and striking plate lock set

17. Double-handed lock		A rim lock which by being turned upside down, can be used on either clockwise or anti-clockwise closing doors. The keyway has a key pin hole at both ends of the slot.
18. Escutcheon (plate)	6.74	A cover for a keyhole – part of the lock furniture (see also Section 6.9.5 – door furniture).
19. Faceplate	6.56	The outer of a double forend – a cover plate fixed to the inner forend.
20. Follower	6.56	Part of the bolt mechanism that withdraws the spring-bolt as it is turned via the spindle. It may have one or more horns.
21. Forend	6.56	The fixing plate on the front of a mortise lock or latch case.
22. Full rebated forend	6.61	Purpose made forend or attachment shaped for use either in a rebated door jamb, or more likely for rebated meeting stiles of double doors.

GENERAL-PURPOSE MORTISE LATCH

TUBULAR MORTISE LATCH

TUBULAR MORTISE CATCH (ROLLER LATCH)

Figure 6.62 Mortice latches: (a) mortice latch, (b) tubular latch, (c) roller latch

Figure 6.63(a)   Rim lock

Figure 6.63(b)   Rim dead lock

Figure 6.63(c)   Rim ratch (with thumb bolt)

Figure 6.63(d)   Cylinder rim lock (cylinder rim night-latch)

23. Handing	6.36	When purchasing specialised types of locks and latches the chances are that they will be handed (see Section 6.4 and Figure 6.36).
24. Hook bolt	6.60(f)	Used for securing sliding doors – as the bolt, which is pivoted enters the keep – it lifts up and then drops over the keep. Also see the 'Claw bolt'.
25. Horizontal lock	6.60d	Mortise or rim lock – they allow a greater distance between the door edge and the spindle or keyhole. It is used with knob furniture.
26. Keep or Keeper		Fixed onto or into the door jamb to retain the bolt of a rim lock or latch.
27. Key	6.59	Means of operating the locking (and in some cases

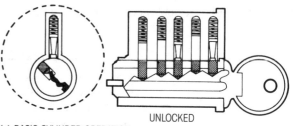

(a) BASIC CYLINDER OPERATION

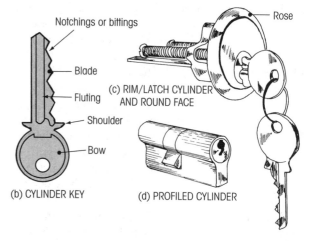

(b) CYLINDER KEY

(c) RIM/LATCH CYLINDER AND ROUND FACE

(d) PROFILED CYLINDER

Figure 6.64(a–d)   Lock cylinders and their operation

Figure 6.64(e)   Examples of how cylinder mechanisms can vary (Union)

latching) mechanism – for key combination (variation) see Table 6.9.

28. Key blank — A key sized to fit a keyhole but without any of the notches (bitting) that would be required to suit a locking mechanism.

29. Knobs   6.74 — Door furniture for a horizontal lock – an alternative to lever handles.

30. Knobset   6.67 — See 'Bored latch set' and 'Bored lock set'

31. Latch   6.62 — Product with only one – bevelled springbolt, or roller bolt used to fasten a door. Only certain types are lockable, for example, nightlatches.

32. Latch bolt   6.56 — see 'Springbolt'

33. Levers   6.58 — Flat shaped plates lifted via the keys bitting until they are horizontal, at which point the deadbolt can be thrown forward. Greater the number of levers the better the locks security – see Table 6.9.

34. Lever handles   6.73 — Door furniture for lock or latch – an alternative to door knobs.

35. Lock — Case containing a latch bolt operated by handles and deadbolt usually operated by a key.

36. Locking latch   6.63 — A latch (springbolt or roller bolt) capable of being locked.

37. Lockset — Matched lock and furniture (handles, spindles, escutcheon, etc.) ready for fixing.

38. Mortise lock or latch — A locking or latching device which is mortised into a door stile.

39. Narrow case lock or latch — A rim lock or latch suitable for a narrow door stile.

40. Nightlatch   6.63(d) — A latch with a bevelled springbolt or a roller bolt that automatically latches as the door is closed. The latch may be withdrawn from its keep by a key from the outside or by a knob or lever from the inside. Latches may also be retained with their case, or deadlocked.

41. Pin tumbler mechanism   6.64 — A cylinder mechanism consisting of a row of spring-loaded tumblers that rise and fall to meet the 'V' notches cut in the key. When all the junctions between pins (the lower part of the tumblers) and the drivers (the upper part of the tumblers) are aligned (by using the correctly cut key), the cylinder plug can be rotated to operate the latch.

Table 6.9   Possible number of variations before the same key pattern is repeated

Lock	Number of levers	Possible number of key variations
Two lever lock	2	12
Three lever lock	3	150
Five lever lock (non BS)	5	500
Five lever lock built to BS3621	5	1000

Note. Cylinder lock variations can be as high as 100,000.

42. Plug   6.64   That part of the cylinder into which the key is put. It can be rotated when the correct key is used to allow the pin and drivers to separate. It may also have a disc tumbler mechanism (not covered in this section).

43. Privacy set   A bedroom or bathroom door lock with an inside turn knob or button capable of locking the knobs or levers. Emergency provision is made to unlock the mechanism from the outside.

44. Rebated lock or latch   6.61   A mortise lock or latch with a fully rebated forend.

45. Rebate set   6.61c   Raises the stile rebate to enable a standard mortise lock or latch to be fitted.

46. Reversible latchbolt   A springbolt which can be turned round to suit the door handing. *Caution: the bolt is under pressure from a compressed spring, and this spring can become dislodged if the case is opened. Therefore if it is necessary to open the case, wear eye protection and carry out the operation within a tall sided open-top cardboard box.*

47. Rim lock or latch   6.63   A lock or latch which is screwed onto the face of the door.

48. Roller bolt   6.60b   An alternative to a bevelled spring-bolt – it does not require handing.

49. Rose   6.56   A plate which fits behind a cylinder or door knob.

50. Sash lock   A two bolt mortise lock commonly referred to as an upright mortise lock (see upright mortise lock).

51. Shoot of a bolt   6.60   The distance a bolt moves.

52. Skeleton key   A key cut to suit a number of older type or low security warded locks – bypass the inner wards. Not possible with modern multi-lever or cylinder locks.

53. Spindle   6.74   A square sectioned bar forms part of the door furniture. It passes through the follower and fits into a lever handle or knob on each side of the door.

54. Springbolt (latch-bolt)   6.56   A bolt that is spring loaded to retain it in the extended position. One edge is bevelled to provide a lead-in to enable it to ride over its striking plate before entering its keep. The handles or knobs operate it.

55. Staple   6.63   A box-shape keep fixed to a door jamb to receive the bolt of a rim lock or latch.

56. Striking-plate   6.56   A metal plate let into and screwed to the door jamb to shroud the mortice holes which receive bolts from mortice locks and latches.

57. Thumbturn   A small fitting on the inside of the door used to operate a deadbolt – not to be used on glass or wood panelled doors.

58. Tubular mortice lock or latch   6.62   A lock or latch mechanism in a tubular case housed within a single round hole bored into the door edge.

59. Upright mortice lock (sash lock)   6.60   Narrow cased mortice lock, its follower hole and key hole are vertically in line.

60. Wards   6.59   Metal rings fixed to the inside of the lock case so that the key must pass around them as the bolt is shot. A lock relying solely on wards for its security is only suitable where security is of little importance (see skeleton key).

Guide to fixing and fitting a mortice lock (Fig. 6.65)

Methods used can vary from joiner to joiner. The object of the exercise is to why a particular method is chosen. If it is because it is quicker, beware! Always make sure that accuracy and quality of work is not being sacrificed. The following example is a proven method of fitting and fixing an upright mortice lock (sash lock):

1. Determine the position of the spindle hole (Figure 6.47). With framed doors, the locks mortice hole must not interfere with any of the tenons of connecting rails, and/or glazing bars to the door stiles.
2. With the door leaf open to a convenient position, gently wedge it from both sides to prevent it swinging. (Over-tightening will cause damage to the door and frame, and possibly to the wall.)
3. Holding the lock forend square up against the doors closing edge lightly mark the upper and lower edges of the lock case onto the outer closing face (Figure 6.65a) – by transferring these marks onto the closing edge will determine the height of the

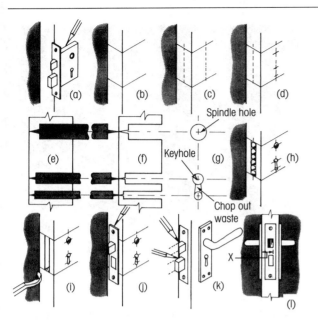

Figure 6.65 Guide to fitting and fixing a mortice lock

mortice hole (Figure 6.65b).

4. With a marking gauge or combination square, gauge the centre of the door leaf and the amount of backset (Figure 6.65c).

5. Measure and transfer to the door the distance the distances down the backset line to the centres of the spindle hole and keyhole (Figure 6.65d). Alternatively mark these distances on to a paper or card pattern, or make a template from a short end of timber. Then simply transfer the dimensions to the door face. *Note*. Some joiners use a pencil pushed through the spindle hole to mark its centre and a bradawl to mark the keyhole – both of these can produce inaccuracies they should therefore be discouraged.

6. Using an auger bit bore a 16 mm (5/8″) spindle hole, and a 10 mm (3/8″) and 6 mm (1/4″) keyhole as shown in Figures 6.65(e, f) – both sides of the door leaf.

7. With a chisel cut out the keyhole from both sides of the door leaf (Figure 6.65g).

    Note: the removal of the key hole waste can be left until the lock case mortice hole has been cut out. In which case a pad or keyhole saw [BENCH AND SITE SKILLS, TABLE 5.1] should be used.

8. Using an auger bit of a diameter slightly larger than the width of the lock case thickness, and a depth stop [BENCH AND SITE SKILLS, SECTION 5.5.4] (adhesive tape wrapped around the shank can be used as a temporary depth stop). Bore a series of holes to a depth equal to the full depth of the lock body plus 3 mm (Figure 6.65h).

9. Chop and clean out the mortice hole square – if a swan necked mortice chisel is available (Figure 6.65i), it is most useful for cleaning out the back of the mortice hole.

10. Position the lock into the hole. Mark around its forend (Figure 6.65j) (using a marking gauge to carefully score along these lines helps to prevent the edges splitting when chopping out the forend housing) chop out a recess until the forend fits flush. Screw the lock to the door.

11. Fit the spindle and fix the handles to the door. Check that the spring bolt operates and that the key activates the dead bolt.

12. With both bolts shot, close the door and mark their position on to the jamb edge (Figure 6.65k).

13. Square across the rebate. Gauge the distance 'X' (Figure 6.65l). Chop out the latch bolt mortice hole in the jamb to form the keep – adjust as necessary to accommodate any seals.

14. Position the striking plate accurately over the latch bolt hole then mark and chop out the dead bolt mortice hole – check it operates fully.

15. Recess the striking plate into the jamb until it is flush with it (the lip may require a deeper bevelled housing) – screw back to the door jamb. *Note*: some joiners prefer to cut out the dead bolt mortice hole after the striking plate has been fixed – in which case extra care must be taken not to damage the honed edge of the chisel.

Guide to fixing and fitting a cylinder rim lock

The amount of backset (distance between the edge of the door and spindle hole) can vary. Standard cylinder rim locks have a backset of 60 mm, with narrow stile doors a 50 mm backset is available but 40 mm is more common.

Illustrated general fitting and fixing guides are usually supplied with each type (pattern) of lock. The fitting and fixing process shown in Figure 6.66 is one example of a Yale cylinder rim lock with a 40 or 60 mm backset.

The following procedure could apply to Figure 6.66:

1. At the pre-determined height, mark a square line onto the outer face of the door leaf from the closing edge of the door leaf by the amount of backset. Then using a 32 mm (1¼″) diameter, centre or auger bit carefully and slowly bore the hole for the cylinder into the door until the point of the bit just protrudes. Remove the bit and finish the hole from the back face of the door (see Figure 6.65f).

2. With the key removed and the rose in place insert the cylinder into the hole.

3. Hold the mounting plate against the inside of the door, and if necessary cut back the flat connecting bar so that it projects 15 mm (9/16″) beyond the surface of the mounting plate.

4. Secure the mounting plate to the cylinder with the connecting machine screws provided. For doors less than 45 mm thick the connecting screws will require shortening. Fix the plate to the door with the wood

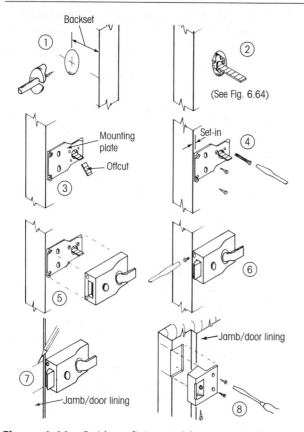

Figure 6.67   Typical arrangement for a knobset

Figure 6.66   Guide to fitting and fixing a cylinder rim lock

Figure 6.67 shows a typical arrangement for a knobset. Different methods of locking and latching are available, for example:

- Non-lockable turnknob operates the latch bolt.
- A turn button on the inside can lock the outside turnknobs.
- A key on the outside and a push button on the inside can lock the outside turnknob.
- A keyed turnknob on the outside and a thumbturn on the inside operate a dead bolt.

6.9.4 Door Closers

Closers are mechanical devices that force or encourage a door to close on its own. Some 'push' others 'pull', rising butts on the other hand (already covered in Section 6.9.1 and illustrated in Figure 6.53) rely simply on the weight of the door leaf to swing closed.

Methods of application vary – from being housed into the floor (see Section 6.7, Figures 6.46e, f) and being attached to the door and/or the frame, some form part of the hinge mechanism (see Section 6.7, Figure 6.46g), while others are concealed.

Figure 6.68 shows detail of an overhead hydraulic door closer together with the different methods of installation, the hydraulic check prevents the door from slamming shut.

screws provided – in this case the mounting plate will need to be set in from the front of the door by 1.5 mm (¹⁄₁₆″).

5. Latch back the lock in the open position. Align the arrow on the back of the case with the arrow on the rotatable slot. Place the lock case over the mounting plate – making sure that the flat connecting bar is inside the slot in the lock case.

   Note. In this case one complete turn of the key from the outside locks the handle inside. So it will not turn even if an intruder brakes a glass panel to reach it, or gains access to the inside by other means.

6. Push the lock case onto the mounting plate – slide it sideways until it is flush with the door edge. With the screws provided secure the case to the mounting plate.

7. With the door closed use the lock case to mark the position of the 'staple' (keep) onto the jamb edge.

8. With the door open us the staple as a template to mark around the flange. Chop out a recess to bring the staple flange flush with the inside face of the jamb, then fix with the screws provided.

6.9.3 Knobsets (bored lock and latch sets)

A combined assembly of a tubular latch or lock and knob furniture.

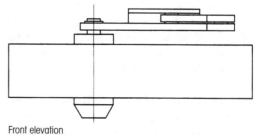

Front elevation
(a) Attached to door without facing arm.

Figure 6.68   'Briton' overhead door closer – three methods of application. *Continued on the next page.*

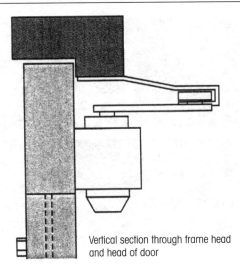

Vertical section through frame head
and head of door

(b) Attached to door with arm parallel to door

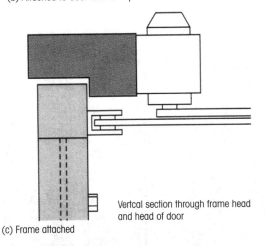

Vertcal section through frame head
and head of door

(c) Frame attached

Figure 6.68 'Briton' overhead door closer – three methods of application.

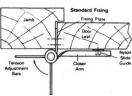

Figure 6.69 'Crompton' door closer

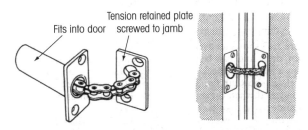

Figure 6.70 'Chasmood' adjuct door closer

The closer shown in Figure 6.69 fixes onto the door jamb's rebate – the fixing is therefore concealed when the door is closed. The tension arm slides over a nylon guide across the face of the door as it opens. It closes under pressure from an adjustable coiled spring housed within the tube. There are several other types on the market all working on a similar coiled spring principle.

Concealed door closers – one example is shown in Figure 6.70 – the tube housing the spring and linked chain mechanism is housed inside the closing edge of the door, the tension retainer plate is let into the closing edge of the door jamb. The required tension is achieved by withdrawing the chain from its retaining tube and linking it to the jamb retainer, this process must however be carried out by following the manufacturer instructions which are supplied with each closer.

There are closers like the one illustrated in Figure 6.71, which come into operation as the door nears its closed position. Once activated they gently close the door without slamming and retain it in the closed position whilst still allowing easy opening.

Figure 6.71 'Chasmood' the dictator door closer

Pairs of self-closing doors with rebated meeting stiles are required to close in the correct sequence. In which case a door selection device will be needed. Figure 6.72 shows such a devise that will ensure that the leading doorleaf will always end up in its final closed position before the trailing door leaf.

Figure 6.72   'Union' door selector

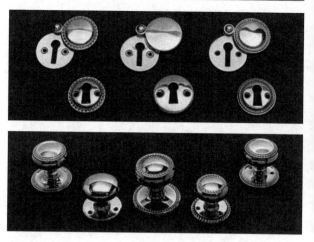

Figure 6.74   Knob furniture (including escutcheons)

6.9.5 Door Furniture

For the purpose of this chapter we are referring to those items that are either fixed directly or indirectly to the face of the door leaf, its lining or frame, as door furniture, for example:

- Lever furniture (Figure 6.73) or lock and/or latch furniture.
- Knob furniture (Figure 6.74) or mortice lock and rim lock furniture.
- Door handles and pulls (Figure 6.75).
- Letter-plates (Fig. 6.76) – or combined letter-plates, knockers and numbers (Figure 6.76).
- Finger and kicking plates (Figure 6.77).
- Security, retainers and fittings – barrel bolts, flush bolts, door chains, indicating bolts, door viewer, (Figure 6.78).
- Sundry items – door stops, clothing hooks, symbol plates (Figure 6.79).

It is important to realise however, that manufacturers may sub-divide such items differently, for example:

Figure 6.73   Lever furniture (Union)

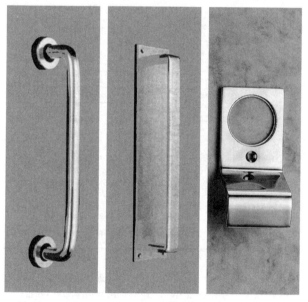

Figure 6.75   Door handles and pulls (cylinder pulls, round bar handles, flat face pull handles)

Door furniture: letter plates, finger plates, kicking plates, knob furniture, door pull handles, door viewer, number plates, cylinder turns and roses, escutcheons.
Lever furniture: lever handles.
Door fittings: door closing devices, indicator bolts, flush bolts, door stops, clothes hooks.

6.10 DOMESTIC GARAGE DOORS

These doors are constructed in much the same way as exterior house doors. The openings being much wider (see Figure 6.26, Table 6.4) will accommodate doors hung in pairs. Unless a large single door is used in which case it would require an 'up-and-over'

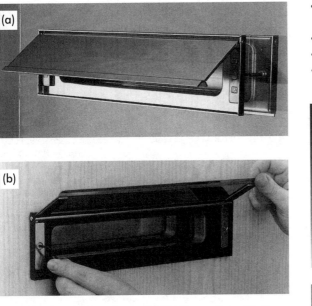

Figure 6.76 Letter plates. (a) RT Fire resistant letter plate. (b) RT Standard letter plate.

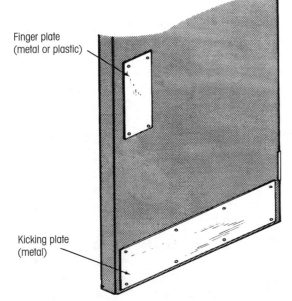

Figure 6.77 Finger and kicking plates (metals and plastics)

mechanism. Horizontal-sliding and/or folding mechanisms can also be used.

Because, with the exception of internal sliding and folding doors, these doors are made to open outwards and are often left open for long periods, provision should be made in their design and construction for all round protection against the weather.

Doors to be considered here, are:

- Ledged and braced battened (matchboarded) double-leaf doors.

- Framed ledged and braced battened (matchboarded) double-leaf doors.
- 'Up-and-over' single door systems.
- Sliding doors.
- Folding doors.

Figure 6.78 Security, retainers and fittings

Figure 6.79 Sundry items

6.10.1 Ledged and Braced Battened (Matchboarded) Double Leaf Doors (Figure 6.80)

Methods of construction for these doors are the same as those shown in Section 6.1.1.

If due to the door's width, brace lines are at an angle less than 45° to the horizontal, then a two-piece single brace via the middle rail can be employed as shown.

Where the two doors meet, a cover lath is fixed (screwed from the inside of the door) to the door leaf which is to receive the lock – this forms the door handing rebate.

Each door is side hung either on scotch tee hinges (Figure 6.84), or cranked bands and hooks (Figure 6.85) which offer longer life and greater security.

General security can be provided by two tower bolts (Figure 6.86) – one at the top and one at the bottom of the inside face of the leading door, and a rim lock (Figure 6.63a) on the closing edge of the trailing door.

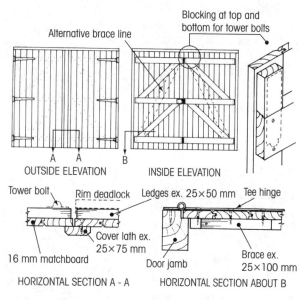

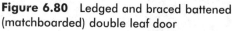

Figure 6.80 Ledged and braced battened (matchboarded) double leaf door

6.10.2 Framed Ledged and Braced Battened (Matchboarded) Double Leaf Doors (Figure 6.81)

As can be seen from the construction details shown in Figure 6.81. These doors are much stronger and heavier than the unframed types. Notice how glazed lights can be incorporated into the design. Meeting stiles can be rebated from the solid, or the doors can be reduced in width to accept two planted beads (the former method is better where security is of special importance).

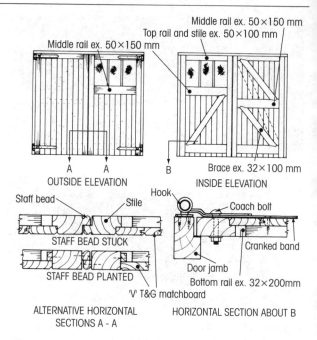

Figure 6.81 Framed ledged and braced battened (matchboarded) double leaf door

Figures 6.87–6.91 shows several suitable bolting mechanisms for securely retaining the leading door to the door frame head and/or floor. A rim lock can be used on the trailing door, and/or a hasp and staple with pad lock may be considered for added security.

Fitting and hanging framed type garage doors (also see Section 6.4)

1. Frame checks:
   - Check the opening for plumb and square.
   - Check the door sizes with the opening (using a timber lath or pinch rod).

2. Fitting the doors (one door at a time):
   - Cut off top and bottom horns – take care not to tear the grain on the back face of the door.
   - Fit the hanging stile – scribe if necessary.
   - Provide a top clearance of 3 mm, and floor clearance of 10 mm.
   - Provide an edge clearance of not less than 3 mm between the hanging stile and jamb, and excluding any 'lead-in' 3 mm between meeting stiles.

3. Hanging the doors:
   - Lay one of the doors flat (with its outer face uppermost) across two timber battens supported on trestles. Position a cranked band (Figure 6.85) (with its hook on plate in place so that the door jamb position can be allowed for) onto the top and bottom rails. Temporary fix the band (use two wood screws per band).

- Place both doors into the opening (use temporary door stops around the frame as required). Pack and adjust with small wedges until correct clearance is achieved.
- Tack a lath across the fronts of the doors and jambs to retain the correct clearance, and brace the doors to restrain any possible movement (this is very important if an assistant is not on hand and/or on windy days).
- Slide each crook (one at a time) into the band eyes – temporary fix in place with two screws.

   Note. Because each hook is a loose fit within the bands folded eye. The top hooks should be lightly pushed away from the door as it is being fixed; conversely the bottom hook should be pushed towards the door. In this way the door should not sag once the packing and restraint laths are removed.

- Lift the doors off their hooks, once again lay each in turn across the trestles, and with a bradawl make the centre position for each of the coachbolts, then remove the bands (each band, and its position should be marked accordingly, for example [LH-TOP. LH-BOT. RH-TOP. RH-BOT]).
- Bore holes to take the coachbolts.
- Paint the backs of the bands and their respective position on the door, then fully fix all four bands back onto the doors – not forgetting to use washers behind the nut or the coach bolt.
- Adopt a similar procedure with each hook plate, but before removing them from the jamb mark around the plate with a pencil, then gently tap the base (collar) of each hook so that an impression is left on the jamb were it has been welded onto the back-plate.
- Any protrusions can now be let into the jamb by either by chiselling back a shallow sloping groove or by boring an upward sloping hole to prevent water lodging behind the crook.
- Using only two screws per hook reaffix them into their respective places.
- Seal the bottom edge of the door by painting and/or preservative – particular attention should be paid to the end grain of the stiles.
- Rehang the doors – check all round for clearance – make any adjustments before adding the remaining wood screws – small amounts twist can be removed by recessing the back-plate of the offending hook or hooks (diagonally opposite) into the jambs.
- Replace temporary door stop with full ones to both door jambs and head.

4. Fit and fix a threshold if required.
5. Fix the remaining hardware as per specification.
6. Glaze as necessary.

6.10.3 'Up-and-over' Single Garage Doors (Figure 6.82)

This type of door is more likely to be made from steel, or GRP (glass-fibre reinforced polyester), than timber, but the great advantage of a timber door is that purpose made sizes are easily made. Nominal door opening sizes are shown in Table 6.4.

Two possible designs for purpose made doors are shown in Figure 6.82. Figure 6.82(a) uses framed, ledged, and braced battened method of construction. Figure 6.82(b) uses an outer solid timber frame with central muntin and mid-rail. The exterior plywood panels are sub-divided by two false muntins.

When the doors are held fully open (overhead), their dead weight makes them inclined to sag across their width. As shown in Figure 6.82, cross stiffeners will need to be in place to stop any deflection – these stiffeners may be available from the door-gear manufacturer (Figure 6.82i) alternatives worth considering are shown in Figures 6.82(ii–iv).

It is important that the over head gear must match the door's weight and its type of construction. Figure 6.83 shows how, when fully open these doors can, depending on the type of operating gear either end up fully retracted (via a top track) into the building – suitable for most wood doors (Figure 6.83a), or left

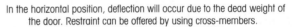

In the horizontal position, deflection will occur due to the dead weight of the door. Restraint can be offered by using cross-members.

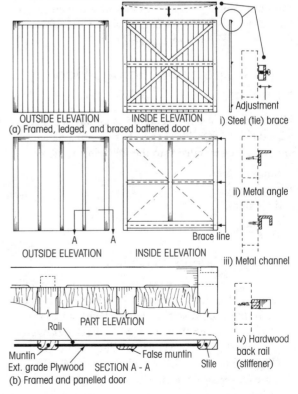

OUTSIDE ELEVATION   INSIDE ELEVATION
(a) Framed, ledged, and braced battened door

i) Steel (tie) brace

OUTSIDE ELEVATION   INSIDE ELEVATION   Brace line

ii) Metal angle

PART ELEVATION

Rail
Muntin   False muntin   Stile
Ext. grade Plywood   SECTION A - A
(b) Framed and panelled door

iii) Metal channel

iv) Hardwood back rail (stiffener)

Figure 6.82 Single 'up and over' garage doors

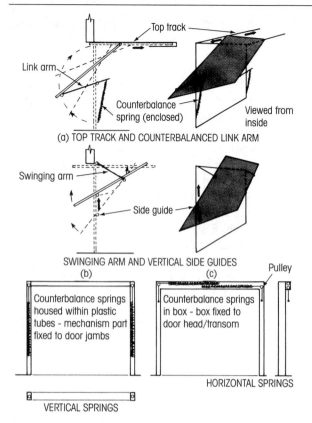

(a) TOP TRACK AND COUNTERBALANCED LINK ARM

SWINGING ARM AND VERTICAL SIDE GUIDES
(b)   (c)

HORIZONTAL SPRINGS

VERTICAL SPRINGS

Figure 6.83 Operating principle for 'up-and-over' garage doors

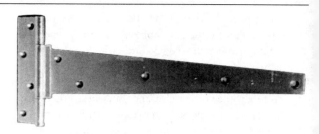

Figure 6.84 Heavy duty (scotch) tee hinges

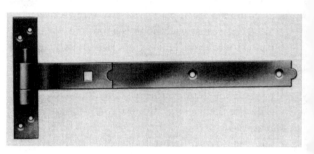

Figure 6.85 Cranked folded eye band with hook on plate

partly projecting to leave a canopy (Figure 6.83b). The counter balancing coil springs may be attached to a link arm (although enclosed), housed within the jambs, or frame head (Figures 6.82 and 6.83).

6.10.4 Side Hung Garage Door Hardware

Hardware for up-and-over doors is usually supplied with the door gear. Hardware for side-hung garage doors can be separated into the following sections:

- hinges
- security
- door stays.

Hinges

Two types of hinges to consider here:

1. *Tee hinges* – light duty tee hinges are more generally associated with shed doors, they are manufactured from thin gauged steel and coated with an epoxy black lacquer (often referred to as Black Japanned) or galvanised (zinc plated). Heavy duty tee hinges (Figure 6.84) – on the other hand are of much stronger construction and available with either a steel or brass pin, and a galvanised zinc protective coating, epoxy black or self coloured.

2. *Cranked folded eye band, with hook on plate* (Figure 6.85) – heavy duty hinge in two parts, with a self colour finish or galvanised. The steel band (strap) in this case is folded to produce a cranked eye at one end to receive a hook that is welded to a back plate. The crank (bend) in the band enables the face of the door to be in line with the front edge of its jamb. When the doors are fully open (90° to the door frame) they can be lifted off their hinges – this useful feature permits easy access for maintenance; for example, all future painting can be carried out under cover. Added security can be achieved by replacing the screws attaching the hooks to the jambs with clutch screws [BENCH AND SITE SKILLS, TABLE 11.2], or removing the Pozidriv keyway – alternatively hinge bolts (Figure 6.55b) could be used.

Security

One door leaf will be held shut by bolts, whilst the other will require a handle and lock mechanism.

Door bolts:

- *Tower bolts* (Figure 6.86a, b) – the most common type of bolt and least expensive. A cranked neck model (Figure 6.68b) is also available. An extended version of this type called a garage door bolt is shown in Figure 6.87. Generally finished in epoxy black, but they can be galvanised. The **garage door bolt** shown in Figure 6.87 is an extended version of the tower bolt.

- *Bow (D) handle door bolt* (Figure 6.88) – strong and heavy extended reach for tall doors suitable at the head and foot of the door. Finished coating of epoxy black.

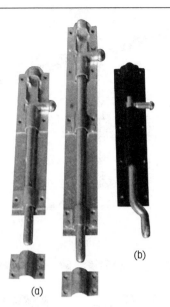

Figure 6.86 Tower bolts: (a) standard tower bolt (b) necked tower bolt

- *Monkey tail door bolt* (Figure 6.89) – strong and heavy extended reach for tall doors suitable at the head and foot of the door. Finished coating of epoxy black.

Figure 6.87 Garage door bolt

Figure 6.88 Bow 'D' handle door bolt

- *Chain bolt* (Figure 6.90) – self-latching top bolt, disengaged by a pull chain.
- *Foot bolt* (Figure 6.91) – a very strong bolt operated by the foot. Finished coating of epoxy black.

Figure 6.89 Monkey tail door bolt

Figure 6.90 Chain bolt

Figure 6.91 Foot bolt

Figure 6.92   Hasp and staple

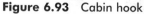

Figure 6.93   Cabin hook

Lock mechanism:
With ledged and braced battened doors a **rim dead lock** (Figure 6.33b) with **pull handle** (Figure 6.75) can be considered. For added security a pad locked **hasp and staple** (Figure 6.92) can be used.

With framed ledged and braced battened doors standard domestic **mortice locks** (Figure 6.63) as well as rim locks can also be used.

Door stays:
Stays provide a means by which garage doors may be safety held open (at least 90° to their opening). Doors already fitted with bolts to the bottom of the door may be provided with an extra keep for when the door is open, and a further bolt provided to bottom of the adjacent door.

If the door hinges back towards a wall or fence **cabin hooks** (Figure 6.93) can be used as door retainers. Failing that, a **Crompton garage door holder** (Figure

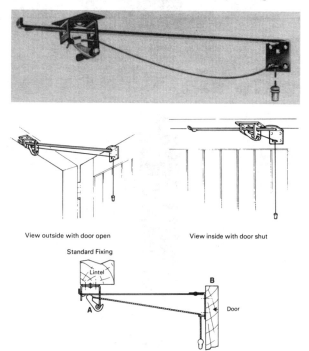

Figure 6.94   Garage door holder

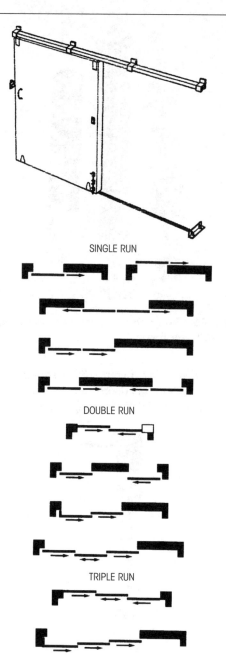

SINGLE RUN

DOUBLE RUN

TRIPLE RUN

Figure 6.95   Straight sliding top hung (Henderson industrial sliding gear

6.94) can be attached to the top of each door – on opening the doors their automatic check secures them in the open position. Doors are closed by pulling an acorn pendant to release the catch and thereby allowing the doors to be closed by hand. At no time should the cords be used to close the doors.

6.10.5 Sliding Door Gear (Section 6.10.4)

Figure 6.95 shows a straight sliding top hung system suitable for both external and internal application to workshops or garages.

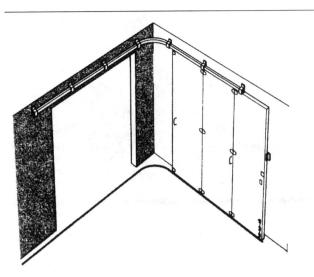

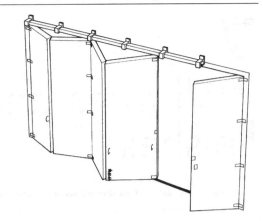

EXAMPLE PLAN DETAILS

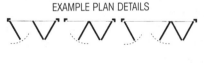

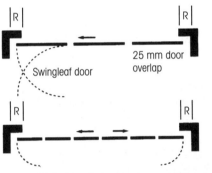

For complete locking, R should not exceed 360 mm, but 115 - 230 mm is ideal. If the swingleaf under the curve is not required, R should be a minimum of 690 mm to accommodate the corner curve.

690 mm min. if swingleaves not fitted

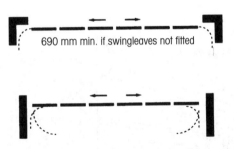

Figure 6.96 Round a corner top hung sliding door (Henderson Industrial Sliding Gear)

Figure 6.97 Folding doors (Henderson Industrial Sliding Gear)

By using either a single, double or triple track system several doors can be slid simultaneously to suite a variety of different sized openings and door combinations – a bottom roller system (not illustrated) is also available.

Alternatively, as shown in Figure 6.96, by link-hinging several doors together they can be slid via a curved track system round a corner or corners.

6.10.6 Folding Door Gear (Industrial and Commercial)

Figure 6.97 shows a top hung track system for a series of external folding doors. This arrangement is suitable for garages, garage showrooms, and workshops and shop fronts. As can be seen, with this system large clear opening are possible were doors can be neatly folded away to the sides.

7
Wall trims and finishes

Wall trims are used either to mask or conceal any gaps which are likely to occur as a result of moisture or thermal movement (Figure 7.1) from dissimilar materials which abut each other. They can also provide a decorative wall feature.

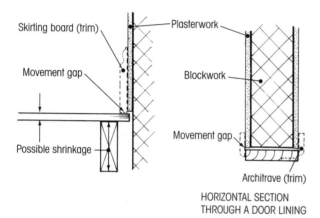
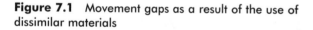

HORIZONTAL SECTION
THROUGH A DOOR LINING

Figure 7.1 Movement gaps as a result of the use of dissimilar materials

Figure 7.2, together with Table 7.1, should be used to identify and locate the most common of these wall trims so as to ascertain whether they have a functional or decorative role, or possibly both.

Functional properties – apart from being used to mask any gaps between the woodwork and wall. Trims can also play a major part in protecting plasterwork.

Wood and wood composite trims are much more resilient to knocks than plasterwork. As shown in Figure 7.3 their position and size reflects were plasterwork is at its most vulnerable against knocks from a vacuum cleaner or sweeping brush, and pushing of furniture against a wall.

Decorative properties – with the exception of a specific 'chair rail', all the other items listed in Table 7.1 will be shaped as a moulding to be in keeping with the design and style of the room.

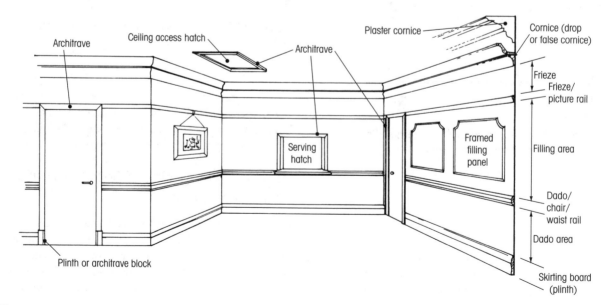

Figure 7.2 Typical location of trims and finishes

Table 7.1 Wall trims and finishes in relation to their use (Figure 7.2)

Wall trims and finishes (mouldings)	Functional	Decorative	Location
Architrave	Yes (P)	Yes	Wall to door/hatch lining. Ceiling to hatch lining
Skirting board	Yes (P)	Yes	Wall to floor.
Dado (chair or waist) rail	Yes/no	Yes (P)	Wall trim (830–900 mm) – above floor level
Picture (frieze) rail	Yes/no	Yes (P)	Wall trim (400–500 mm) – below ceiling level
Delft (plate) shelf	Yes (P)	Yes	Wall trim (400–500 mm) – below ceiling level
Cornice (drop or false)	Yes (P)	Yes	Wall trim – can be used to house concealed lighting units
Infill panel frame	No	Yes (P)	Wall trim – mid-way between picture rail & dado rail

P = primary use.

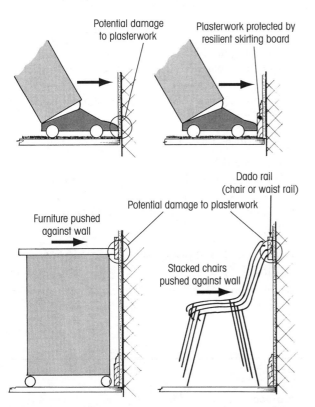

Figure 7.3 Wall trims used to protect plasterwork

7.1 MOULDINGS (SHAPED SECTIONS)

Unlike moulded sections used for doors and windows [BENCH AND SITE SKILLS, SECTION 1.9.4, FIGURE 1.89, TABLE 1.15] the shaped edge or edges of moulded trims do not always reflect their end use. Some do however have recognisable profiles, which carry a common name recovered from a bygone age of the classical periods. It is these classical lines [BENCH AND SITE SKILLS, SECTION 1.9.4, FIGURE 1.93] that we tend to copy when reproducing the various shapes we call mouldings.

7.1.1 Designing Moulding Profiles

Figure 7.4 shows how some of the classical shapes based on the Roman and Grecian styles can be reproduced using simple geometrical means.

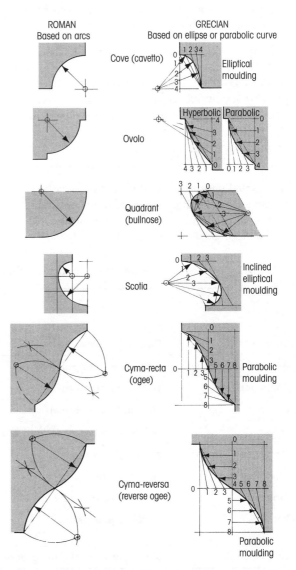

Figure 7.4 Forming moulding shapes geometrically

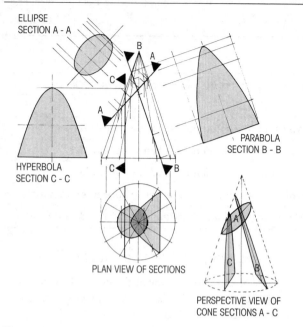

Figure 7.5   Grecian mouldings are derived from conic sections (sections through a cone)

Roman styles are based on the arc's of a circle [BENCH AND SITE SKILLS, SECTION 13.5.3], whereas the Grecian styles are based on the softer curves of the ellipse, parabola and hyperbola formed as a result of conic sections (Figure 7.5 shows from where these are derived).

7.2 MATERIAL

If a painted finish is required it can be most cost effective to use clear grades of softwood, preferably Redwood (dead and large live knots must be avoided on the moulded portion). Medium density fibreboard

(MDF) [BENCH AND SITE SKILLS, SECTION 2.5.6] moulding is a more modern alternative providing an unblemished surface usually pre-painted as part of the manufacturing process. MDF profiles are also available covered in a variety of different films, including most of the decorative hardwoods.

Where a natural finish is required both softwoods and hardwoods are used. A large range of hardwoods are used, some for their good clear close grain machining properties, others for their grain characteristics [BENCH AND SITE SKILLS, SECTION 1.10.3, TABLE 1.20, FIGURE 1.106]. Table 7.2 lists some of the hardwoods used in the manufacture of mouldings.

7.3 FORMING THE PROFILES

Propriety mouldings are produced either by:

- moulding machine
- spindle moulder (vertical spindle or shaping machine)
- embossing (pressing)
- machine carving (using computer numeric control, CNC)

Purpose-made mouldings are produced either by:

- spindle moulding
- portable powered hand router [BENCH AND SITE SKILLS, SECTION 8.15]
- combination and multi planes (universal plane) – Figure 7.8.

Hand tools (including moulding planes – Figure 7.7)

If you purchase pre-formed mouldings from a supplier you expect uniformity throughout that particular

Table 7.2   Selection of hardwoods used as mouldings

Species	Colour (Heartwood)	Density kg/m^3) average @ 15%mc	Hardness	Remarks
Aspen (Popular)	Grey, whitish to pale brown	450	Soft–medium	Medium working properties
Basswood	Whitish to pale brown	420	Soft–medium	Good working properties
Beech	Whitish to pale brown	720	Hard	Good working properties
Lime	Yellowish, or white to light brown	560	Medium	Excellent for carving
Mahogany (American)	Reddish brown	560*	Medium	Good working properties
Oak (Red Oak)	Pink to reddish brown	790	Hard	Good grain characteristics
Obeche	White to pale yellow	390	Soft	Pins without splitting
Ramin	White to pale yellow	670	Medium–hard	Good machine carving properties
Tulipwood (Poplar)	Straw brown to slightly greenish.	480*	Soft–medium	Good working properties

*Density can vary.

pattern. Each pattern will carry a specific style name or batch code number to enable you to reorder at a later date. However, if you change your supplier and ask for the same style of moulding don't be surprised if the profile does not match your previous order – product cutter profiles for the same style commonly vary between suppliers. If on the other hand your company has the machinery (spindle moulder) to produce small batches of moulding this problem should not arrive.

Almost any moulding can be produced with a portable powered router providing the appropriate cutters are available. For large profiles the router will be hand held and passed over the work piece via the appropriate fence or jig. For smaller moulding profiles of a short run off, the router may be inverted into a purpose made table and used as a small spindle moulder [BENCH AND SITE SKILLS, SECTION 8.15.4].

Working mouldings by the use of hand tools can be a slow process, but there are occasions that warrant this. Usually as a result of when a short length non-standard moulding is required, or when it is impracticable to start making up a machine cutter to suit. Hand working means using all the tools at our disposal, this may include using portable powered hand tools such as the router and planer – not only can they speed up the process they can, used correctly, add to detailing accuracy. Figure 7.6 shows how a simple parabolic ovolo moulding profile often used on the edge of skirting boards can be reproduced using hand tools, for examples:

- Ensure the sample is clear of paint build-up (Figure 7.6a) – remove as necessary.
- Make a template of the profile out of card (Figure 7.6b), or use a proprietary needle template former (Figure 7.6c).
- Mark off the end profile of the section to be copied (Figure 7.6d).
- Mark lines along all the square edges to be formed on the face side and face edge (Figure 7.6e) – this will help when setting width gauges of the moulding tools.

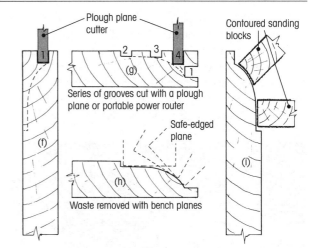

Figure 7.6(f–i) Copying a preferred order mould using hand tools (continued)

- With the stock (base material, for example a length of timber) held vertically in the vice and using a plough plane or portable powered hand router cut a groove to form the quirk (Figure 7.6f).
- With the stock held securely flat form as many grooves as practicable to the contour depth of the prescribed moulding profile (Figure 7.6g).
- Gradually remove remaining waste with bench planes (Figure 7.6h) whilst constantly checking for the finished profile with the template (Figure 7.6b).
- Using step down grades of abrasive papers and shaped sanding blocks (Figure 7.6i) remove all the flats left by the planing.

This method can be adapted to suit many different profiles, but there will be occasions when inward curves (coves) will need to be formed. In these cases special wooden moulding planes are used (Figure 7.7).

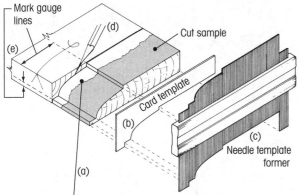

Figure 7.6(a–e) Copying a preformed ovalo mould using handtools

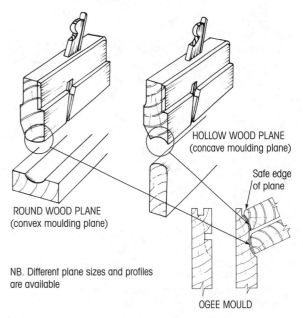

Figure 7.7 Wooden moulding planes

Figure 7.8 shows a plane and its range of cutters enabling it to form a variety of moulding shapes.

For very small detailing and beading a purpose scratch stock (Figure 7.9) could be considered – it is most suitable for scratching a profile into dense hardwoods. The blade made from an old saw blade is shaped by grinding or filing with a cutting (scraping) angle of about 80° to

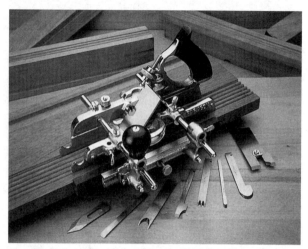

Figure 7.8   Clifton multiplane, one of the many specialist handplanes manufactured by Clioco (Sheffield) Tooling Ltd

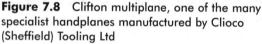

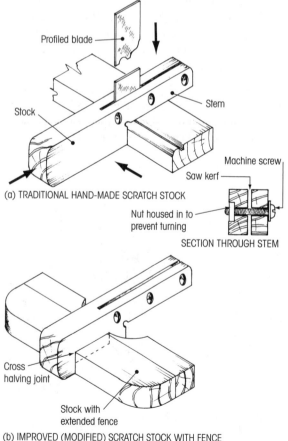

(a) TRADITIONAL HAND-MADE SCRATCH STOCK

SECTION THROUGH STEM

(b) IMPROVED (MODIFIED) SCRATCH STOCK WITH FENCE
Note: Made from hardwood – usually beech

Figure 7.9   A purpose scratch stock

suit a required profile. In use the scratch stock (fence) must be held square against the edge of the timber whilst pushing the cutter forward and at the same time lowering it very slowly into the wood.

Fixing mouldings

We will be covering the following areas of work:

- framing openings with architrave
- skirting boards
- rails around a room
- forming panels.

7.4 ARCHITRAVES

Architraves are more usually associated with forming a frame around internal door linings, but in fact architraves are also used to frame wall and ceiling hatches and some window linings.

Figure 7.10 shows several methods used to frame internal door linings, but before describing the different fixing methods we should consider which joint to use and why.

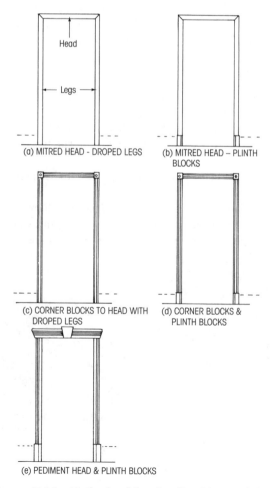

(a) MITRED HEAD - DROPED LEGS

(b) MITRED HEAD – PLINTH BLOCKS

(c) CORNER BLOCKS TO HEAD WITH DROPED LEGS

(d) CORNER BLOCKS & PLINTH BLOCKS

(e) PEDIMENT HEAD & PLINTH BLOCKS

Figure 7.10   Methods of framing lined internal door openings

As shown in Figures 7.10(a, b) the simplest method of joining the head to the legs is by using a mitre joint. The only problem here is you cannot conceal any gaps which more often as not will as shown in Figure 7.12 occur as the timber looses yet more moisture – chances are you will have noticed the effects of this in your own home without realising why.

Corner blocks – another way of joining the head to the legs is by using a corner block (as shown in Figures 7.10c, d). Figure 7.11 features a Yorkshire Rose to show how these may be personalised, but as shown in Figure 7.12 care should be taken to avoid any potential block shrinkage by either housing the architrave into the block or using a material of minimal shrinkage such as MDF. Figure 7.10(e) shows the use of an ornate pediment head, again it would be advisable for this to house the tops of the legs to conceal any shrinkage.

Plinth blocks – when as shown in Figures 7.10(b, d, e), 7.11 and 7.16, the thickness of the architrave is greater than the thickness of the skirting board, legs can extend to the finished floor level and butted against it. If on the other hand the architrave is thinner than the skirting board or is of an ornate nature liable to become damaged at about floor level, it is advisable to incorporate a solid block (although usually shaped with a chamfer to follow the outline of the architrave and allow the door to open beyond 90°) of timber or MDF known as a plinth or architrave block – three examples are shown in Figures 7.10(b, d, e) (see also Figure 7.11 and Section 7.4.4, Fixing plinth blocks).

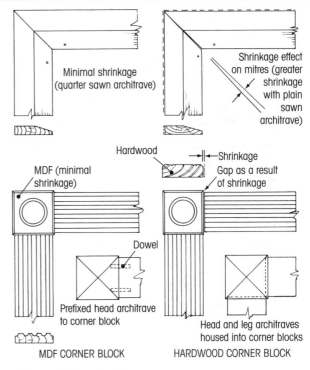

Figure 7.12 Shrinkage at corners

7.4.1 Ordering Architrave

Several common profiles are show in Figure 7.13 all of that are fixed in a similar way. The amount (linear metres) required for each door opening will depend on:

- The opening size (this can vary between door sizes – whether they are imperial or metric sized. So beware of the term 'Standard Opening Size' – **always check before ordering.**
- Width of the architrave.
- Condition of the architrave.

Timber merchant can supply architrave as 'sets' (2 legs 1 head) with a small provision of waste – again you must state the door's size, for example:

An imperial door 6′6″ × 2′6″ (1981 mm × 762 mm) using 50 mm wide architrave would require:

- 2 legs @ 2.1 m
- 1 head @ 0.9 m

Would require a standard 5.1 m length.

Beware – wider architrave would require a longer head length and possibly a longer leg length.

7.4.2 Fixing Architraves (Three-sided Openings, Figure 7.14)

Figure 7.14 shows two methods and sequences of fixing architrave around a rectangular door opening with mitred head and dropped legs. For example.

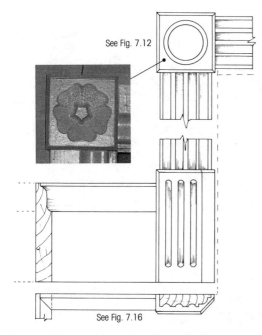

Figure 7.11 Incorporating a decorative feature within corner and plinth blocks

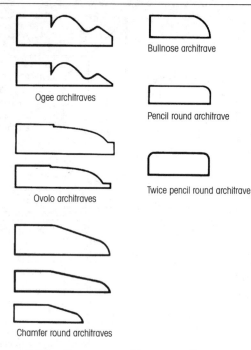

Figure 7.13   Common architrave profiles

Method 'A'

Stage 1. First architrave leg

Mark the margin distance – usually about 4–5 mm at intervals around the door opening – the thickness of one leg of your 4-fold rule [BENCH AND SITE SKILLS, SECTION 5.1.2] is usually adequate. Narrow margins with non-tapering architraves can restrict the amount the door is allowed to swing open without catching it. Narrow margins can also restrict the knuckle of the hinge housing, and the lead-in required for the 'lip' of a lock striking plate (Figure 6.56). Large margins on the other hand reduce the width required for nailing particularly with rebated casings, and can change the proportioned features of a decorative architrave beyond traditionally accepted levels.

Position the leg of the architrave against the wall as shown and mark the point where the internal margin for the leg meets the head margin.

This is the 'heel' of the mitre that will be cut at 45° from the 'toe'.

Use a mitre box [BENCH AND SITE SKILLS, SECTION 5.11, FIGURE 5.113] and tenon saw to cut the mitre – avoid using a mitre block as this can be less accurate. Alternatively use a proprietary hand mitre saw [BENCH AND SITE SKILLS, SECTION 5.11, FIGURE 5.114].

Using 38–45 mm oval nails fix (leaving nail heads to draw) to the door lining.

Note. If there is any doubt as to the squareness of the opening, use a short end of architrave as a gauge to pencil in cross lines as shown at each corner. This will allow you to mark a true bisected mitre for an accurate heel and toe angle. Any minor adjustments to the mitre

joint can be made by taking fine shavings off with a block plane [BENCH AND SITE SKILLS, SECTION 5.4.1, FIGURE 5.30].

Stage 2. Head architrave

• Mitre the adjoining end of the head architrave – adjust by planing if required.

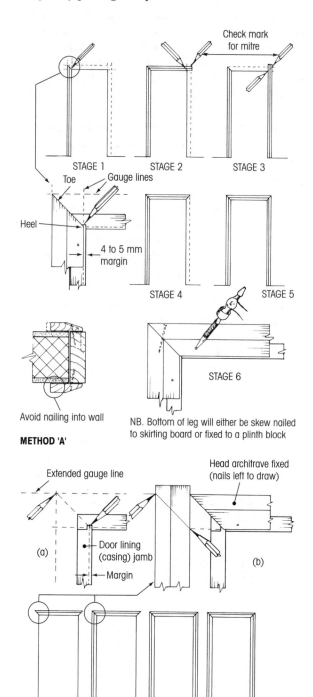

Figure 7.14   Sequence of fixing an architrave around a door opening

- Once fitted mark the position of the heel to toe at the other end.
- Cut the mitre **but do not fix in place**.

Stage 3. Second leg
- Position and mark the top mitre as before.
- Cut the mitre.

Stage 4. Head architrave
- To the margin lines fix to the door lining using 38–45 mm oval nails fix (leaving nail heads to draw).

Stage 5. Second leg
- Reposition architrave leg – adjust by planing if necessary.
- To the margin lines fix to the door lining using 38–45 mm oval nails fix (leaving nail heads to draw).

Stage 6. Final stage
- Complete the nailing process at about 300 mm centres.
- Ensuring that the outer edges of the mitre are in line, from the top of the mitre drive a nail at an angle across the mitre as shown.
- Punch all the nails just below the surface.
- The bottom outer edge will eventually secured by skew nailing to the skirting board.

Method 'B'

Stage 1. Head architrave
- Mark the margin distance – usually about 4–5 mm at intervals around the door opening – the thickness of one leg of your 4-fold rule [BENCH AND SITE SKILLS, SECTION 5.1.2] is usually adequate. Narrow margins with non-tapering architrave's can restrict the amount the door is allowed to swing open without catching it. Narrow margins can also restrict the knuckle of the hinge housing, and the lead-in of a lock striking plate. Large margins on the other hand reduce the width required for nailing particularly with rebated casings, and can change the proportioned features of a decorative architrave beyond traditionally accepted levels.
- As shown at 'a' – using a short end of architrave mark cross over gauge lines at each corner.
- Using a mitre box and tenon saw cut a mitre at one end.
- Whilst holding the head architrave in position against the gauge lines on the lining and plaster work check the heel and toe of the mitre for fit – adjust as necessary.
- Whilst holding the head architrave in position mark the heel and toe of the opposite end – check for 45° and cut as before.
- Using 38–45 mm oval nails fix (leaving nail heads to draw) to the door lining.

Stage 2. First architrave leg
- Held plumb against the head architrave – mark the toe and heel from the extended head gauge lines as shown at 'b'.

- Check the bisection line with your mitre or combination try-square – if satisfactory cut as before using a mitre box.
- If it is found to be off 45° cut by hand and if required fit to head mitre using a block plane.
- Using 38–45 mm oval nails fix (leaving nail heads to draw) to the door lining.

Stage 3. Second leg. Repeat as stage 2.

Stage 4. Final stage
- Complete the nailing process at about 300 mm centres
- Ensuring that the outer edges of the mitre are in line, from the top of the mitre drive a nail at an angle across the mitre as shown.
- Punch all the nails just below the surface.
- The bottom outer edge will eventually be secured by skew nailing to the skirting board.

7.4.3 Scribing Architraves

Where an architrave has to abut a wall and it spans a gap narrower than its width, and its abutting surface is either uneven or out of plumb. It will require shaping to ensure a close fit against its abutment.

Figure 7.15(a) shows a situation, which will require both one leg and part of its head scribing. Figure 7.15(b) show the finished arrangement – collectively Figures 7.15(a–e) show a method of achieving this, for example:

1. Cut the left architrave to full height 'h' (Figure 7.15a).
2. Determine the scribing width 'x' (Figure 7.15c) by measuring that portion which overhangs the lining when held plumb with it against the wall, plus the margin distance.
3. Either cut a scribing block or use dividers, or a compass to width 'x' (Figure 7.15c).
4. Temporary tack the architrave plumb to a distance as shown in Figure 7.15(c) then mark the scribe line down the full length of the architrave.
5. Detach the scribed leg.
6. Remove the waste wood either by planing or sawing – slightly undercut as shown in Figure 7.15(e) to ensure a good fit.
7. Reposition, check the fit, then mark and cut the mitre.
8. Cut the head architrave is required to length 'w' plus 20 mm (Figure 7.15a).
9. Position the head architrave at a distance as shown in Figure 7.15(c) – temporary tack in place.
10. Scribe to the bulk head ceiling line as previously described at (2).
11. Remove the waste wood either by planing or sawing – slightly undercut as shown in Figure 7.15(e) to ensure a good fit.
12. Reposition, check the fit, then mark and cut the mitre.
13. Repeat the procedures as described in Section 7.4.2.

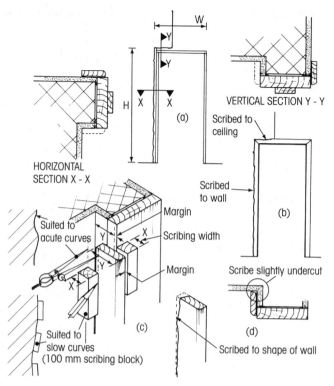

Figure 7.15 Scribing architrave to an uneven wall or ceiling

7.4.4 Fixing Plinth Blocks (Also know as 'Architrave Block')

Where as shown in Figure 7.16(a) the architrave is thicker than the skirting board, unless the lower corner of the architrave needs protection, there is no need for a plinth block. If on the other hand the skirting board is thicker than the architrave and or protection is required (Figures 7.16b, c) a plinth block is usually the answer.

Traditionally plinth blocks are (as shown in Figure 7.16d) first fixed to the architrave via a bare-faced tenon glued and screwed to the back of the block – the legs are fixed in the normal way. The block will then form an end stop for the skirting board, the profile of which is usually housed into the block. In this way any shrinkage gaps which may occur are shrouded.

Alternatively both the architrave and the skirting board can be butted against the block. In this way the block is fixed first, followed by the architrave, then the skirting board.

The relationship between the architrave and the plinth block can vary (one example is shown in the horizontal section, Figure 7.16e).

7.4.5 Fixing Architraves (Four-sided Openings)

Figure 7.17 shows a method and sequence of fixing architraves around a wall serving hatch or ceiling access trap.

Stage 1. Marking out
- Mark a margin distance of about 3–5 mm around the opening.
- If there is any doubt as to the squareness of the opening (pay particular attention to ceiling openings), use a short end of architrave to gauge the width at each corner to enable the true mitre angle to be formed.

Note. This procedure can be used to find the mitre angle of any angle (both acute and obtuse).

Stage 2. Left-hand architrave (a)
- Cut the bottom mitre.
- Hold firmly into position whilst marking the heal point of the top mitre – cut the mitre.
- Nail into position – leave nail heads to draw.

Stage 3. Head architrave (b)
- Cut and fit the left-hand mitre.
- Hold firmly into position whilst marking the heal point of the RH mitre – cut the mitre and put to one side.

Stage 4. Bottom architrave (c)
- Cut and fit LH mitre.
- Hold firmly into position whilst marking the heal point of the RH mitre – cut the mitre.
- Nail into position – leave nail heads to draw.

Stage 5. Right-hand architrave (d)
- Cut and fit bottom mitre.
- Hold firmly into position whilst marking the heal point of the top mitre – cut the mitre.

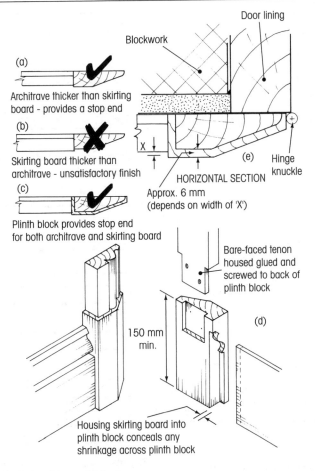

Door lining

Blockwork

(a)
Architrave thicker than skirting board - provides a stop end

(b)
Skirting board thicker than architrave - unsatisfactory finish

(c)
Plinth block provides stop end for both architrave and skirting board

X

HORIZONTAL SECTION

Approx. 6 mm
(depends on width of 'X')

(e)   Hinge knuckle

Bare-faced tenon housed glued and screwed to back of plinth block

150 mm min.

(d)

Housing skirting board into plinth block conceals any shrinkage across plinth block

Figure 7.16   Providing a plinth (architrave) block

Stage 6. Final assembly

- Hold (b) and (d) in position – check mitres (a–b, b–d, and c–d) for fit – adjust if required.
- Nail 'd' into position, then 'b'.
- Ensuring that all the mitres are fair faced carefully nail across each mitre.
- Complete the nailing pattern then punch all nail heads just below the surface in preparation to receive any filler.
- Using a block plane remove a small arris (sharp corner) all-round the framework.

7.5 SKIRTING BOARDS

We have already seen how skirting boards form a decorative finish between the wall and floor and how they offer protection to the wall against everyday activities like vacuuming the carpet or sweeping the floor.

The skirting board size (in particularly its depth) and moulding details usually reflects the period that the property was built.

A selection of commonly available sections is shown in Figure 7.18. If wood is to be used, available lengths are as for processed timber, for example: lengths will start at 1.8 m then upwards in increments (multiples of) of 300 mm up to a maximum of 6.3 m.

7.5.1 Treatment at Corners

External and internal corners are dealt with differently. External corners are mitred (an example is shown in Figure 7.19). Internal corners scribed (an example is shown in Figure 7.20) – the main reason for using a scribed joint is to minimise the shrinkage gap across the thickness of the skirting board. Remember that the greatest amount of shrinkage across a section of timber takes place tangentially followed by radially and then finally the least is longitudinally. Figure 7.21 shows how this can affect an internal mitre. The amount of shrinkage will depend on the moisture content of the wood on installation, and the position of the growth rings. For example are they predominantly across the width as with quarter-sawn timber, or down its depth as with tangentially sawn timber [BENCH AND SITE SKILLS, SECTION 1.4.7].

The mitre – very few corners are perfectly square, but just like those angles which are acute (less than 90°) or of obtuse (greater than 90°) the mitre angle as shown in Figure 7.22) can be easily found, by bisection of the

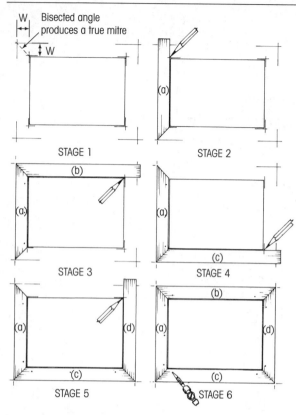

Figure 7.17　Sequence of fixing architraves around wall and ceiling hatch opening

angle. This is best done by using a mitre bevel board, which needs to be between 75 and 100 mm wide and about 900 mm in length.

Stage 1.
Position a the mitre board flat on the floor and up against one wall and overlapping the corner by as much as its own width – mark the floor at its outer edges.

Stage 2.
Re-position the board to touch the other wall and mark floor as before.

Stage 3.
By marking a line from the corner point of the wall to where the outer lines intersect, the bisecting line will produce a true mitre line.

Stage 4.
The mitre line can now be marked on to a bevel board – ready to be transferred to a sliding bevel.

Stage 5.
This angle can be transferred direct to the skirting board, or used as the angle for marking onto a mitre box.

The scribe – as we have seen internal corners are scribed to reduce the effect of shrinkage. As can be seen from Figure 7.20. cutting a scribed joint is not as straightforward as using a mitre, it involves several processes for example:

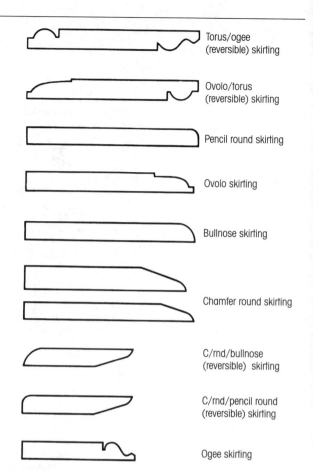

Figure 7.18　Common skirting sections

Stage 1.
Using a mitre box cut a true mitre as if to fit an internal corner, this in effect will produce a front view of the skirting boards profile (as shown in Figure 7.20a).

Stage 2.
Remove the waste portion (Figures 7.20b, c) – with the exception of the top cut, cuts may be slightly under-cut, this will ensure a close fitting joint.

Stage 3.
Before fixing offer the scribe up to the face of a short end of skirting board to check the fit.

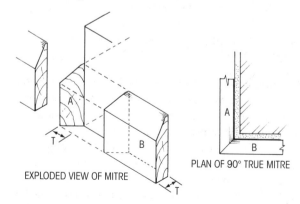

Figure 7.19　Cutting a true mitre to a 90° external corner

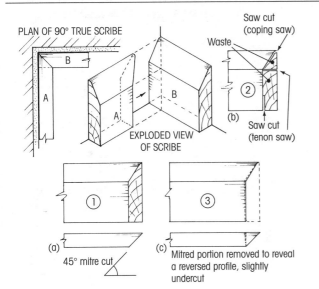

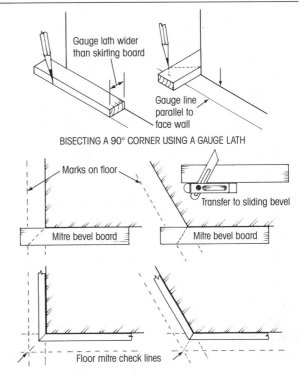

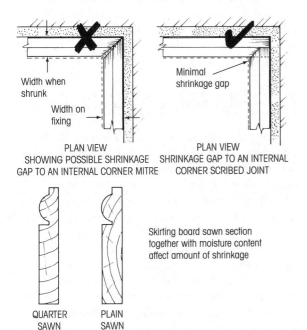

Figure 7.20 Cutting a true scribe (reverse profile) to a 90° internal corner

Note. This method of cutting a reverse profile can be used on any moulded profile – curved sections are generally cut away with the use of a coping saw.

7.5.2 Stop Ends

Whenever a moulding comes to a stop without an abutment its mould is either returned to the wall it is fixed to or, the floor it sits on.

Returned to wall – Figure 2.23 shows two methods of returning the moulded profile, in Figure 2.23(a) the profile is cut across the end grain. This method is known as a mason's mitre. The second method shown in

Figure 7.21 Possible effect of shrinkage on internal corner joints

Figure 7.22 Bisecting angles for a mitre bevel board

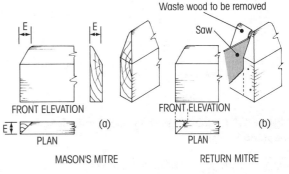

Figure 7.23 Returning a skirting board towards the wall

Figure 2.23(b) is used with complex moulds and quality work – a true mitre is formed with a short end of skirting board that is glued and pinned together. When set, the waste end is carefully cut back leaving a true return without any end-grain exposure.

Returned to floor – Figure 2.24 shows two methods of returning a moulded profile back towards the floor. The first example (Figure 2.24a) is only suitable for a skirting board with a chamfered face, whereas Figure 2.24(b) can be used with any shape of mould – even the most complex.

7.5.3 Lengthening Skirting Board (Heading Joints)

Heading joints are only used where the run is too great for a standard length of skirting board – with good planning heading joints can often be avoided.

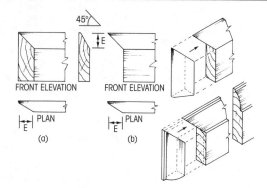

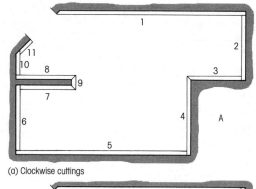

(a) Clockwise cuttings

Figure 7.24   Returning a skirting board towards the floor

If such a joint is required as shown in Figure 7.25 it is formed by making two accurate parallel 45° cuts using a mitre box.

Great care is required for this joint as any in accuracy will be difficult to conceal. The overlapping portion should face the main natural light source (window) to reduce the effect of any shadow line. The joint is secured by skew nailing.

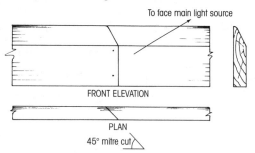

Figure 7.25   Joining skirting board in its length (heading joints)

7.5.4 Fixing Skirting Boards

Before any fixing takes place it is important to plan the sequence you intend to follow – not only for economic reasons but to avoid cutting mitres and or scribes at both ends. Two guidance fixing sequences are shown in Figure 7.26.

In Figure 7.26(a) the sequence follows a clockwise direction, whereas Figure 7.26 follows an anti-clockwise direction – only lengths numbered three, seven, and nine, have either a mitre or scribe at both ends.

Skirting boards are generally fixed to walls by nailing through the face of aboard, as shown in Table 7.3 the methods used can very according to the different types of base material used for walls.

Fixing centres can vary – with long lengths 600 mm centres is usually adequate.

More often as not the skirting board will be slightly sprung in its length. It will therefore need to be pushed to the floor before being fixed. As shown in Figure 7.27 a short end of board (kneeler board) can be used for this purpose.

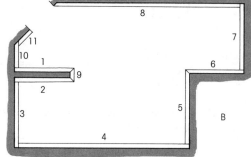

(b) Anti-clockwise cuttings

Figure 7.26   Sequence of fixing skirting board

7.5.5 Scribing Skirting Boards

If as shown in Figure 7.28(a) the floor is uneven and a close fit is required then it can be scribed similar to architrave. However, a word of warning, you should always bear in mind that if you reduce the depth of the skirting board at the ends (Figure 7.28b, c) the lengths of the abutting skirting boards should be reduced to bring them in line. The alternative is to increase the depth of the skirting board to be scribed by gluing a strip of timber to the under side (Figure 7.28c). Then by using an off cut of the strip as a scribing block scribe the skirting board to the floor.

7.6 RAILS AROUND A ROOM

As shown in Table 7.1 the two rails commonly fixed around a room are the dado and picture rail. The dado rail can have two main functions, one to divide the height of the room into two horizontal areas, and secondly as a means of protecting the wall from being damaged from chair backs and wall furniture. Hence the reason for calling it a chair wall rail, or possibly a waist rail because it is usually situated at about waist height. Figure 2.29 shows four typical dado sections.

The primary function of the picture rail (Figure 7.30) was to act as a bracket for a picture hook from which a picture was hung – these could be very large and heavy so their fixings were very important. Today, picture rails just like the dado rail are often provided more as a decorative feature.

Table 7.3 Fixings [BENCH AND SITE SKILLS, SECTIONS 11.1, 11.2, 11.5–11.8]

Wall base material	Fixing medium	Fixing	Comments
Brickwork	Wood plugs	*Oval brad nails	Usually found in old properties and renovation work
Brickwork	Plastics plugs	Screws	
Brickwork	Direct	Masonry nails	Goggles must be worn
Breeze (clinker) blocks	Direct	Masonry nails	Goggles must be worn
Breeze (clinker) blocks	Plastics plugs	Screws	
Aerated concrete blocks	Direct	Cut clasp nails	
Aerated concrete blocks	Plastics plugs	Screws	

*As a general rule the length of the nail is usually 2½ times the thickness of the material to be fixed.
†Masonry nails should penetrate the base material by not less than 25 mm.
Note. Always ensure that any concealed services (electric cables, etc.) are detected (see Section 2.1.7) before any fixings are made. In some situations panel adhesive can be used as a fixing.

7.6.1 Setting Out

Figure 7.31 illustrates the importance of taking a datum line around the room. There are several methods of striking this line described in BENCH AND SITE SKILLS, SECTION 6.7, for example:

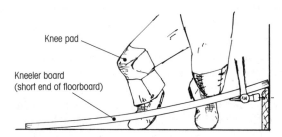

Figure 7.27 Holding 'sprung' skirting board down to an uneven floor

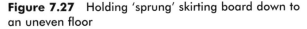
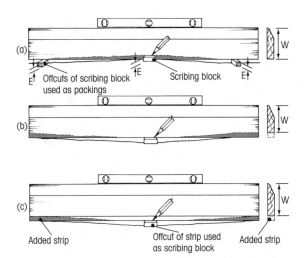
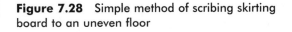

(a) Offcuts of scribing block used as packings — Scribing block

(b)

(c) Added strip — Offcut of strip used as scribing block — Added strip

NB. In most cases the end depth, 'W', **should not** be altered (cut away) as at (b).

Figure 7.28 Simple method of scribing skirting board to an uneven floor

- spirit levels
- water levels
- Cowley automatic level.
- Laser levels (not referred to in this book).

Once this level line is struck – ideally as the top line of the dado rail, this can be repeated at the picture rail height. It can be easier to use a 'gauge rod' cut to the filling height – this should be kept and marked 'template' as it will be useful at a later stage when, assuming that a dado rail is to be fixed first before the picture rail.

7.6.2 Treatment at Corners

The same principles as those used for the skirting board will apply (Section 7.5.1). When corners have to be bisected the methods shown in Figure 7.22 can still be used, but if there are wall units covering the floor at this

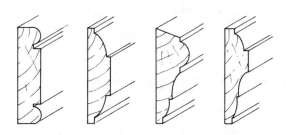
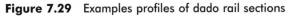

Figure 7.29 Examples profiles of dado rail sections

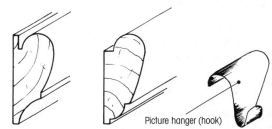

Picture hanger (hook)

Figure 7.30 Examples profiles of picture rail sections

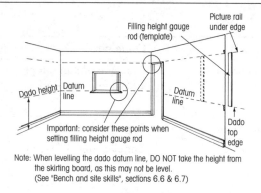

Figure 7.31   Setting out for dado and picture rails

point use two short ends of board hand-held across the corner instead of laying them on the floor.

7.6.3 Stop Ends

The same methods as those shown in Figure 7.23 can be used. If however a complicated shape is to be returned in the solid, then as shown in Figure 7.32 the profile can be marked from the back. Then very carefully cut away the waste wood with a coping saw, with its blade set to cut on the pull stroke, ragged edges on the face mould can be avoided. The shape can then be finished with a sharp chisel and abrasive paper held round an appropriate former.

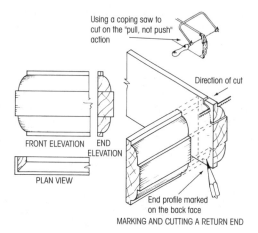

Figure 7.32   Scribing a returned stop end

7.6.4 Lengthening Dado and Picture Rails (Heading Joints) (see also Section 7.5.3)

Lengthening joints should always be avoided, the picture rail should never be joined in its length. Dado rails can be joined by using adjacent 45° cuts as shown in Figure 7.25.

7.6.5 Fixing Rails

The sequence described in Section 7.5.4 for fixing skirting boards will still apply – similarly the types of fixing used as shown in Table 7.3. Concealed propri-

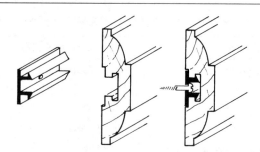

Figure 7.33   Proprietary concealed fixing devices used with a dado rail

etary snap fixing suitable for dado rails are available – a typical example is shown in Figure 7.33 they should be fixed according to the manufacturer's instructions.

Special regard should be given to fixing picture rails as the occupants could decide to use them as they were originally intended. That is to say for hanging pictures and mirrors from, these could be extremely heavy therefore much care must be taken at the fixing stage.

7.7 DELFT (PLATE) SHELF

These are usually associated with hallways or dining rooms to support ornamental plates or ornaments out of harms way. They can very often take the place of a picture rail.

Figure 7.34(a) shows a modern design, whereas Figures 7.34(b, c) illustrates two classical designs – all have dentil bracket supports. Sizes range from 100 mm by 100 mm to 150 mm by 150 mm, they are usually fixed to the wall by using screws and plugs, Figure 7.34(d) shows how the fixings can be concealed behind a bracket.

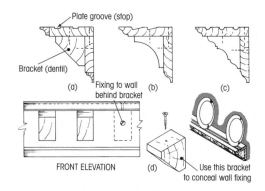

Figure 7.34   Delft shelf (plate shelf)

7.8 CORNICE (DROP OR FALSE CORNICE)

Cornices are more usually associated with plasterwork where wall meets ceiling, or the finish around the top of a wall or cupboard unit. There are however occasions where

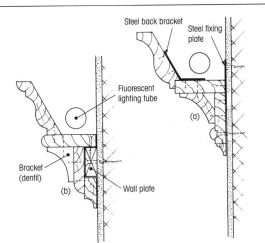

Figure 7.35 Wood cornice with concealed lighting

we can use a cornice to head wall panelling or independently as a feature, and as a means of housing concealed lighting to illuminate the freeze area of a room.

Figure 7.35 shows a vertical section through to simple cornice capable of concealing fluorescent tubes. The illustration in Figure 7.35(a) is made up of two runs of moulding and a shelf. Whereas Figure 7.35(b) is built up off a wall plate with a horizontal board, dentil brackets and moulded capping boards.

7.9 INFILL PANEL FRAME

This type of framework is purely decorative. Traditionally this work was undertaken by the plaster using a fibres plasterwork technique (a skill still used today). However, a less expensive alternative is to use a wood moulding. Figure 7.36 shows examples of the type profile that could be used for this purpose. There is a great variety in proprietary decorative mouldings on the market, which also include those that have been embossed or even carved.

The infill panel arrangements can vary greatly in their design, usually to suit the decor of the room. Figure 7.37 shows how corner details may vary, in Figure 7.37(a) a basic 45° mitre cut is made, were as in Figure 7.37(b) the mitre is formed by bisecting the angle as shown in Figure 7.22. The mitre for the quadrant at Figure 7.37(c) is more complex. In this case as shown in Figure 7.37(d) the mitre line is developed using

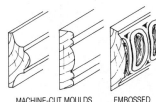

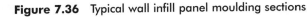

Figure 7.36 Typical wall infill panel moulding sections

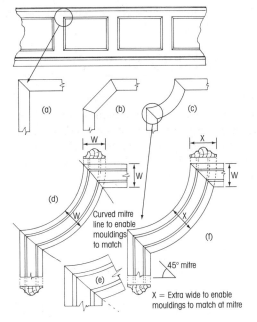

Figure 7.37 Infill panel corner details

geometry to produce a curved mitre line. Had a straight cut been made the abutting moulding detailing would not have met up, illustrated in Figure 7.37(e). Alternatively, as shown in Figure 7.37(f) the curved moulding is widened to meet the requirements at 'X'.

7.9.1 FIXING PANEL MOULDINGS

Methods of fixing the moulding to the wall really depends on their section size length and conditions (any slight distortions). For example the light sections of moulding may be stuck to the wall using a suitable

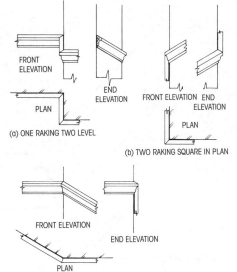

Figure 7.38 Possible arrangement for raking moulding

panel type of adhesive. Heavier or longer sections will require mechanical fixings.

7.10 RAKING MOULDS

Moulding trims inclined at an angle (raking moulds) are used to eventually joined up to one or more different levels to provide continuity of the moulding. In this case we are concerned with raking moles which continue around a corner. Three examples are given below and shown in Figure 7.38.

- One raking and two level – square in plan (Figure 7.38a).
- Two raking – square in plan (Figure 7.38b).
- One raking one level – non square in plan (Figure 7.38c).

Figure 7.39 illustrates a method of finding the bevel for the back of the moulding, the top edge bevel for the inclined mitre, and the required profile of the inclined moulding if the overall profile detailing is to meet at the mitre joint.

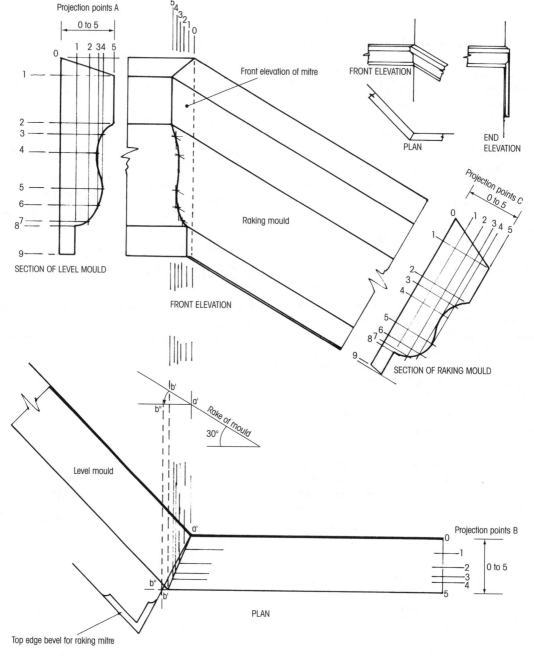

NB. Projection points at A, B and C are the same distance apart

Figure 7.39   Raking mouldings – bevel and section development

8 Windows

8.1 INTRODUCTION

Windows consist of glass panels held in a surrounding frame of timber (Figure 8.1a), or other material, such as plastic or metal, which are set into the walls and roofs of buildings to allow day light to enter, those inside to see out, and, in certain cases, to allow for ventilation (Figure 8.1b). If possible the windows should be of sufficient size to allow the level of illumination inside the building to be between 1% and 5% of the brightness of the sky outside.

As windows are subject to wind pressure the area of the panes of glass are limited (Table 8.1) and the glazed surface if large would be reduced by dividing up the area with horizontal and vertical members and/or sub-frames (casements and sashes) set into the main frame (Figure 8.1c).

The improved quality of timber windows and increased length of life in use, has been made possible by the use of water and weather proof adhesives and the preservation treatment given to timber.

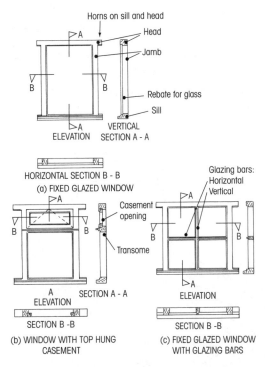

Figure 8.1 Window glazing arrangements

The manufacture of windows is regulated by the following Codes of Practice (CoP), British Standard (BS), International Standards Organisation (ISO), and Building Regulations Approved Documents (BR).

Manufacturing	*[ISO 9002]*
[BS644 Part 1 1989 and BSS 1186 Parts 1 & 2 1988]	
Weather resistance	[BS6375 Part 1 1989]
Timber preservation	[BS5589 Table 5 Category B]
Means of escape	[BS5588 Part 1
	Section 3 Para 11.5]
[BR 1991 Approved Document B paragraph 1.8]	
Glazing	[BS6262 1982.
	BS8000 Part 7 1990]
Glazing in critical locations	[BR Approved Document N]
Ventilation	[BR 1992 Approved Document F]

Table 8.1

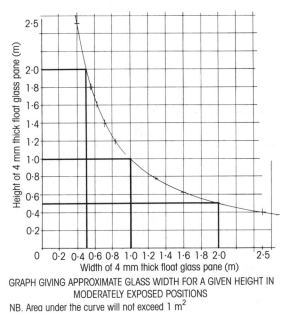

GRAPH GIVING APPROXIMATE GLASS WIDTH FOR A GIVEN HEIGHT IN MODERATELY EXPOSED POSITIONS

NB. Area under the curve will not exceed 1 m^2

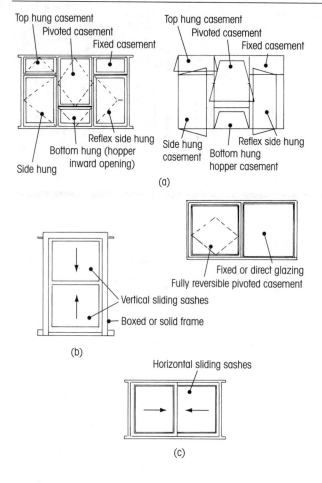

(a)

(b)

(c)

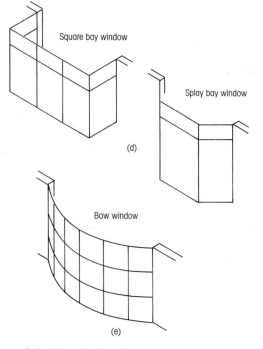

(d)

(e)

Figure 8.2   Hanging and positioning of opening lights and window types

8.2 GENERAL CLASSIFICATION OF WINDOWS

(1) Fixed glazing to   Glass fixed directly into the
    frame (Fast sheets)   frame (Figure 8.1a).
(2) Sash or box framed:
    (i)   Vertical sliding (Figure 8.2b).
    (ii) Horizontal sliding (Yorkshire light)
          (Figure 8.2c).
(3) Casement. Window Casements are fixed, hinged or pivoted and rebated into the main frame (Figure 8.2a).
(4) Bay window. A window being square or polygonal in plan forming an outward projection from a room (Figure 8.2d).
(5) Bow window. As a bay window, but segmental in plan (Figure 8.2e)

8.3 PARTS OF A WINDOW

The names and parts of a window will vary with its type.

8.3.1 Casement Window (Structural Frame). (Figure 8.3a)

- Sill   The bottom horizontal member of the frame.
- Head   The top horizontal member of the frame.
- Jambs   The outer vertical members of the frame.
- Mullion   Intermediate vertical member between the jambs.
- Transom   Intermediate horizontal member between head and sill.

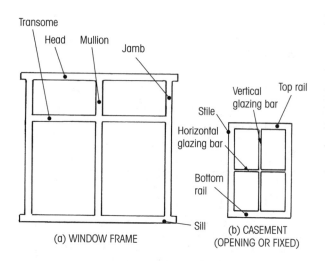

(a) WINDOW FRAME

(b) CASEMENT (OPENING OR FIXED)

Figure 8.3   Members of casement window

8.3.2 Casement Members (Casements Fixed, Hinged or Pivoted). (Figure 8.3b)

- Top rail   The top horizontal member of the casement.
- Bottom rail   The bottom horizontal member of the casement.
- Stiles   The outer vertical members of the casement.
- Glazing bars   Intermediate vertical and horizontal members.

Casements can be:

(a) Fully rebated into the frame (Figure 8.4a).
(b) Partially rebated into and lipped onto the face of the frame (Figure 8.4b).

In this case the gap between the faces of the rebate in frame and casement is not so critical. This wider gap is masked by the over lipping of the casement onto the frame, allows for a greater opening tolerance, and the casement is not so likely to bind on the frame if there is expansion of the timber.

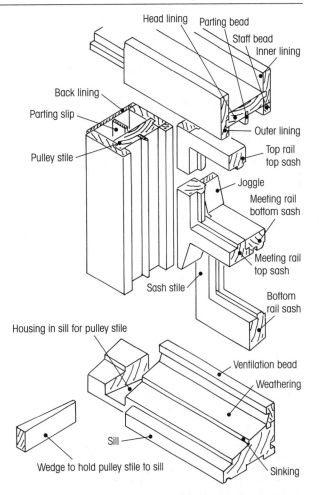

Figure 8.5   Components of a sliding sash window

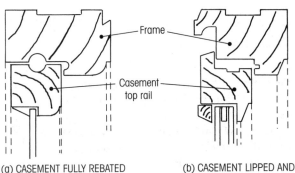

(a) CASEMENT FULLY REBATED INTO FRAME

(b) CASEMENT LIPPED AND REBATED INTO FRAME

Figure 8.4   Rebating casements into a frame (vertical section)

8.4 BOXED FRAMED OR SLIDING SASH WINDOW

In this type of window the opening lights (sashes) slide up and down within the vertical plane of the window. (Figure 8.2b).

Note. Vertical sliding sash windows can also be made with solid rather than boxed pulley stiles if helical springs rather than sash weights are used to support the sashes (Figure 8.32b, c).

8.4.1 The Boxed Frame. (Figure 8.5)

- Back lining: A vertical board, which forms the back of the box in which weights

are housed.

- Head: The top horizontal member which carries the parting bead, inner and outer top linings.
- Inner and Outer linings: Vertical and horizontal boards attached to the pulley stiles and head to form the inner and outer sides of the boxed frame.
- Parting bead: A slip housed into the pulley stiles and head to act as a separator between the sliding sashes.
- Pulley stiles: The vertical members of the frame which form a guide for the sashes.
- Sash weight: Cylindrical solid iron weights used to counter balance the weight of the sashes for ease of opening are attached to them by cords or chains running over pulleys housed in the top of the pulley stiles (Figure 8.32a).
- Sill: The bottom horizontal member into which the pulley stiles and

vertical inner and outer linings are housed.

- Staff bead: Fixed to the inner lining to keep the lower and inner sash in place.
- Wagtail, parting, slip or mid-feather: Used to keep the sash weights apart within the vertical boxes.

8.4.2 Sashes. (Vertically sliding) (Figure 8.5)

- Bottom rail: The bottom horizontal member of the bottom sash.
- Glazing bars: Intermediate horizontal and vertical members between stiles and top and bottom rails of the sash.
- Meeting rails: The top horizontal member of the bottom or inner sash and the bottom horizontal member of the top or outer sash.
- Stiles: The outer vertical members of the sashes.
- Top rail: The top horizontal member of the **top** outer sash.

8.5 THE RANGE OF STANDARD CASEMENT WINDOWS AVAILABLE FROM MANUFACTURERS

All windows can be made to any required size using one off or small batch production techniques. This is a relative expensive way of manufacturing windows and it is the usual practice to purchase standard windows from a large volume manufacturer, as mass production methods gives a standard product of known quality at a reasonable price. Joinery manufacturers produce a wide range of windows of various sizes, seeking to make their own range of windows unique while at the same time meeting the requirements of the relevant standards. Figure 8.6(a) gives a representative sample of the range and size of timber casement windows available from a manufacturer.

8.5.1 Preparing a Window Schedule

From the architect's drawing of a building which has the windows identified by a reference code a window schedule (list) of the windows required can be prepared as shown in Table 8.3. From the schedule an order for the windows can be prepared. A schedule is also of help to an estimator in estimating the cost of a building.

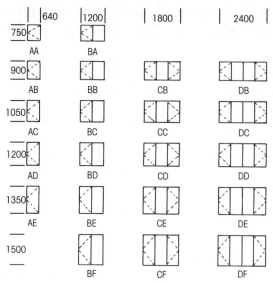

Note: Windows prefixed 'A' or 'B' can be ordered right- or left-hand

Figure 8.6 Example of a page from the catalogue of a timber window frame manufacturer

8.5.2 Receipt and Storage of Standard Windows on Site

On delivery the windows should be checked against the order for the correct number type and size. Check also for any damage [BENCH AND SKILLS, SECTION 14.3.1].

Before the windows arrive, on-site provision should be made for their receipt and storage.

Stacked flat, the windows are laid on bearers set in a flat plane and kept clear of the ground (Figure 8.7a) or vertically as (Figure 8.7b). The windows should be stored under cover, or if outside covered with a waterproof sheet so arranged that air can circulate round the windows. Figure 8.7c is an example of how windows should not be stored.

8.6 CASEMENT WINDOWS

The full rebating of a casement into the frame (Figure 8.4a) is no longer standard practice as it calls for accurate sizing and fitting of the casement into the frame if it is not to bind (catch) on the frame when being opened. Also you cannot have a mixture of fully rebated casements and direct glazing to a frame without changing the rebate size in the frame (Figure 8.8a), or using glazing beads larger than necessary. Alternatively the frame and casement can be rebated so that the casement overlaps the face of the frame (Figure 8.4b). The rebates in the frame can be the same size to accommodate both the casement and the direct glazing.

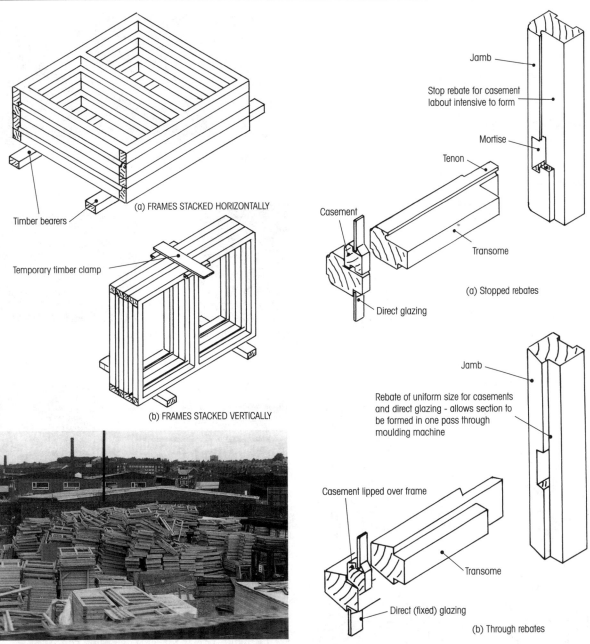

(c) How windows should not be stored

Figure 8.7 Site storage of windows.

Figure 8.8 Stopped and through rebates in window frames

This makes the working of the frame sections much simpler as the same profile can be worked along its complete length (Figure 8.8b).

8.6.1 Size and Profile of Casement Window Sections

BS644 gives recommended profiles and minimum sizes for members in casement windows, but most manufacturers vary the size and to some extent the section profiles by using more robust sections than those in the British Standard. Figure 8.9 shows the profiled sections used by some manufacturers.

8.6.2 Material Used in Window Frame Manufacture

The majority of timber used in the manufacture of windows is European Redwood (*Pinus sylvestris*) [BENCH AND SITE SKILLS, TABLE 1.2], which is non-durable and must be treated with preservative after the joints and sections have been worked, using a double vacuum process [BENCH AND SITE SKILLS, SECTION 4.4.2].

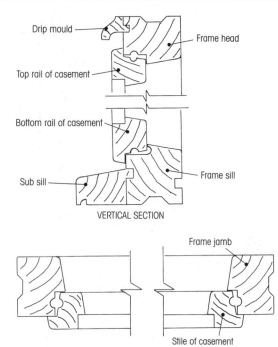

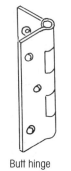

Figure 8.9   Storm-proof window sections

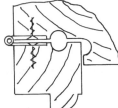

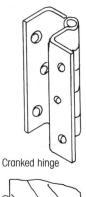

Figure 8.10   Hinging casements to window frames

Alternatively hardwoods can be used, but again if not classed as durable they must be treated with preservative.

8.6.3 Ironmongery for Windows

Factory produced casement windows come complete with hinges, casement stays and fasteners.

(a) Side opening mechanism with friction stay

(b) Fully reversible pivoting mechanism

(c) Casement stays and fasteners

Figure 8.11   Hardware (ironmongery) for casement windows

The type of hinge used will depend on the opening movement of the casement:

- Opening about one edge using butts or cranked hinges (Figure 8.10a, b).
- Open and slide about one edge using concealed sliders and pivots (Figure 8.11a).
- Pivot and fully reversible casements using concealed sliders and pivots (Figure 8.11b).

Casement stays and fasteners come in a variety of profiles and many incorporate locking mechanisms for added security (see Figure 8.11c).

8.7 FIXING TIMBER CASEMENT WINDOWS INTO THE WALLS

Two methods of building in windows into walls are used.

(a) Building in the window as the wall is being built.
(b) Forming an opening in the wall and then fixing in the window after the wall is built.

Hardwood and pre-glazed windows are usually fixed into the openings after the walls are completed, as this saves the hardwood being stained by cement and mortar splashes, and from other damage that may occur during the construction process.

8.7.1 Building in Windows as the Walls are Constructed

(a) The window sill and head are cut and notched for building into the opening depending on the following conditions:

   (i) When mortise and tenon joints are used the ends of the sill and head project beyond the jambs of the frame to form horns [BENCH AND SITE SKILLS, SECTION 7.5.7] (Figure 8.12a).

   (ii) When the head and sill are combed jointed to the jambs of the frame and the projecting part of the sill is notched round the brickwork (Figure 8.12b).

   (iii) The sill is cut to the width of the opening (Figure 8.12c).

(b) When the wall is built to sill height the widow is placed on the wall and adjusted for correct height

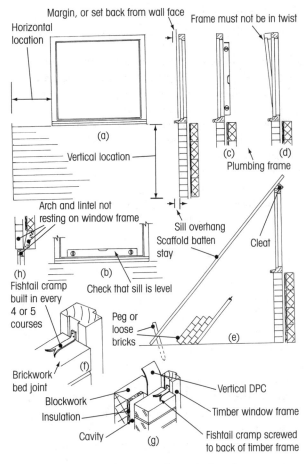

Figure 8.13 Building in timber window frames

and horizontal position. The sill is then checked for being level and the window set back the correct amount from the face of the wall (Figure 8.13a, b).

(c) The frame is checked to make sure it is not in twist and then plumbed in the vertical plane of the wall (Figure 8.13c, d).

(d) An inclined strut or struts are used to hold the window frame in position while being built into the brickwork, being secured to the window head by a cleat and fixed to a peg at the bottom, or weighted in place by a number of bricks or concrete blocks (Figure 8.13e).

(e) As the brick work is built up against the window frame galvanised mild steel, or non-ferrous metal (bronze) fish tail cramps are screwed to the back of the jambs of the frame and built into the bed course of the brickwork at every four to six courses (Figure 8.13f).

(f) A vertical damp-proof course (DPC) is also built in (Figure 8.13g).

(g) Any lintel placed above the window should be fixed clear of the head of the frame which should be considered as non-load-bearing (Figure 8.13h).

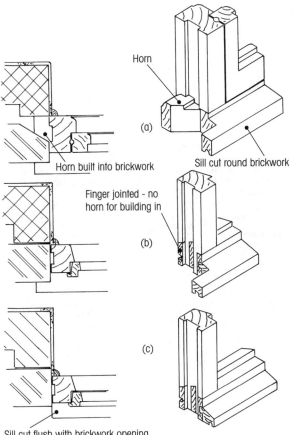

Figure 8.12 Cutting window sills for building in

8.7.2 Fixing Timber Windows in Prepared Openings in Masonry Walls

(a) To help the bricklayer form the opening for the window a timber template or profile is made, and braced square with the height equal to that of the window and a width equal to that of the window plus 6 mm to give a fixing tolerance of 3 mm at each side. The vertical tolerance is obtained by setting the profile on 6 mm thick blocks. The profile is made in such away that it can be taken out without damage to the surrounding wall (Figure 8.14a).

(b) The profiles are positioned, levelled, plumbed and stayed as previously shown in (Figure 8.13e).

(c) When the opening is completed and the window is ready for fixing, the profile is eased out by removing the bottom packings and carefully tapping out the profile. If the profile is tight it may be necessary to collapse it by removing the horizontal members, but usually the timber dries and shrinks sufficiently for the profile to be tapped out without damage to the surrounding brickwork.

(d) The window frame is placed and secured in the preformed opening in the wall by the use of frame fixing sleeve plugs [CARPENTRY AND JOINERY BENCH AND SITE SKILLS, SECTION 11.7.2] (Figure 8.14c) or galvanised mild steel fixing plates are screwed to the back of the window frame jambs and secured by screws to the masonry after inserting the frame into the opening (Figure 8.14b).

Note. Figure 8.14(d, e) shows the positioning of the window frame in relationship to the outer face of the wall. If the window is set 110 mm in recess a tile, brick or stone sub-sill will be required.

8.7.3 Sealing the Gap Between Window Frame and Surrounding Brickwork

With the development of a range of extrudable sealants which retain their elasticity for long periods of time even under the effects of ultra violet light from the sun, it is common practice to seal the gaps between wall and window frame with a mastic using a mastic pointing trowel, or gun dispenser depending on the type of sealant used.

(a) Trowel applied sealants consist of an inert filler mixed with a slow drying viscous medium which is applied to the gap and pointed with a mastic pointing trowel (Figure 8.15a). Over a period of time the oil dries out and the sealant looses its elasticity and consequently, its ability to maintain

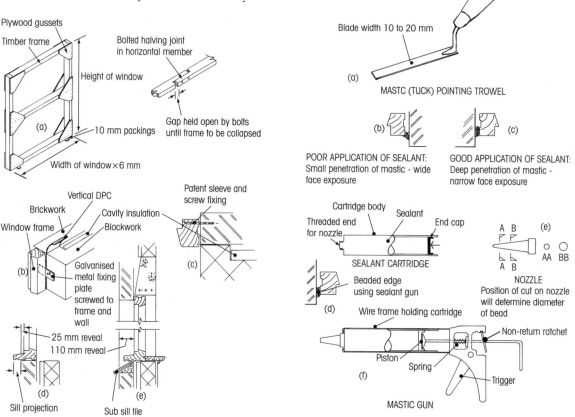

Figure 8.14 Fixing casement windows in prepared openings

Figure 8.15 Application of sealing mastic by gun and knife

Table 8.2 An example of a companies product information and a Health and Safety data sheet

Product	Flexible frame sealant	Silicon frame sealant
Identification	A high quality acrylic sealant. Waterproof, flexible and durable	A high quality neutral cure, low modulus silicon rubber sealant cures to a permanently flexible seal. Water proof and mould resistant.
Uses/applications	Interior and exterior use. Good adhesion, sealing gaps and draughtproofing.	Good adhesion ideal for sealing joints in wood metal and upPVC frames
Description	Water based filled acrylic polymer	Low modulus silicon sealant.
Physical properties		
Appearance	White/brown smooth paste	White/brown paste.
Odours	very slight when wet.	Slight whilst curing.
Performance		
Fire.	Non-flammable	Non-flammable
Water	Wet-water will disperse product	
Dry-water resistant	Water resistant	
Chemicals	resistant to petrol, oils and weak alkalis. Not resistant to acids	Resistant to attack by dilute acids and alkalis
Biological	Contains a non-mercurial preservative. No bacteria growth	Does not support the growth of bacteria or fungi
Thermal	Serviceable from −25°C to +40°C	−50°C to + 120°C
Durability reliability	Approximately 10 years	Twenty years.
Compatibility. With other products	Compatible with most building surfaces	Good adhesion to most building surfaces
Working characteristics		
Rheology	Easily spread viscous paste	Easily spread viscous paste
Working time	Approximately 20–30 minutes	Approximately 15 minutes
Cure time	Approximately 48 hrs.	Approximately 24 to 48 hrs
Cure type	Water evaporation	Neutral curing system
Temperature range	Do not apply below 5°C	Do not apply below 0°C
Coverage	Approximately 11 m using a 6 mm bead	Approximately 11 m using a 6 mm bead
Shelf life	At least 12 months in unopened cartridge	At least 12 months in unopened cartridge
Health and safety data sheet		
Identification	Flexible frame sealant in white plastic cartridge	Window and door frame sealant.
Company	Manufacturers Ltd., Makesville. MV1 3OD. Tel: 1234 5678901	Manufacturers Ltd., Makesville. MV1 3OD. Tel: 1234 5678901
Composition	No hazardous ingredients that require notification	Neutral Cure Silicone.

Table 8.2 *Continued.*

Product	Flexible frame sealant	Silicon frame sealant
Methyltrimethoxysilane. Methanol		
Hazards		
Inhalation	Low toxic hazard.	Low toxic hazard. Use in well ventilated area.
Ingestion	Do not swallow. Eat or drink while using product	Do not swallow. Eat or drink while using product
Skin & eye contact	People with sensitive skin may develop dermatitis use gloves and overalls. Eye irritation, wear goggles	People with sensitive skin may develop dermatitis use gloves and overalls. Eye irritation, wear goggles during application
Chemical and physical	No hazards associated with product	No hazards associated with product
Environmental	No adverse effects	No adverse effects
First aid		
Inhalation	Relieve by breathing fresh air	Relieve by breathing fresh air. Medical attention if recovery is delayed.
Skin contact	Wash off with soap and warm water	Remove with paper towel before washing with detergent and water
Ingestion	Do not induce vomiting. Seek medical attention	Do not induce vomiting. Seek medical attention
Eye contact	Flush immediately with water for several minutes and seek medical attention	Flush immediately with water for several minutes and seek medical attention
Fire fighting	Non flammable but will support combustion. Normal fire fighting media are suitable	Non flammable but will support combustion. Normal fire fighting media are suitable
Accidental release	Soak up with sand, earth, sawdust place in metal or plastic containers	Soak up with sand, earth, sawdust place in plastic or metal containers. Notify local authority if product enters sewer or other water ways.
Handling and storage	Store in dry location 5°C–30°C. Keep away from food animals and children	Store in dry location 5°C–25°C. Keep away from food animals and children
Exposure. Personal protection	Wear protective gloves if the skin is sensitive.	Use in well ventilated areas, wear eye and hand protection and well fitting clothing

Note. There is other information contained in this documentation which is of use to other specialists, but what is mainly illustrated is for the benefit of the end user.

Table 8.3 Window frame schedule

Drawing No Descriptions	Windows												
	W01	W02	W03	W04	W05	W06	W07	W08	W09	W10	W11	W12	Totals
2400 × 1200 mm high four light. Left and right end side hung casements. All complete with beads and rebated for double glazed units. Framed in softwood with hardwood sill. Suppliers cat Ref DD.	1			1			1				1		4
1800 × 1200 mm high three light. Left and right end side hung casements. All complete with beads and rebated for double glazed units. Framed in softwood with hardwood sill. Suppliers cat Ref CD.			1										1
640 × 900 mm high single light. Left side hung casements. All complete with beads and rebated for double glazed units. Framed in softwood with hardwood sill. Suppliers cat Ref AB.		1			1						1		3
1800 × 1050 mm high three light. Left and right end side hung casements. All complete with beads and rebated for double glazed units. Framed in softwood with hardwood sill. Suppliers cat Ref CC.							1						1
1200 × 1200 mm high two light. Right hand side hung casement. All complete with beads and rebated for double glazed units. Framed in softwood with hardwood sill. Suppliers cat Refs BD.								1	1				2
1200 × 900 mm hugh two light. Left hand side hung casement. All complete with beads and rebated for double glazed units. Framed in softwood with hardwood sill. Suppliers cat Ref BB.										1			1

Note: This schedule to be read in conjunction with Figure 8.6 and the drawing in the appendix.

the seal. When using this type of mastic it is necessary to force it as deep as possible into the gap (Figure 8.15b, c).

(b) Gun applied sealants may be silicone or acrylic based (Table 8.2) supplied in cartridges (Figure 8.15d) with a plastic nozzle (Figure 8.15e). A bead of sealant is forced from the cartridge via the nozzle by a trigger-activated piston in the applicator gun (Figure 8.15f), into the gap that requires sealing. Although low modulus silicone sealants retain their elasticity, the surface will harden and eventually become brittle under the prolonged effects of sunlight. For the best results and prolonged service life, force the sealant as far back as possible into the gap (Figure 8.15c).

8.8 WINDOW BOARDS AND THEIR FIXING

The window board forms the finish or trim to the top of the inner leaf of the cavity wall where it finishes at the window opening (Figure 8.14e). Window boards can be formed from solid timber, blockboard, plywood or plastic (Figure 8.16). Window boards are fitted at the first fixing stage, before the walls are plastered.

SOLID TIMBER

Solid timber lipping

BLOCKBOARD

Timber bearer and lipping

PLYWOOD

Plastic foam filling

PLASTIC EXTRUSION

Figure 8.16 Window board types

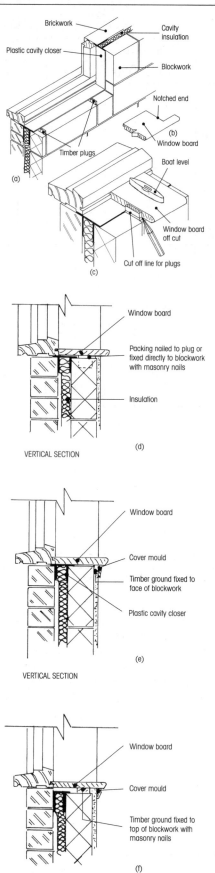

VERTICAL SECTION (d)

VERTICAL SECTION (e)

VERTICAL SECTION (f)

Figure 8.17 Fixing window boards

8.8.1 Fixing Window Boards

(a) The cavity is closed by the bricklayer using a course of cut blocks and patent cavity closers (Figure 8.17a)

(b) If the cavity has been closed with bricks on edge and a patent cavity closer the mortar joint of every fourth brick is raked out and a wooden plug inserted [BENCH AND SITE SKILLS, SECTION 11.6.1] care being taken not to loosen the bricks.

(c) The ends of the window board are notched to fit round the masonry (Figure 8.17b).

(d) Using an off-cut of window board the rebate is placed in the groove in the sill and the board is levelled using a boat or pocket level. The plugs are marked as shown in Figure 8.17(c) and cut off level. Once the plugs are cut to leave the required projection the window board is pushed home making sure the tongue is pushed fully into the groove in the window sill, before nailing through the window board with oval nails directly into the projecting plugs.

(e) Alternative methods of fixing the window boards.

(i) Cut the plugs off flush with the surface of the brickwork and make up the level to the underside of the window board with timber packings, then nail through the window board and packings into the plugs (Figure 8.17d).

(ii) Fix a timber ground to the face of the wall at the correct level for the window board, and nail the window board to the ground using oval nails, finally cloaking the ground with a cover mould (Figure 8.17e).

(iii) If the cavity is closed with cut course of blocks and a patent cavity closer had a timber ground to the top of the blocks (Figure 8.17f) and nail the window board using oval nails directly into the timber ground.

8.8.2 Cover Moulds

Because of the slight differential movement between plaster and timber a gap will eventually appear between

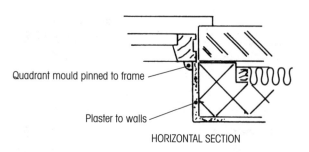

HORIZONTAL SECTION

Figure 8.18 Cover moulds to internal window reveals

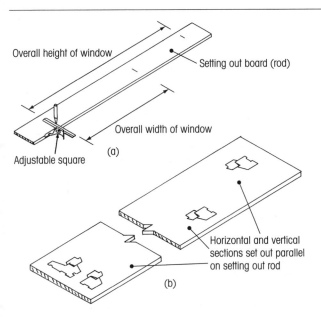

Figure 8.19 Setting out a casement window

the abutting plastered wall and the window frame. To mask this gap a small cover mould, ovolo or cavetto, can be pinned to the timber window frame where it makes contact with the plastered wall (Figure 8.18).

8.9 CONSTRUCTING A NON-STANDARD CASEMENT WINDOW

All windows go through the same stages of manufacture, except that in the case of mass and batch production, some stages, such as marking out can be dispensed with by the use of jigs, templates and specially adapted woodworking machines. In the one-off production of a window all the stages of manufacture must be followed.

8.9.1 Setting out the Window Frame and Casements

From the architect's drawings, specification and window schedule [BENCH SKILLS, SECTION 12.9.2], a setting out rod is prepared. This is a planed timber board with a width of from 150 to 250 mm and about 20 mm thick. The length will depend on the longest dimension of the joinery item being set out. As an alternative, plywood or MDF (medium density fibreboard) may be used. The advantage of using a board for setting out, is that it is dimensionally stable along its length [BENCH AND SITE SKILLS, SECTION 1.7.5] and as the vertical (height) and horizontal (width) sections of the window (Figure 8.19a) are set out along the length of the board the dimensions will remain stable.

(a) The overall height of the vertical and width of the horizontal sections are marked out full size on the rod using an adjustable square, rule or tape (Figure 8.19a). Carefully check that these dimensions are correct.
(b) Full size sections of the window members are drawn in their appropriate positions (Figure 8.19b). [CARPENTRY AND JOINERY BENCH AND SITE SKILLS, SECTION 5.2.2].

8.9.2 Cutting List

Once the window frame and casements are set out full size on the setting out rod a cutting list can be prepared, giving in table form the number, length and cross-sectional sizes for the members of the window [BENCH AND SITE SKILLS, SECTION 5.2.1] (Table 8.4). From this list the timbers are cut to length and machined into rectangular sections of the correct size.

Note. Lengths of timber profiled (moulded) to the sections of window frame members, can be obtained, and if this is the case the various members can simply be cut to length.

Table 8.4   A cutting list for casement windows

BETTER BUILDERS LTD.

Description of the work: No.2 single light casement windows 1 m high 0.9 m wide. Finish for paint

Job number: JS123   *Client*: C. Brown   *Date*:

No.	Description		L	W	T	Finished size W	T	Timber	Remarks
4	Frame	Jambs	1010	75	63	70	58	Softwood	Paint finish
2		Heads	1000	75	63	70	58	ditto	ditto
2		Sills	1000	125	75	120	70	Hardwood	ditto
4	Casement	Stiles						Softwood	ditto
2		Top rails						ditto	ditto
2		Bottom rails						ditto	ditto

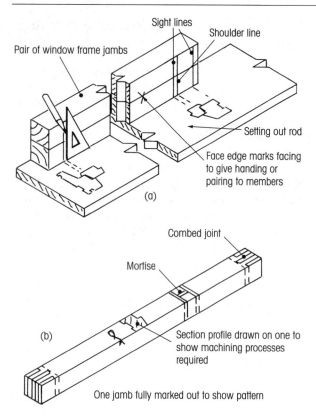

Figure 8.20 Marking out window frame members

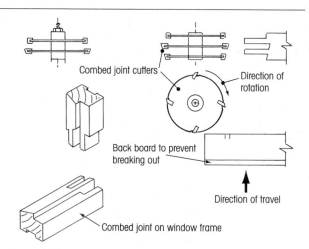

Figure 8.21 Cutting combined joints on window frame members

mortise and tenon. Combed joints give a greater surface area when gluing two members together.

(b) The grooves, moulds and rebates could then be worked on the sections by hand, but more realistically by a router [BENCH AND SITE SKILLS, SECTION 8.15], or using a spindle moulder in the case of a single window.

8.9.5 Assembly of the Window Frame and Casements

(a) Joints would be checked for a fit and the exposed inner surfaces of the members which are not easy

8.9.3 Marking out the Window Frame and Casement Members

(a) Then wrot and squared sections of the window members would be paired (handed) by drawing on face side and face edge marks with a pencil.
(b) The paired window jambs and intermediate mullions, and casement stiles would be placed by the vertical section on the setting out rod and the position of all joints and shoulder lines would be transferred onto the squared stock using a set square or an adjustable square (Figure 8.20a). The sill, head and transom of the frame, and rails of the casement would have lines transferred to them from the horizontal section.
(c) After the lines are transferred to the squared stock the position of the joints and shoulder lines would be clearly identified, and the finished cross section shape of the member would be drawn to show the position of grooves, mouldings and rebates (Figure 8.20b).

8.9.4 Cutting the Joints and Moulding the Sections

(a) The mortise and tenons would be worked either by hand [BENCH AND SITE SKILLS, SECTION 7.5.7] but more usually by machine. Combed joints cut by specialist woodworking machinery on external angle joints (Figure 8.21), have largely replaced the

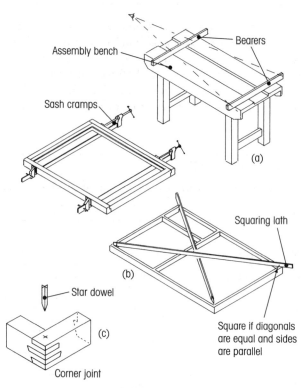

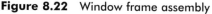

Figure 8.22 Window frame assembly

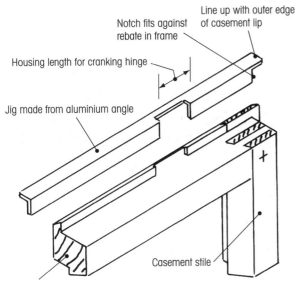

Line up with outer edge of casement lip

Notch fits against rebate in frame

Housing length for cranking hinge

Jig made from aluminium angle

Casement stile

Top rail casement

Figure 8.23 Jig for marking out hinge housings on casement window frames

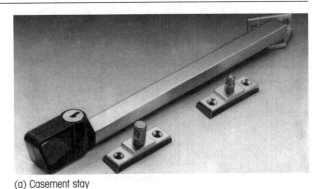

(a) Casement stay

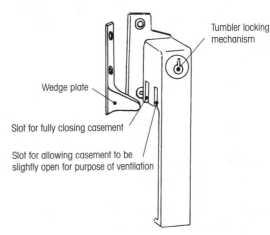

Tumbler locking mechanism

Wedge plate

Slot for fully closing casement

Slot for allowing casement to be slightly open for purpose of ventilation

(b) Casement fastener

Figure 8.24 Lockable casement stays and fasteners

to get at when the frame and casements are assembled, are cleaned up using a smoothing plane [BENCH AND SITE SKILLS, SECTION 5.4] and a hand held electric sander of the disc or belt type [BENCH AND SITE SKILLS, SECTION 8.11] to remove the machining cutter marks and prepare the surface to receive wood preservative paint or stain.

(b) Suitable bearers would be attached to the assembly bench and a check made to see that the top edges lie in a flat plane (Figure 8.22a).

(c) Adhesive would be applied to the joints and the main frame assembled and placed on the bearers.

(d) The frame would be cramped up and the diagonals checked with a squaring lath (Figure 8.22b).

Note. The window could be assembled in a mechanical or hydraulic cramping frame in which locating stops are incorporated to hold the window square.

(e) Non-ferrous metal (aluminium) star dowels are driven through the joint to hold it in place until the adhesive sets (Figure 8.22c).

8.9.6 Cleaning up the Faces of the Window Frame and Casements

This can again be done by hand, using a smooth plane and an electrical orbital or belt sander to obtain flat and mark free surfaces ready for painting or staining.

8.9.7 Hanging the Casements and Fixing the Ironmongery

(a) Check that the casements fit into the frame with the correct clearances.

(b) Mark the position of the hinges on the frame and casements and house into the lip of the casement (Figure 8.23) [BENCH AND SITE SKILLS, SECTION 7.6.4]. Screw the hinges to the casement and then to the rebate in the frame. Check that the casements open without catching on the frame.

(c) Casement fasteners (Figure 8.24b) and casement stays (Figure 8.24a) are then screwed to the frame and casements (see also Figure 8.11).

8.9.8 Casement Window Security

Small window bolts operated with a key can be incorporated in the stiles of the casements [BENCH AND SITE SKILLS, SECTION 12.11, FIGURE 12.32] or casement fasteners and stays incorporating a locking mechanism (Figure 8.24) can be fixed.

8.10 BAY AND BOW WINDOWS

(a) As indicated in Section 8.2 bay windows, square or semi-polygonal in plan (Figure 8.25a, b), project outwards beyond the face of the external wall. They can be purpose made (Figure 8.25a) or formed by joining a number of standard windows together using special corner posts (Figure 8.25b). A

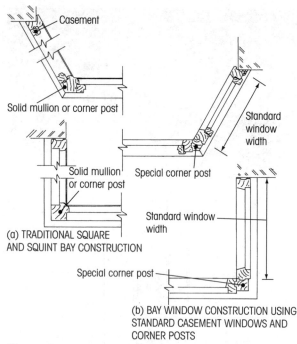

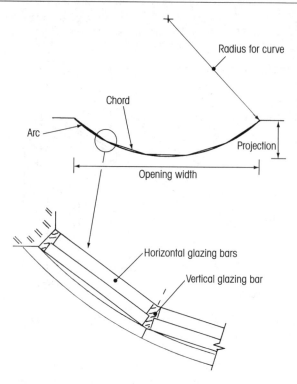

Figure 8.25 Horizontal sections through square and semi-polygonal bay windows

Figure 8.26 Bow window outline

traditional non-standard bay window can be constructed as a complete frame into which casements can be fitted (Figure 8.27b).

Note. If the projection of the curve is greater than a fifth of the width of the window the bow window is often referred to as a segmental or circular bay window.

(b) Bow windows have a sill, which is a segmental curve in plan on its outer edge while the horizontal glazing bars form short chords to the curve [BENCH AND SITE SKILLS, SECTION 13.5.3] (Figure 8.26).

8.10.1 Making a One Off Bay Window

(a) The window will be shown in the architect's drawings, and although detailed construction information may not be given, it will at least be possible to determine the width, height and projection of the bay, and its shape in plan (Figure 8.27a).

(b) From the dimensions obtained, set out full size horizontal and vertical sections of the bay window on a sheet of plywood, or medium density fibreboard (MDF), paying particular attention to the three critical dimensions of height width and projection (Figure 8.27b).

(c) From information on the setting out board a template giving the out line of the face of the brickwork would be made for use by the bricklayer (Figure 8.27c).

(d) A cutting list as Table 8.4 would be prepared and the timber members worked as described for a casement window. Figures 8.27(d, e) show the general construction details.

(e) The bay would be assembled in the vertical position as follows:
  (i) Sill and head sections are jointed together, using handrail bolts to join the sill (Figure 8.27e).
  (ii) Jambs and through mullions are placed into the sill while at the same time fitting the transoms between the jambs and mullions.
  (iii) The head would then be secured to the tops of the jambs and mullions usually using nails.

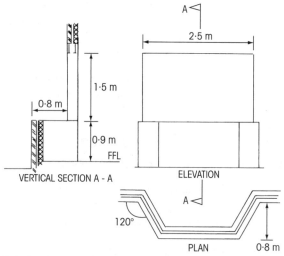

Note: Check setting out of brickwork under bay with bay window manufacturer

(a) Baywindow opening in wall

Figure 8.27 *Continued opposite*

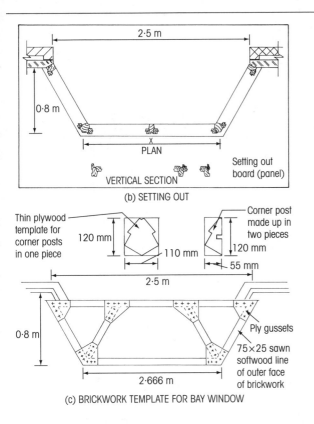

(b) SETTING OUT

(c) BRICKWORK TEMPLATE FOR BAY WINDOW

Figure 8.27   (a) Architect's brickwork detail for a bay window opening; (b) and (c) setting out

Note. If the bay window supports a roof or additional framing above it, the mullions run the full height of the bay with the transom members being cut between mullions and jambs.

(iv) The casements are made, fitted and hung as previously describe in Section 8.8.7.

8.10.2 Bow Window Construction

Setting out is as for a bay window except that the outer edge of the sill is curved in plan and the mullions and transoms are replaced by vertical glazing bars between which the horizontal bars are cut. Although bow windows are normally fixed glazed it is possible to incorporate opening casements.

(a) If the width of the opening and the projection of the bow is known, the radius of the curve for setting out the sill line can be found by using a geometrical method or using the formula (Figure 8.28a):

$$\text{Radius} = \frac{\left(\dfrac{(0.5 \text{ opening}^2)}{\text{projection}} \right) + \text{projection}}{2}$$

Figure 8.28(b) shows the setting out for a bow window and its general construction.

Note. The slope of the weathering on the exposed curved sill, if at a constant angle would give a scalloped effect when viewed in elevation. To keep the line of the top edge of the sill horizontal, the slope angle of the weathering needs to be varied along the segment of the sill (Figure 8.29).

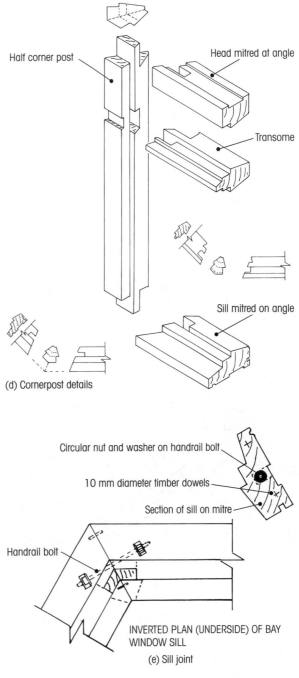

(d) Cornerpost details

(e) Sill joint

Figure 8.27   Details of bay window construction

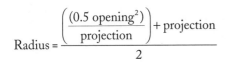

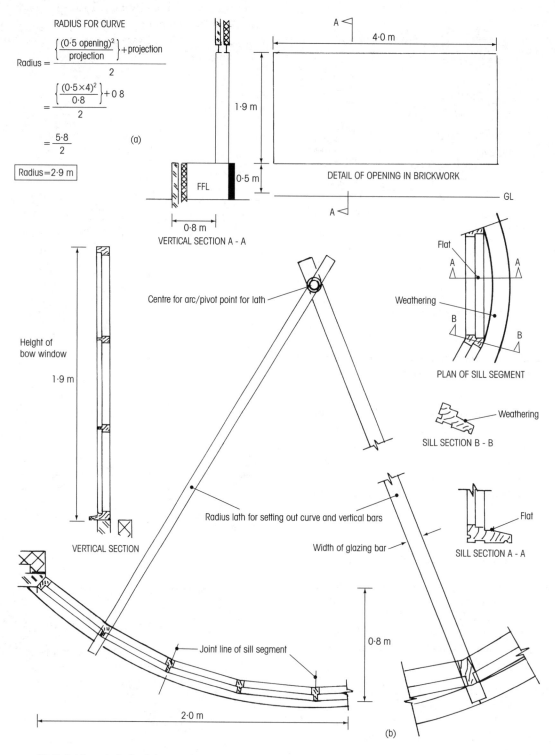

RADIUS FOR CURVE

$$\text{Radius} = \frac{\left\{\dfrac{(0.5 \text{ opening})^2}{\text{projection}}\right\} + \text{projection}}{2}$$

$$= \frac{\left\{\dfrac{(0.5 \times 4)^2}{0.8}\right\} + 0.8}{2}$$

$$= \frac{5.8}{2}$$

(a)

Radius = 2.9 m

A

4.0 m

1.9 m

0.5 m

FFL

0.8 m

VERTICAL SECTION A - A

DETAIL OF OPENING IN BRICKWORK

GL

A

Flat

A   A

Weathering

B   B

PLAN OF SILL SEGMENT

Weathering

SILL SECTION B - B

Height of
bow window

1.9 m

Centre for arc/pivot point for lath

Radius lath for setting out curve and vertical bars

Width of glazing bar

Flat

SILL SECTION A - A

VERTICAL SECTION

Joint line of sill segment

0.8 m

2.0 m

(b)

NB. To find the centre for the arc by geometry see Figure 5.60a.

Figure 8.28 Setting out a bow window

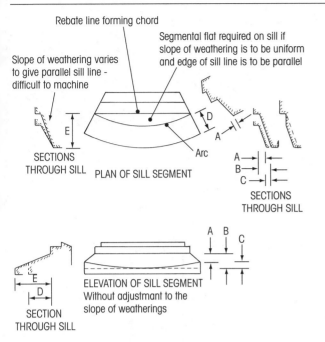

Figure 8.29 Method of weather a curved sill when the rebates form a chord to the curve

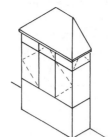

HIPPED ROOF OVER SQUARE BAY   LEAN TO ROOF OVER SQUARE BAY

FLAT ROOF OVER SEGMENTAL BAY

HIPPED AND GABLE END ROOF OVER SPLAYED END BAY

Note: Pitched roofs can be formed over curved bays, as can flat roofs over square and splayed bays

Figure 8.30 Roof configurations to bay windows

8.10.3 Semi-circular Bay Windows

Their construction is similar to that of a semi-polygonal bay except that the outer line of the sill is curved and the casements form short chords to the curve [BENCH AND SITE SKILLS, SECTION 13.5.3] (see Figure 8.29).

8.10.4 Roofs Over Bay Windows

These can be pitched and covered in tiles as a house roof, or flat and covered with lead or three-layer felt. Large manufacturers of bay windows also supply GRP (glass fibre reinforced plastic) flat roofs to fit over the bay windows they produce. The general construction is the same as for flat and pitched roofs (see Chapters 3 and 4). Figures 8.30 and 8.31 show the general configuration and construction of the roofs.

8.11 VERTICALLY SLIDING SASH WINDOWS

The window consists of a boxed or solid frame into which are fitted two vertically sliding sashes, one offset above the other. Their vertical movement is controlled by helical spring balances, or weights suspended on cords running over pulleys, to balance the weight of the sashes. Figure 8.32(a) shows the principle on which the weights work in counter balancing the weight of the sash and glass so that they can be moved up and down with the minimum of

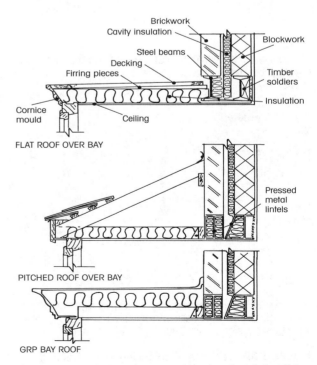

FLAT ROOF OVER BAY

PITCHED ROOF OVER BAY

GRP BAY ROOF

Figure 8.31 Bay roof construction (sectional details)

physical effort. For the weights to hold the sashes in the closed position, the combined weight of the weights holding the top sash should be slightly heavier than the glazed sash, and the weights holding the bottom sash slightly lighter in weight. The same principle also applies when helical spring sash balances are used (Figures 8.32b, c).

8.11.1 Setting Out and Marking Out a Sliding Sash Window

(a) The vertical and horizontal cross-sections of the window are drawn full size on the setting out rod. (Figure 8.33).

(b) From the setting out rod a cutting list of the number and size of the components is prepared (Table 8.5).

(c) Components are cut to length and machined to their rectangular sections.

(d) The members are paired by the use of face and edge marks and marked out from the setting out rod to show all joints, and a cross-section is drawn on the components to show what and where

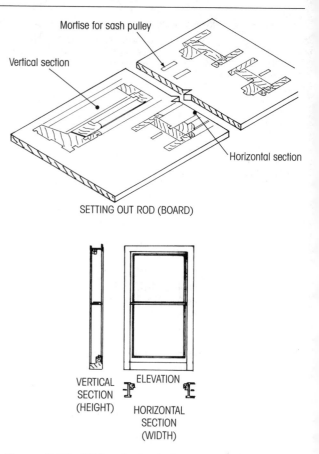

SETTING OUT ROD (BOARD)

VERTICAL SECTION (HEIGHT)   ELEVATION   HORIZONTAL SECTION (WIDTH)

Figure 8.33   Sliding sash window setting out rod

grooves, rebates, weatherings and mouldings have to be worked on the individual members of the window (Figure 8.20b). General jointing arrangements for the sashes are shown in Figure 8.34, but it should be noted that the joints between stiles, bottom and top rails can be combed as an alternative to the use of mortise and tenon joints.

8.11.2 Forming Pockets in the Pulley Stiles

As the weights are encased in the boxed frame and the cords holding the weights to the sashes occasionally require renewing, it is necessary to form a pocket in the pulley stile to give access to the weights. The pockets are formed on the inside part of the pulley stile in which the lower sash slides, and in such a position that the weights can be lifted out without binding on the opening. The top edge of the pocket is the length of the longest weight plus 50 mm above the sill, and the pocket length is usually no longer than 225 mm (Figure 8.35a). For the majority of the time the pocket remains masked by the lower sash when in the closed position.

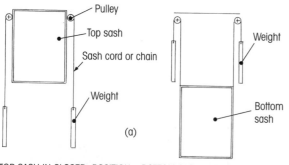

TOP SASH IN CLOSED POSITION (Weights slightly heavier than glazed sash)

BOTTOM SASH IN CLOSED POSITION (Glazed sash slightly heavier than weights)

(a)

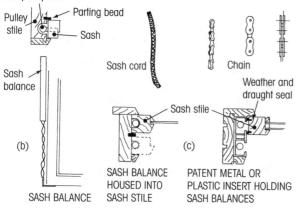

(b)

SASH BALANCE

(c)

SASH BALANCE HOUSED INTO SASH STILE

PATENT METAL OR PLASTIC INSERT HOLDING SASH BALANCES

Figure 8.32   Principle of balancing vertical sliding sashes

Table 8.5   A cutting list for a sliding sash window

BETTER BUILDERS LTD.

Description of the work: No.1 Sliding sash window 700 mm wide 1200 mm high. Finish for paint

Job number: JS234   *Client*: D .Cress   *Date*:

No.	Description	L	Sawn size		Finished size		Timber	Remarks
			W	T	W	T		
	FRAME							
2	Pulley stiles	1100	125	32	120	27	Softwood	
1	Head lining	700	125	32	120	27	ditto	
1	Sill	700	150	100	145	95	Hardwood	
2	Outer linings	1200	100	25	95	20	Softwood	
1	Outer lining head	550	100	25	95	20	ditto	
2	Inner linings	1200	90	25	85	20	ditto	
1	Inner lining head	550	90	25	85	20	ditto	
2	Parting beads	1100	25	16	20	10	ditto	
1	ditto	550	25	16	20	10	ditto	
2	Staff beads	1100	25	19	20	14	ditto	
1	ditto	550	25	19	20	14	ditto	
1	Ventilation bead	550	40	19	35	14	ditto	
1	Parting slip	1100	45	12	45	12	ditto	Sawn
8	Glue blocks	400	50	50	45	45	ditto	Triangular
2	Wedges	100	75	50	70	45	ditto	Tapered
	SASHES							
4	Sash stiles	630	50	50	45	45	ditto	
1	top rail	700	50	50	45	45	ditto	
1	bottom rail	700	75	50	70	45	ditto	
2	meeting rails	700	63	50	57	40	ditto	

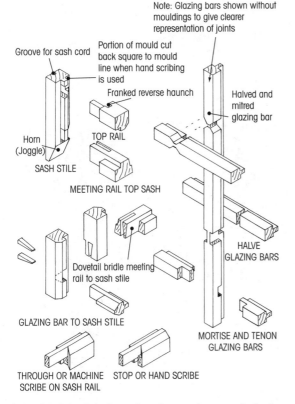

Figure 8.34   Jointing of sash members and glazing bars

To form the pockets:

(a) Mark the position of the pocket on the pulley stile. (Figure 8.35b).

(b) Drill a hole through the back of the groove that takes the parting bead (the diameter of the hole no greater than the width of the groove), and using a Jig saw form a saw cut the length of the pocket on the inside edge of the groove (Figure 8.35c). This will allow the pocket to be removed with out disturbing the parting bead.

(c) Using a pocket chisel (Figure 8.35d) cut across the grain at points 1 and 2, turn the pulley stile over and make cuts at 3 and 4 (Figure 8.35e). Take a hammer and smartly tap the back of the pocket at the point shown in Figure 8.35(f). This will cause the short grain between the chisel cuts to shear and the pocket to fall out.

(d) The pocket is then secured in place by using a cup and counter sunk screw [BENCH AND SITE SKILLS, SECTION 11.2.7] (Figure 8.35a).

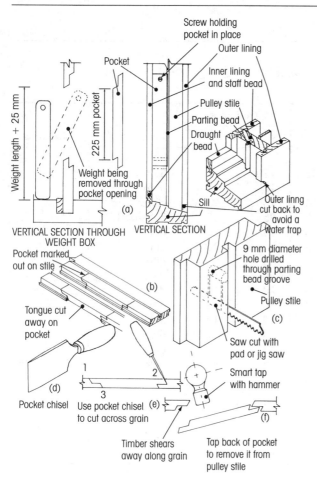

Figure 8.35 Forming pockets in the pulley stiles

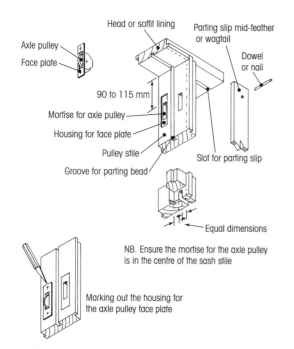

Figure 8.36 Fixing axle pulleys to the pulley stiles

8.11.3 Fitting the Axle Pulleys

(a) Mark the positions for the mortises for the axle pulleys and mortise.

   Note. The distance down of the axle of the pulley from the head lining should be between 90 and 115 mm.

(b) Fit the pulley in the mortise, mark round the face plate of the pulley, and house in flush, before screwing in place (Figure 8.36).

8.11.4 Forming the Sliding Sashes

The sashes are made in much the same way as casements, except that where the two vertical sashes overlap, meeting rails dovetailed to the stiles are used (Figure 8.37a), or horns (joggles) are left at the lower ends of the upper sash stiles, and at the upper end of the bottom sash stiles and a mortise and tenon joint is used (Figure 8.37b).

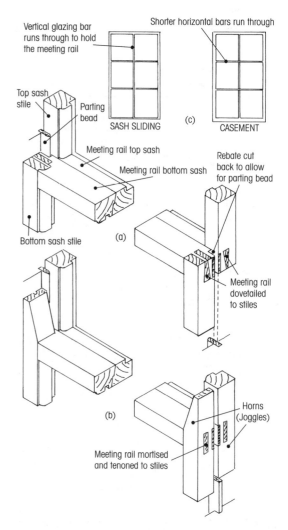

Figure 8.37 Jointing of meeting rails to stiles of sash

Note. If glazing bars are incorporated in the sashes the vertical bars connecting the top and bottom rails of the sashes run through as they help the meeting rails carry the weight of the glass. Otherwise it is normal practice to let the bars running in the shorter direction be the through bars (Figure 8.37c).

8.11.5 Assembling the Boxed Frame

(a) When the various components have been machined to the correct profile, the pockets cut, the axle pulleys fitted and the housings cut in the head and sill; the pulley stiles are secured to the head and sill by wedging and nailing into the sill and head housings (Figure 8.38).

　Care must be taken to ensure that the pulley stiles are not in "wind" (pronounced wined) but lie in a flat plane (Figure 8.39a). To assist in this a jig is set up on the assembly bench (Figure 8.39b) to hold the assembled head, sill and pulley stiles square and flat. It also leads to peace of mind if the diagonals of the frame are checked with a squaring rod (Figure 8.39c).

(b) Once the pulley stiles, sill and head are secured square and true, the inside linings are cut, fitted and nailed (Figure 8.39d). The inner head lining is carefully cut square and to accurate length between the fixed linings and nailed in position (Figure 8.39d).

(c) The frame is then turned over and the outer linings are cut, fitted and nailed, but as the outer linings project past the face of the head and pulley stiles; the outer head lining has to be mitred to the vertical outer linings (Figure 8.39e).

(d) The parting beads are put in, the head bead first then the two side beads. Staff beads are cut and mitred at the corners and tacked to the edge of the inner linings and sill (Figure 8.39f).

　Note. A draught (ventilation) bead, which is a deeper staff bead, is sometimes fixed to the inner edge of the sill (Figure 8.39f).

(e) Glue blocks are tacked inside the hollow head. The parting slips (wagtails) are attached to the slots in the head lining and the back lining nailed in place (Figure 8.39f). Finally the sashes are fitted to check that they will slide and then out for glazing and weighing.

8.11.6 Hanging the Sliding Sashes

If the window is to leave the workshop complete with hung sashes, the glazed sashes are weighed and pairs of weights approximately equal to the weight of the sashes are selected.

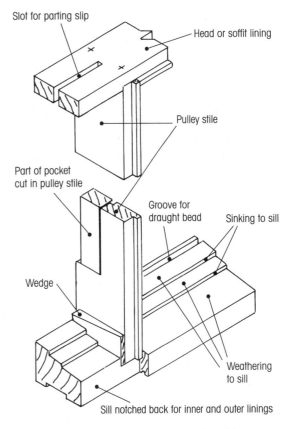

Figure 8.38　Securing pulley stile to head lining and sill

Note. The pair of weights for the top sash should be about a kilogram heavier than the weight of the sash, while those for the bottom sash about a kilogram lighter.

　The weights are connected to the sashes by cords or chains and the sashes are hung as follows:

(a) The pockets, vertical parting beads and staff beads are removed.

(b) To assist in threading the sash cord over the pulleys a mouse (Figure 8.40a) is used. The mouse is passed over the pulley and falls down the box until it can be pulled out through the open pocket. The string attached to the sash cord, is pulled until the cord passes over the sash pulley down and out through the open pocket. (Figure 8.40b).

(c) Take the sash cord which is purchased by the Knot (about 10.75 m) and using the mouse start at the inside left hand pulley (1). Thread the cord over the left hand pulley, down and out through the pocket then up and over the inside right hand pulley (2), down and out of the pocket, up and over the outer

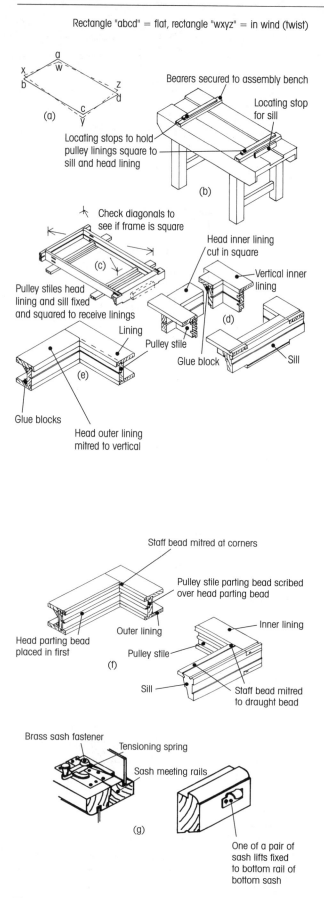

Rectangle "abcd" = flat, rectangle "wxyz" = in wind (twist)

(a)

Bearers secured to assembly bench

Locating stop for sill

Locating stops to hold pulley linings square to sill and head lining

(b)

Check diagonals to see if frame is square

(c)

Pulley stiles head lining and sill fixed and squared to receive linings

Head inner lining cut in square

Vertical inner lining

Lining

Pulley stile

Glue block

Sill

(d)

(e)

Glue blocks

Head outer lining mitred to vertical

Staff bead mitred at corners

Pulley stile parting bead scribed over head parting bead

Head parting bead placed in first

Outer lining

Pulley stile

Inner lining

Sill

Staff bead mitred to draught bead

(f)

Brass sash fastener

Tensioning spring

Sash meeting rails

(g)

One of a pair of sash lifts fixed to bottom rail of bottom sash

Figure 8.39　Assembly of a boxed frame

left-hand pulley (3), down and out of the pocket, and finally up and over the outside right hand pulley (4) and out of the right hand pocket (Figure 8.40c).

(d) Tie the leading end of the cord to one of the weights for the top sash (Figure 8.40d) put the weight inside the box, and pulling on the cord raise the weight 30 or 40 mm above the sill (Figure 8.40e). Cut the cord to a mark on the pulley stile, which gives sufficient length of cord to fix to the edge of the sash. Be sure to tie a knot in the cut end of the cord so that it does not slip back over the pulley into the box. Tie the other weight for the top sash to the other cut end of the cord, pull the weight inside the box and repeat the process. When it comes to determining the length of the cord required to attach the weights to the bottom sash, pull the weight up to within 30 or 40 mm of the pulley and cut to the mark on the pulley stile that will give sufficient cord for fixing to the sash stile (Figure 8.40f).

(e) Attach the top sash to the cords coming from the pulleys at the outer edge of the frame and place against the outer linings. Replace the pockets and tap the vertical parting beads in place. Finally attach the bottom sash to the remaining pair of cords and place the sash against the parting beads, then screw or nail the staff beads in place, making sure there is sufficient clearance for the sash to slide easily (see Chapter 13, Figure 13.16).

8.11.7 Ironmongery to the Sliding Sashes

House in and screw with brass countersunk screws the brass sash fastener to the sash meeting rails and make sure the two meeting rails are pulled close together as the sash fastener is closed. Screw a pair of sash lifts to the inside of the bottom rail of the bottom sash (Figure 8.39g).

8.11.8 Solid Frame Sliding Sash Windows

The development of tensioned helical spring sash balances (Figure 8.32b) has eliminated the need for boxed frames, and a solid frame can now be used to carry the sashes. Figure 8.32(c) shows the general construction details of a solid frame sliding sash window. Profiled to accommodate the balancing mechanism which should be installed as the manufacturer's instructions.

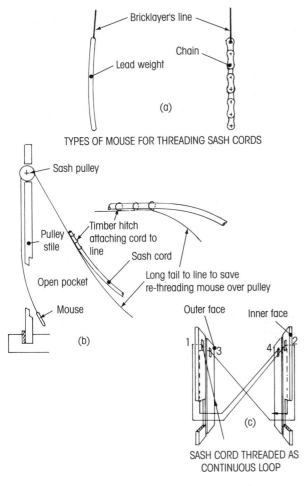

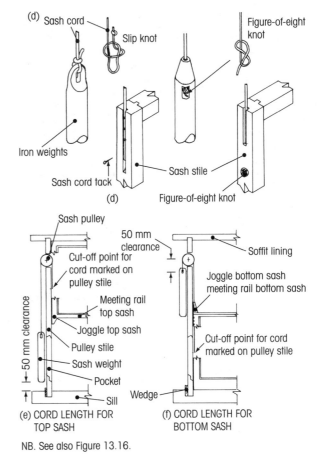

NB. See also Figure 13.16.

Figure 8.40   Threading and fixing sash cords

8.12 TIMBER SUB-FRAMES TO METAL WINDOWS

Metal windows can be built directly into the surrounding masonry, and with the use of sealants, which retain their elasticity and adhesion, a weather-proof seal between masonry and metal window can be achieved. As the profile of metal window frame sections used in domestic dwellings is relatively small (Figure 8.41a) a timber sub-frame is used as otherwise it is difficult to make a satisfactory weather proof joint. Ferrous metal (steel) window frames are given protective coatings by various methods and their contact under damp conditions with timber sub-frames containing acid, such as oak, cedar or Douglas fir is not so critical [BENCH AND SITE SKILLS, SECTION 1.10.35]. To keep the joints between timber and metal, draught and weather proof, and to allow for differential movement a mastic compound is used to bed the metal frames into the timber sub-frame. Figure 8.41(b) shows the general construction, and Figure 8.41(c) shows a spring wire glazing clip used for holding the glass in a metal window frame.

8.13 GLAZING TO WINDOWS

Glass for glazing windows is manufactured as clear, obscure (patterned) or wired. The properties of glass can be modified in the manufacturing process to give certain other properties of light and heat absorption and toughness. Glass produced without specialist treatments and not wired, will when broken, produce pointed shards which have razor sharp edges.

Most, if not all, clear sheet glass is now produced by the float method, which gives a glass that does not distort the light as it passes through. Glass is identified by thickness, 3 and 4 mm glass being the norm for general domestic glazing. Glazing in vulnerable areas, i.e. low panels in glazed doors and storey frames have to be glazed as specified by the *Building Regulations Approved Document "N"* (Figure 8.42).

When measuring for glass the dimensions into the rebate are taken, then 4 mm is deducted from the height and width to give a glass size with a clearance gap of 2 mm all round (Figure 8.43a).

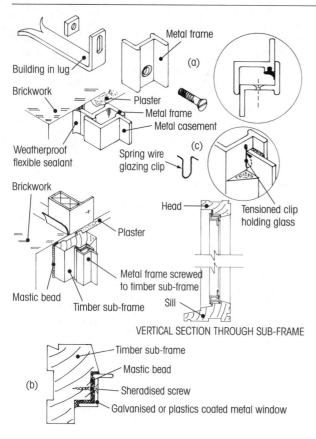

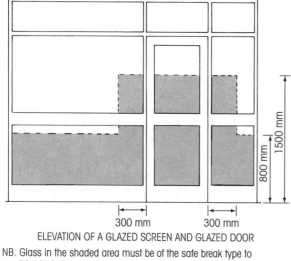

ELEVATION OF A GLAZED SCREEN AND GLAZED DOOR

NB. Glass in the shaded area must be of the safe break type to BS 6206/91. The glass, when broken, must disintegrate into small, detached particles with no sharp edges or points. Only the glazed panels above the door can be glazed with non-safe break glass.

Figure 8.42 Areas where safe breaking glass must be used

8.13.1 Single Glazing to Painted Casements and Sashes

(a) Check that the glass is the correct size and all the timber rebates are primed (Figure 8.43a).

(b) Run a bead or bed of linseed oil putty (timber glazing mastic) against the back of the rebate (Figure 8.43b).

(c) Position plastic seating blocks in the bottom rebate (Figure 8.43c).

(d) Lift the glass pane and place the bottom edge onto the seating blocks and gently push the glass into the top and side rebates and press evenly round the edges using a rag to bed the glass into the putty till a uniform bed thickness of between 1 and 2 mm is formed (Figure 8.43d).

(e) Place in other locating blocks along the top and sides of the glass forcing them into the putty in the space between the edge of the glass and rebate. This will help keep the side hung casements square (Figure 8.43e).

Figure 8.41 Timber sub-frames for metal windows

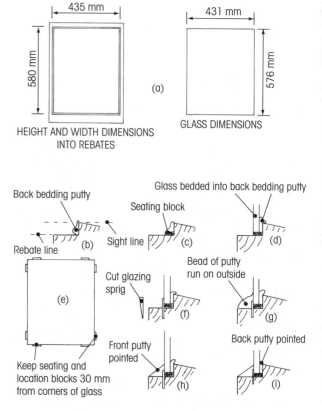

Figure 8.43 Glazing using linseed oil putty (timber glazing mastic)

(f) Drive in glaziers sprigs (small flat cut nails). Round pins produce pressure points where they come in contact with the glass. Drive in the glaziers sprigs using a small hammer or firmer chisel which is slid over the glass to strike the sprig (Figure 8.43f).

(g) Run a fillet of putty all round the exposed part of the rebate in front of the glass (Figure 8.43g).

(h) Using a putty knife or chisel point the putty to give a bevelled fillet on the outer edge of the glass. The line of this fillet should be just below that of the rebate sight line. Point first along the top edge of the pane of glass then down the sides and finally across the bottom. The putty should be of such a consistency that the pointing is smooth and does not break up as the putty knife is drawn along the surface of the putty in the pointing process (Figure 8.43h).

(i) Point the bedding putty on the inside of the glass (Figure 8.43i).

8.13.2 Double Glazing to Timber Casements and Sashes

Double glazing is used to:

(a) Reduce heat loss from the room and reduce the amount of sound entering the room from the outside (Figure 8.44a,b).

(b) The greater mass of the double glazed unit having two thicknesses of glass will marginally reduce the passage of sound.

(c) Reduce the incidence of condensation on the room side surface of the glass by maintaining the temperature of this inner surface above dew point (the temperature at which moisture condenses out of warm air when the air makes contact with a cold surface) (Figure 8.44c).

The double glazed units consist of two sheet of 4 mm float glass held apart by a 6 to 12 mm wide spacer running round the perimeter of the glass and adhered to them to form an air tight seal. The cavity between the glass panes being filled with dry air or argon gas. The width of the glass and spacers can be varied for special conditions of glazing, but usually for domestic glazing the glass thickness and spacing are as indicated in Figure 8.44(d).

8.13.3 Fixing Double Glazing Units into Casements and Sashes

Linseed oil putty must not be used when microporous paint and timber stains are used, or double glazing units are to be fixed.

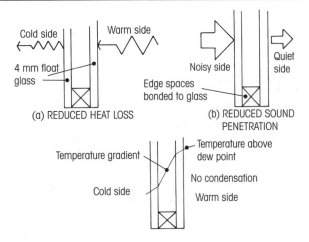

(a) REDUCED HEAT LOSS

(b) REDUCED SOUND PENETRATION

(c) ELIMINATES CONDENSATION ON WARM SIDE

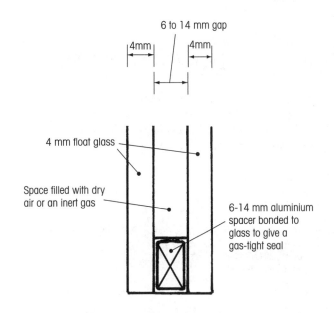

(d) SECTION THROUGH A DOUBLE GLAZED UNIT

Figure 8.44 Benefits of double glazing

(a) Bed the self-adhesive sealing strip along the back of the rebate. If a non hardening sealant is being used, 2 mm thick stop blocks must be placed against the back of the rebate to stop the mastic being squeezed out by the unit (Figure 8.45a, b).

(b) Position the seating blocks along the bottom rebate, lift the glazing unit onto the seating blocks and press into position, putting in the spacing blocks along the top and down the edges of the unit.

(c) Run a self adhesive glazing strip, or bedding mastic round the outside perimeter of the unit and press the glazing beads in place, screwing or nailing them in position (Figure 8.45a, b).

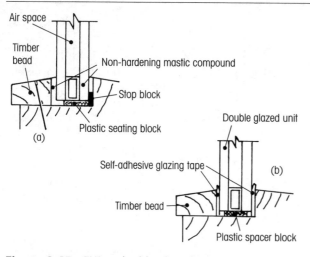

Figure 8.45 Fixing double glazed units into deep rebates

8.14 WINDOW VENTILATION

The improved design of windows and the general use of draught exclusion strips give an almost hermetically sealed unit, and with the upgrading of draught exclusion from rooms where some circulation of air is necessary the *Building Regulations Approved Document "F" 1995* requires that ventilation in windows must be provided. It is practice to incorporate ventilators in:

(a) The head of the window frame.
(b) The top rail of the casement or sash.
(c) Within the glazed area of the window.

Figure 8.46 shows this arrangement of ventilators.

8.15 WINDOWS WITH SHAPED HEADS

All windows with the probable exception of bay windows may be constructed with shaped heads, which can take one of the following curved shapes:

(a) Segmental (Figure 8.47a).
(b) Semi-circular (Figure 8.47b).
(c) Semi-elliptical (true or approximate) (Figure 8.47c–d).

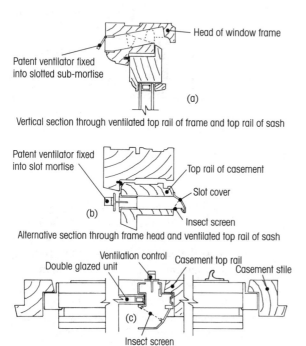

Figure 8.46 Window ventilators

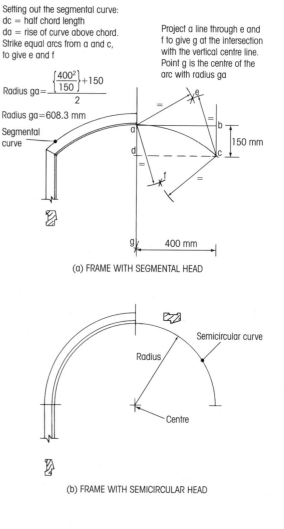

Setting out the segmental curve:
dc = half chord length
da = rise of curve above chord.
Strike equal arcs from a and c, to give e and f

$$\text{Radius } ga = \frac{\left\{\frac{400^2}{150}\right\} + 150}{2}$$

Radius ga = 608.3 mm

Project a line through e and f to give g at the intersection with the vertical centre line. Point g is the centre of the arc with radius ga

(a) FRAME WITH SEGMENTAL HEAD

(b) FRAME WITH SEMICIRCULAR HEAD

Note: Elliptical inner and outerlines are not parallel

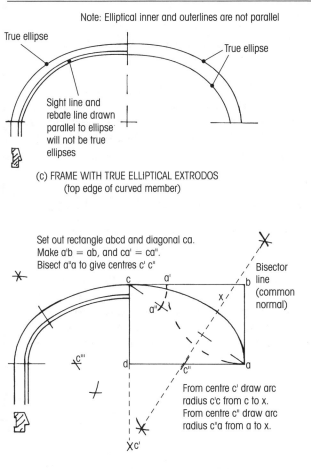

(c) FRAME WITH TRUE ELLIPTICAL EXTRODOS
(top edge of curved member)

Set out rectangle abcd and diagonal ca.
Make a'b = ab, and ca' = ca".
Bisect a"a to give centres c' c"

Bisector
line
(common
normal)

From centre c' draw arc
radius c'c from c to x.
From centre c" draw arc
radius c"a from a to x.

(d) APPROXIMATE ELLIPSE WITH THREE CENTRES

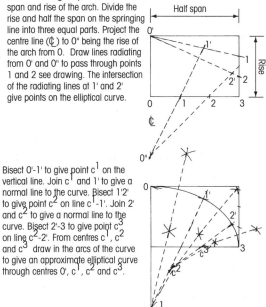

Draw a rectangle equal to half the span and rise of the arch. Divide the rise and half the span on the springing line into three equal parts. Project the centre line (₵) to 0" being the rise of the arch from 0. Draw lines radiating from 0' and 0" to pass through points 1 and 2 see drawing. The intersection of the radiating lines at 1' and 2' give points on the elliptical curve.

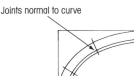

Bisect 0'-1' to give point c^1 on the vertical line. Join c^1 and 1' to give a normal line to the curve. Bisect 1'2' to give point c^2 on line c^1-1'. Join 2' and c^2 to give a normal line to the curve. Bisect 2'-3 to give point c^3 on line c^2-2'. From centres c^1, c^2 and c^3 draw in the arcs of the curve to give an approximate elliptical curve through centres 0', c^1, c^2 and c^3.

Joints normal to curve

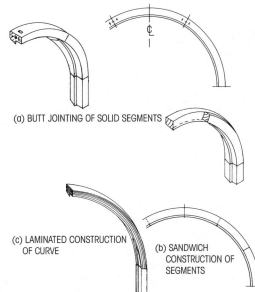

(e) SETTING OUT FOR AN APPROXIMATE ELLIPSE WITH FIVE CENTRES

Figure 8.47 Window frames with curved heads

Note. The use of a true ellipse is very labour intensive and therefore it is more usual to use an approximate ellipse built up from a series of segmental curves. (Figures 8.47d, e).

With the development of effective preservative treatments and of adhesives, which are weather proof, curved window heads can be formed in the following ways:

(a) Solid segments of the curve butt jointed together (Figure 8.48a).
(b) Sandwich construction of the curve (Figure 8.48b).
(c) Laminated construction of the curve (Figure 8.48c).

Joints between the jambs and curved heads of the frame can be formed as follows:

(a) Handrail bolts and dowels (Figure 8.49a).
(b) Bridle joints (Figure 8.49b).
(c) Halving joints (Figure 8.49c).
(d) Mortise and tenon joints (Figure 8.49d).

(a) BUTT JOINTING OF SOLID SEGMENTS

(c) LAMINATED CONSTRUCTION OF CURVE

(b) SANDWICH CONSTRUCTION OF SEGMENTS

Figure 8.48 Method of forming curved window heads

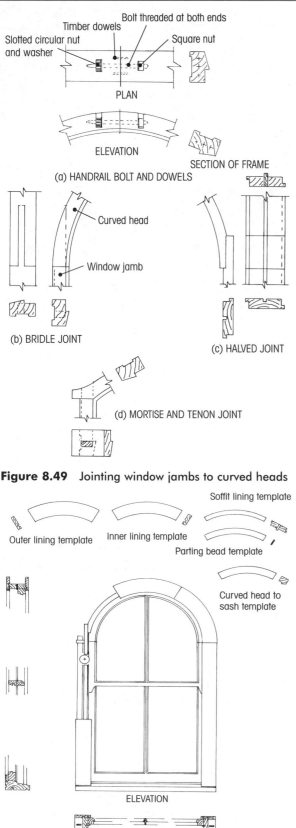

Figure 8.49   Jointing window jambs to curved heads

8.15.1 Setting out the Curved Head of a Sliding Sash and Casement Window

The setting out is done full size using a sheet or sheets of plywood, or medium density fibreboard (MDF). Figure 8.51 shows the setting out for a casement and Figure 8.50 shows that for a sliding sash window with a semicircular head.

8.14.2 Shaping the Curved Head

If solid segments are used for the curve, a template (Figure 8.50) is marked out and cut to the required shape. This template is then used to mark out the curved blanks on a board planed to the required thickness (Figure 8.52a) before bandsawing out the curved segments (Figure 8.52b) [BENCH AND SITE SKILLS, SECTION 9.7.5]. The sawn blanks are then machined to the required profile using a spindle moulder and ring

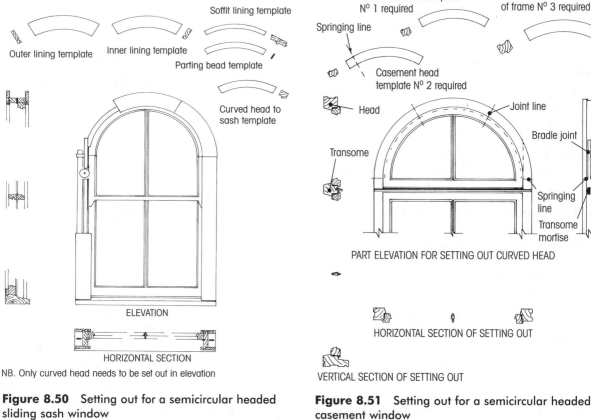

Figure 8.50   Setting out for a semicircular headed sliding sash window

Figure 8.51   Setting out for a semicircular headed casement window

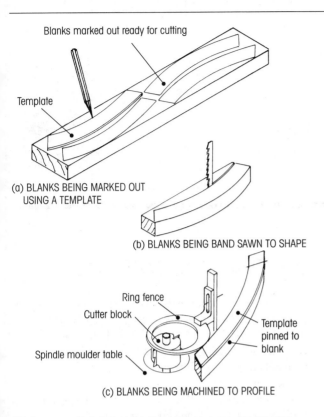

Blanks marked out ready for cutting

Template

(a) BLANKS BEING MARKED OUT USING A TEMPLATE

(b) BLANKS BEING BAND SAWN TO SHAPE

Ring fence

Cutter block

Spindle moulder table

Template pinned to blank

(c) BLANKS BEING MACHINED TO PROFILE

NB. For purposes of clarity, helmet guard is not shown. Moulding the blanks is a specialist operation and must only be undertaken by a competent wood machinist

Figure 8.52 Forming the curved segments for semicircular window head

fence (Figure 8.52c) before being butt jointed together using handrail bolts [BENCH AND SITE SKILLS, SECTION 11.4] or other jointing techniques. Curved heads formed by lamination or sandwich construction would be assembled, and then machined to the correct profile.

9
External cladding

A non-load-bearing weather protective covering (skin) to a wall or roof of a building. We are only concerned here with the external cladding of walls with timber and wood-based products. It is worth noting that the American term for a timber cladding is 'siding'.

The primary function of a cladding is to protect the main structural elements from the weather, however, when choosing a cladding many other factors must be considered – these can include:

- Aesthetics (appearance).
- Durability [BENCH AND SITE SKILLS, SECTION 1.10.4e].
- Strength to resist impact.
- Resistance to abrasion.
- Fire performance.
- Thermal properties.
- Minimum maintenance.
- Compliance with current Building Regulations (restrictions are in force).

Figure 9.1 shows the effect of how (regulations permitting) vertical and horizontally placed cladding can appear when fully or partly covering an elevation of a building. Elevations at (a) and (c) give the impression of lifting the height of the building by making it appear taller than it is. Just as elevations (b) and (d) appear to stretch and shorten the building by making it look longer.

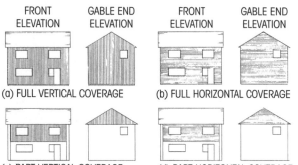

Figure 9.1 Possible effect of using full or partial vertical or horizontal cladding

9.1 CLADDING MATERIALS

The range of cladding materials is vast – however, some have restrictions put on their use, particularly wood-based products. Table 9.1, which should be read in conjunction with Figure 9.2, lists several of these materials together with their form, type, and application.

For the purpose of this chapter the term 'masonry' is given to include stonework, brickwork, and blockwork.

In the UK brickwork is the most popular cladding associated with timber-framed buildings.

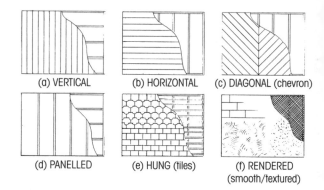

Figure 9.2 Cladding formats on timber grounds

9.2 TIMBER BOARDS AS CLADDING

Timber in its many forms has been used as cladding for hundreds of years, its service life can, depending on species and climatic conditions, be as short as 2 years or as long as 50 plus. The simple fact has been that if, or when, any deterioration to the cladding has occurred it has been an easy task to replace any defective boards with little or no disturbance to the building structure.

Table 9.1   Cladding materials, their form, type, form and application

Material	Form/type	Application	Figure reference
Timber (various species of preservative treated SW and some HW) [BENCH AND SITE SKILLS, SECTION 1.3.1]	Square edged Profiled weatherboard Matchboard	Vertical Horizontal, or Diagonal Vertical or Diagonal	9.2a 9.2b, c 9.2a, c
Western Red Cedar (SW) – preservative treated or untreated heartwood [BENCH AND SITE SKILLS, TABLE 1.18]	Square edged. Profiled weatherboard. Shingles and Shakes.	Vertical Horizontal, or Diagonal Vertical hung	9.2a 9.2b, c 9.2e
Plywood (WBP) [BENCH AND SITE SKILLS, SECTION 2.1.3]	Strips Strip and Full sheets	Horizontal Panel	9.2a 9.2d
Hardboard (tempered) [BENCH AND SITE SKILLS, SECTION 2.5.5]	Strips Strip and Full sheets	Horizontal Panel	9.2a 9.2d
Cement bonded particle board [BENCH AND SITE SKILLS, SECTION 2.3.4]	Strip and Full sheets	Panel	9.2d
Cement based non-combustible boards (non-asbestos reinforcing fibres such as calcium silicate). [BENCH AND SITE SKILLS, SECTION 2.7]	Strip and Full sheets	Panel	9.2d
Plastics (PVC-U) – Section 9.6	Profiled weatherboard	Horizontal, Vertical or Diagonal	9.2a–c
Clay	Tiles (square edged and patterned)	Vertical hung	9.2e
Concrete	Tiles (square edged, shaped and patterned)	Vertical hung	9.2e
Slate (natural and synthetic)	Square edged and shaped	Vertical hung	9.2e
Steel (treated and coated)	Strip and Full sheets (profiled)	Vertical, Horizontal, or Diagonal	9.2a–c
Aluminium	Strip (profiled)	Vertical	9.2a
Cement based rendering on steel mesh (rust proofed expanded metal), or proprietary metal lath	Smooth or textured finish	Surface covering	9.2f
Masonry	Stonework, Brickwork, Blockwork	Erected from a concrete foundation	

Key:
SW – Softwoods.
HW – Hardwoods.

Whereas today irrespective of the wood species used as cladding, with modern preservatives and fixing techniques life expectancy is usually guaranteed for a minimum of 25 years against fungal and insect attack.

9.2.1 Restrictions on the Use of Timber and Wood Based Cladding

There are two main areas to consider here:

* durability, and
* fire performance.

Durability – [BENCH AND SITE SKILLS, SECTION 1.10.4(e)] the natural resistance of a wood to fungal decay, is accessed by the performance of its heartwood when it is subjected to conditions that would be ideal for the development of fungal decay. The sapwood of all wood species has little or no resistance to fungal decay.

Timber from softwoods, with the possible exception of Western Red Cedar (providing it does not contain any sapwood) will require preservative treatment [BENCH AND SITE SKILLS, SECTION 4.3].

Timber cut from the heartwood of durable hardwoods such as European Oak, American mahogany, Afrormosia, Iroko, or Teak, to name a few, can in some circumstances be used without the added protection of a wood preservative.

Fire performance – timber and wood-based products can only be used as external cladding on areas defined as not requiring non-combustible materials. Unless, it can be proved that they can be adequately protected by a non-leaching approved flame retardant suitable for exterior use. For example, restrictions are in place on areas of timber cladding in close proximity to a boundary, for fear of fire spreading from another building or another source by way of radiant heat of flying embers, etc.

Before boundary distances can be determined in order to satisfy the Building Regulations and any relaxations can be put in place. The amounts of unprotected areas must be calculated, these areas can be made up of door or window openings, and the surfaces occupied by any combustible cladding.

9.2.2 Board Profiles

As you can see from Figure 9.3 that with the exception of square edged boards, sections used as weatherboarding are usually fixed horizontally, were as match-

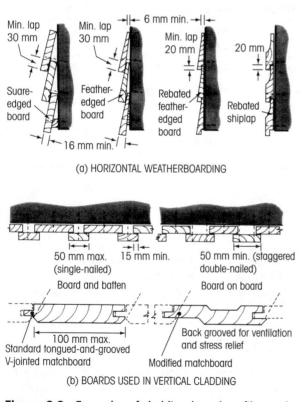

NB. Nails should not penetrate lapped board, to allow moisture movement

(a) HORIZONTAL WEATHERBOARDING

(b) BOARDS USED IN VERTICAL CLADDING

Figure 9.3 Examples of cladding board profiles and their fixing

boarding is fixed vertically or diagonally. In all cases cladding profiles should be designed, positioned, and fixed in such a way that water is shed from its surface as quickly as possible to limit the amount of moisture intake at joints and or overlaps.

9.2.3 Fixing Cladding Boards

We need to consider three things here, namely:

- Timber grounds to which the cladding is to be fixed.
- Board positioning and fixing.
- Fixings.

Timber grounds – these will as shown in Figure 9.2 run in the opposite direction to the boards, unless a counter batten has been used (Figure 9.5c). Grounds may be required to be pressure treated with a wood preservative [BENCH AND SITE SKILLS, SECTION 4.3]. Their section size should be not less than 38 mm wide, and their thickness should be at least $1\frac{1}{2}$ times the thickness of the cladding board. Distance apart will depend on their thickness, but in any case no greater apart than at 600 mm centres. Grounds will either be nailed directly to timber stud work (timber-framed construction), or screwed to the wall via a suitable plug [BENCH AND SITE SKILLS, SECTION 11.6].

Board positioning and fixing – all boards should be fully supported in their length at not less than 600 mm centres and at their ends. Where it is practicable provision should be made for any potential moisture movement.

Figure 9.3(a) shows how horizontal boards need only be single nailed to their support. Not only does this reduce the risk of the board splitting as it shrinks, but the board is aloud to move whilst still being trapped by the one above.

Figure 9.3(b) shows three examples of boards used as vertical cladding. Probably the one most favoured is 'board or batten on board' types. This is because they shed water quickly and provide initial back ventilation to the face boards. Positioning the boards in relation to their end grain section can affect the way they move in the event of any moisture movement distortion. Figure 9.4 illustrates the possible effect of not using quarter-sawn boards.

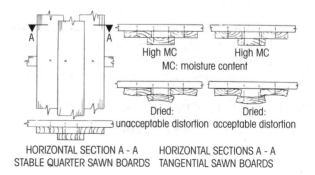

Figure 9.4 Possible effect of vertical boards drying out

Back venting can be a very important aspect of cladding; firstly it ensures that any moisture vapour passing through the external wall can be harmlessly dispersed with out any adverse effect on the cladding. Secondly any moisture taken in by the boards as a result of wind driven rain, etc., can be evenly dried out without any significant increases in moisture content. Thereby avoiding conditions that could otherwise render the boards liable to fungal decay.

Figure 9.5(a) shows an arrangement for back venting the bottom of horizontal cladding, whereas Figures 9.5(b, c) show how vertical cladding can be dealt with. Venting provision must also be made at the top, under window sills (Figure 9.9) and soffit for a through movement of air. All these gaps should be covered with a plastic or stainless steel fly-proof mesh.

By venting the back of the cladding, a vented cavity has been formed which the local authority may regard as a fire hazard. If this is the case, a cavity barrier of wired mineral wool (see Figure 9.5) may be acceptable to resist the movement of smoke or flame.

Nailed boards – [BENCH AND SITE SKILLS, SECTION 11.1] boards should be fixed at centres not greater than 600 mm. Nails should penetrate the substrate by about 25 mm – length should be not less than 2½ times the cladding board thickness. However, **annular ring nails** need only be twice the board thickness.

Nailing patterns as shown in Figure 9.3 should help to counter the possible effect of moisture movement without the boards splitting. Face fixing with round rust proof wire nails is common with weatherboard and board on board cladding. Matchboard on the other hand can be secret nailed (Figure 9.3b – where nail heads can be partly concealed by the groove) or fixed with proprietary metal clips.

Nails should either be galvanised steel, or made from stainless steel, phosphor bronze, or silicon bronze. If the cladding boards have been pre-pressure treated with a wood preservative containing CCA salts [BENCH AND SITE SKILLS, SECTION 4.3.2], then aluminium nails should not be used (because of the risk of chemical reaction).

Note. If Western red cedar cladding boards are to be used the same conditions will apply as for shingles (Section 9.4.1).

9.2.4 Detailing

The object to detailing in this case is to ensure that the finished edges, corners, junctions, or abutments are constructed in such a way as to prevent the ingress (entry) of water to that part of the structure it protects. Good detailing should also enhance the appearance of the cladding as a whole.

Figure 9.5 Fixing and back venting boarded cladding

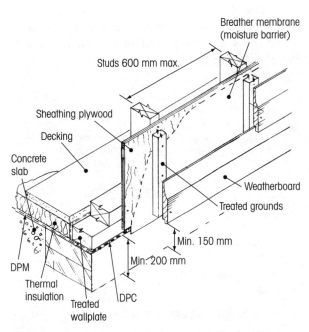

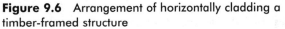

Figure 9.6 Arrangement of horizontally cladding a timber-framed structure

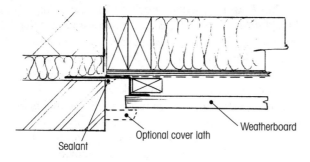

Figure 9.7 Horizontal section showing an open ended abutment against a wall

Figures 9.5 and 9.6 show how the foot so the cladding can be dealt with – notice that a minimum distance of 200 mm has been left between the finished ground level and the bottom edge of the cladding. This should avoid problems with low ground vegetation, and any incidence of water splashing would be reduced to a minimum.

Butting horizontal cladding up against masonry should be avoided – as shown in Figure 9.7 a suitable gap can be left so that end grain treatment can be maintained with ease. Alternatively a cover lath could be used.

Corner detailing is very important – not only for appearance, but also to cover to cover any endgrain exposure. Figure 9.8 illustrates two horizontal sections through 'board-on-board' cladding, and two though horizontal cladding.

Figure 9.9 shows how 'board-on-board' cladding can be detailed around a window opening. Extra ventilation under the sill of a wide window can be achieved by inserting a grill into the under-sill board, or by planting a weatherboard onto the under-sill board and widening the ventilation gap.

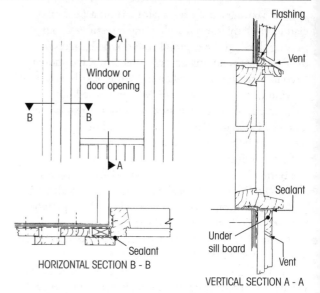

Figure 9.9 Board on board details around a window door opening

9.3 PLYWOOD CLADDING [Bench and Site Skills, Section 2.1.3]

Plywood must be of an exterior grade classified as suitable for external cladding and not less than 10 mm thick. It is also available for this purpose with a pre-finished surface of heavy duty coating synthetic resin – possibly covered with a layer of white or coloured mineral chippings.

Plywood can be used in strip form as horizontal weatherboard or in vertical panels.

Timber grounds will be fixed and used in similar way as boarded cladding – back venting will still be required.

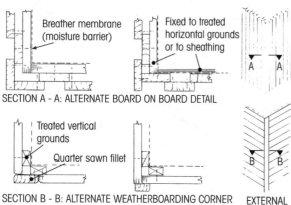

Figure 9.8 Horizontal sections of corner details

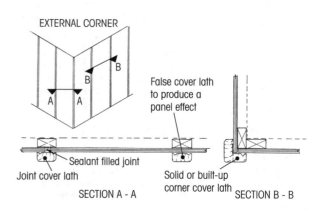

Figure 9.10 Face and corner details for plywood cladding

Figure 9.10 shows details of how panels are vertically abutted in their width and how an external corner joint can be made. Notice how false cover laths are used to give the panel appearance.

Note. Tempered hardboard, cement-bonded particleboard, and cement based non-combustible boards can all be fixed in a similar way to plywood.

9.4 SHINGLES AND SHAKES

Shingles and shakes are best described as wood tiles that have been manufactured from Western Red Cedar. Where the Building Regulations allow this tiles can be used for tiling a pitched roof or as in this case cladding a wall.

As shown in Figure 9.11(a), shingles are thin slices of wood with relatively smooth faces. Produced as a result of ripping down a cross sectioned portion of a log that had previously been crosscut into lengths to suit the length of the shingles. Shakes on the other hand, are as shown in Figure 9.11(b) hand split (cleft) by using a cleaver and mallet. By following the grain in this way this will produce rough textured surfaces. These blocks can then be diagonally sawn to produce two tapered shakes.

Note. The rest of this section is concerned only with shingles.

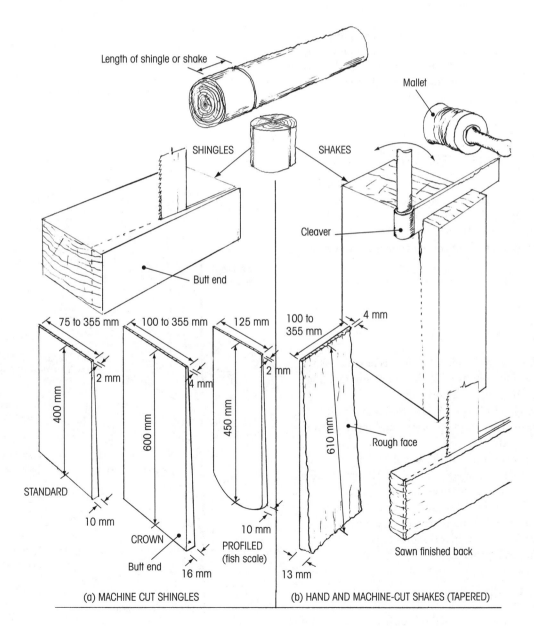

Figure 9.11   Shingles and shakes

9.4.1 Restriction on the Use of Western Red Cedar as a Cladding Material

As with any timber cladding (see Section 9.2.1) the main restrictions are going to be imposed by Building Control (Building Regulations) regarding:

- durability and,
- fire performance.

Durability – Western Red Cedar contain its own natural preservative against fungal decay [BENCH AND SITE SKILLS, SECTION 1.10.3g], but life expectancy can, depending on region and climatic conditions, vary considerably. Therefore most shingles these days are pre-pressure treated with a non-leaching preservative to give them an expected live life span of not less than 50 years.

Fire performance – shingles are available pre-pressure treated with a non-leaching fire retardant. However, as previously explained in Section 9.2.1 the restrictions on the use of shingle will unless instructed otherwise from 'Building Control' (overseeing the Building Regulations) will still apply.

9.4.2 Grades, Sizes and Shapes

Cedar shingles are produced in four grades which reflect their quality with regard to the presence of knots, sapwood, or defects. For example No 1. Grade (Blue labelled) should be 100% knot free, 100% free of sapwood and cut off the edge grain (quarter sawn).

They are sized and shaped as either:

- standard
- crown
- profiled.

Standard shingles (Figure 9.11) – rectangular, 400 long, in various widths of 75–355 mm, and tapering from 2 to 10 mm at the butt end.

Crown shingles (Figure 9.11) – rectangular, 600 long, in various widths of 100–355 mm, and tapering from 4 to 16 mm at the butt end.

Profiled shingles (Figure 9.11) – 450 mm long and 125 mm wide, and tapered from 2 to 10 mm at the butt end which is shaped to give a patterned effect (the one illustrated is a fishtail pattern) when laid in mass. They can either be laid on there own or mixed with other shaped and/or rectangular shingles to produce interesting wall patterns.

9.4.3 Application

Figure 9.12 shows two methods of fixing shingles using a single coursing method. In method (a) shingles are fixed direct via a moisture barrier (breather membrane) to the sheathing plywood. In method (b) tile laths (sometimes called slate lath) have been used as fixing grounds. The gauge (distance from the top of the lath to the top of the one above or below it) of these laths can vary, for example using a two layer covered standard shingle it could be between 140–180 mm.

Single coursing provide for at least two layer s of shingles at ever point. The exposed portion should never be greater than half the shingle length minus 13 mm. For example, using a standard shingle 400 mm long would mean that the exposed portion should not be greater than 187 mm. The starting course at the base should as shown in Figure 9.12 should be doubled. Side joints between shingles should be offset by not less than 40 mm over joints in any adjacent courses. Provision should always be made for potential moisture movement – shingles should therefore be 6 mm apart.

Nailing – pre-drilling for nails is not required providing the nails are positioned not less than

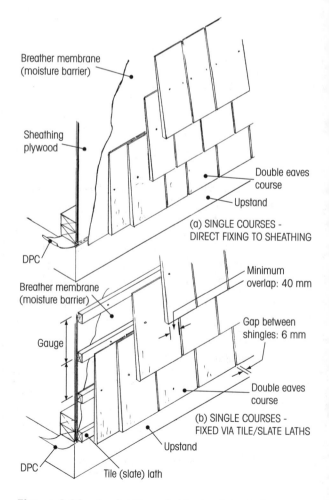

Figure 9.12   Application of shingles to a wall

20 mm in from the edge. Two nails per shingle should be used not less than 40 mm above the 'butt' of the next course.

Nails – due to the chemical contents (natural extracts) within the wood cells of Western Red Cedar (see notes below) and the preservatives used, special nails are available from either the shingle manufacturer or supplier. These will more than likely be made from stainless steel, phosphor bronze, or silicon bronze.

Note. Because Western Red Cedar contains natural extractives [BENCH AND SITE SKILLS, SECTION 1.10.3g] which are water soluble and could therefore leach out and have a harmful effect on some metals. Flashings made out of metal can either be protected with a polyester finish, or coated with a thick layer of bitumen paint.

9.5 FINISHES TO WOOD AND WOOD BASED CLADDING

Wood finishes should have both decorative and weather protective properties, yet be permeable (micro-porous) to water vapour. That is to say, if water vapour happens to enter the cladding via the structure it is protecting (most should have been vented away via the back cavity) it can safely permeate through the wood and its surface finish without becoming trapped. The same principle would apply should water get behind the cladding via surface joints, etc.

Conversely if an impermeable paint or varnish film had been used, and moisture had entered the wood via small cracks, or abrasions, etc., it could become trapped, leading to increased moisture content, and cause a breakdown of the surface film. Examples of film breakdown include bubbling, flaking, and splitting.

With clear transparent finishes (vanishes), any wood surfaces exposed to sunlight (ultraviolet [UV] light) can discolour. Unlike finishes containing a pigment which are usually UV resistant. The effect of trapped moisture behind a clear finish usually appears as blue stain fungus [BENCH AND SITE SKILLS, SECTION 3.2.1].

Two suitable groups of breather (microporous) finishes are:

- exterior emulsion-type paints, and
- exterior wood stains.

Exterior emulsion-type paints – water-based acrylic paints with an opaque coloured 'silk or satin' finish. Application methods vary, some manufacturers recommend two brush coats, others three. Maintenance expectancy can be between 5 and 8 years; however this will largely depend on the amount and type of exposure, for example atmospheric pollution.

Exterior wood stains (a solvent or water-based product) – water-repelling non-opaque (translucent) matt to semi-gloss finishes available in many light-fast colours. Usually contain an inhibitor against algae, and blue stain fungi. Application usually involves a minimum of two brushed coats.

Note. It is well known that Western Red Cedar cladding doesn't require any surface protection because of its natural durability, but it soon looses its initial red colouring as it becomes weathered to take on a silver grey colour. However, the reddish colour we associate with Western Red Cedar can be retained to some degree by applying a pigmented wood stain.

9.6 PVC-u CLADDING

Made from foamed PVC that has been extruded to leave a smooth durable profile with an impervious skin. Apart from the standard white, various colours such as brown, beige, and green are available.

9.6.1 Profile and Trims

The shape of the profile more often as not reflexes its use, for example shiplap weatherboard (Figure 9.13a) is used horizontally. Just as 'V' types can be used vertically (Figure 9.13b) and horizontally and diagonally.

A range of different trims is available to suit differing situations, for example:

Starter trims (Figure 9.14):

(a) base starter
(b) drip starter for over doors or windows.

Edge trims (Figure 9.15):

(a) channel
(b) drip trim

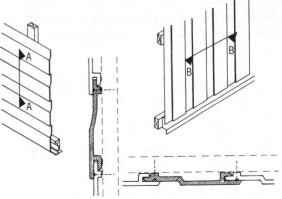

(a) SHIPLAP HORIZONTAL CLADDING VERTICAL SECTION A - A   (b) OPEN "V" VERTICAL CLADDING HORIZONTAL SECTION B - B

Figure 9.13 PVC-u cladding

HC = Horizontal cladding

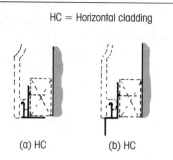

(a) HC   (b) HC

Figure 9.14   Start trims for plastics cladding

VC = Vertical cladding
HC = Horizontal cladding

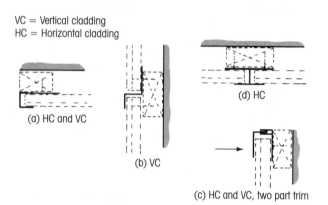

(a) HC and VC

(b) VC

(d) HC

(c) HC and VC, two part trim

Figure 9.15   Edge trims for plastic cladding

(c) head trim (two parts)
(d) butt trim (joint cover).

Corner trims (Figure 9.16):

(a) external corner trim (horizontal cladding)
(b) external two part corner trim (horizontal cladding)
(c) internal two part corner trim (horizontal cladding).

9.6.2 Fixing

As with all plastics components provision should be made for any thermal movement. Therefore fixing should only be carried out when temperatures are below 30°C and above freezing point 0°C. Gaps of 4 mm should be left at each end of the boards (planks) to allow for any movement – some fittings (cover plates, etc.) will have built-in spacing lugs or marks.

Treated timber grounds will be arranged similar to those used with timber cladding. Provision for back venting should be made (back venting may not be allowed in certain areas) – where required, particularly if coloured cladding is used.

Fixing centres for grounds should not exceed 600 mm for buildings up to two storeys. For cladding higher than two storeys and up to five storeys ground centres must not be less than 400 mm. All coloured cladding should be fixed to grounds at not less than 400 mm centres.

Boards are secret nailed through the top edge fixing groove with stainless steel nails. Colour co-ordinated plastic headed nails are available for exposed fixing.

Note. If, as shown in Figure 9.17, thermal insulation is to be incorporated a clear airway of not less than 25 mm should (regulations permitting – see Section 9.2.3) be provided behind the cladding.

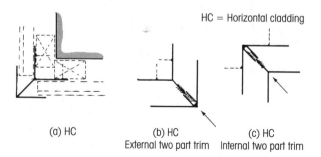

HC = Horizontal cladding

(a) HC   (b) HC   (c) HC
                  External two part trim   Internal two part trim

Figure 9.16   Corner trims for plastics cladding

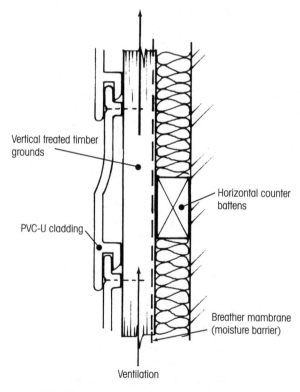

Vertical treated timber grounds

Horizontal counter battens

PVC-U cladding

Breather membrane (moisture barrier)

Ventilation

VERTICAL SECTION THROUGH HORIZONTAL CLADDING

Figure 9.17   Thermal insulation behind PVC-U cladding

10
Casing-in and panelling

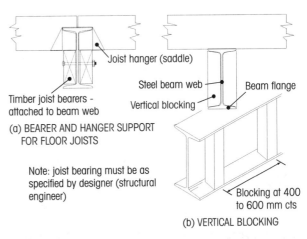

10.1 CASING-IN STRUCTURAL MEMBERS

This is an area of work that provides an internal finish to items that would otherwise look unsightly, and if necessary provides fire protection in accordance with the Building Regulations.

10.1.1 Steel Beams (Binders)

Beams projecting below a floor slab need encasings to protect them from fire (Figure 10.1), and to enhance their appearance.

Encasement support (Figure 10.2) will require either:

(a) a series of cradles fixed at centres of 450–600 mm, depending on the sectional size of any counter battens or panel thickness.

Joist hanger (saddle)

Timber joist bearers - attached to beam web

Steel beam web
Vertical blocking
Beam flange

(a) BEARER AND HANGER SUPPORT FOR FLOOR JOISTS

Note: joist bearing must be as specified by designer (structural engineer)

Blocking at 400 to 600 mm cts

(b) VERTICAL BLOCKING

Figure 10.2   Encasement support of steel beams

(b) framed grounds to both sides of the beam and its underside.

Both of these means of support will require fixing points. These may be provided by:

- Joist bearers that would have been installed to support the lower edge of the floor joists – these may be connected to the beams via a bolt set through the web of the beam (Figure 10.2a).
- Vertical blocking (soldiers), these should fit tight between the flanges of the beam (Figure 10.2b).

Cradles (Figure 10.3a) consist of a series of 'U' shaped pre-fabricated frames constructed by using any suitable framing joint [BENCH AND SITE SKILLS, SECTION 7.5]. The top of each of the cradle legs is fixed to the sides of every joist, and the lower portion fixed to bearers.

Framed grounds (Figure 10.3b) consist of a light pre-fabricated timber framework using a choice of framing joints [BENCH AND SITE SKILLS, SECTION 7.5]. Side grounds will be attached to the vertical blocking; soffit grounds will be fixed to the side grounds.

Unframed grounds (Figure 10.3c) consist of an assembly of timber runners attached to vertical blocking.

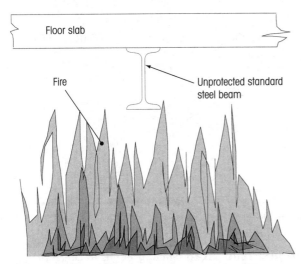

Floor slab

Fire

Unprotected standard steel beam

Figure 10.1   Steel beam at risk from fire

NB. For sake of clarity, joist hangers are not shown in this view

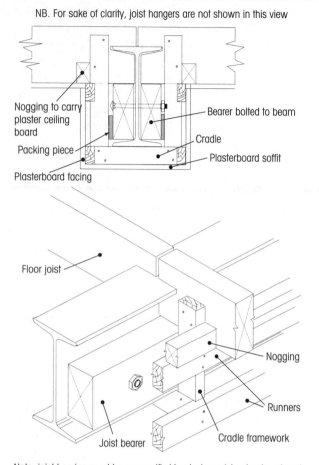

Note: joist bearing must be as specified by designer (structural engineer)

(a) USING CRADLES AROUND A STEEL BEAM

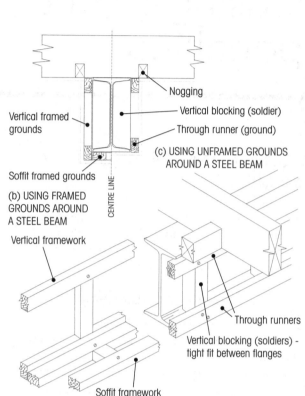

(b) USING FRAMED
GROUNDS AROUND
A STEEL BEAM

(c) USING UNFRAMED GROUNDS
AROUND A STEEL BEAM

Figure 10.3   Fixing timber cradles and grounds to steel beams

Facings – finally the grounds will be clad with a fire-resistant material that must be within the accepted limits of the '*spread of flame*' class to comply with current building regulations.

10.1.2 Concrete Beams

In this case, before any grounds are assembled it is common practise to check for any discrepancies in beam size. Timber grounds will be attached to the beam using a method approved by the building designer, this may involve using a cartridge-operated fixing tool [BENCH AND SITE SKILLS, SECTION 8.19] to drive hardened steel pins into the concrete.

Figure 10.4 shows an optional arrangement for fixing the grounds. In Figure 10.4(a) two side frames and an under-beam frame would be used. In Figure 10.4(b) the under-beam framework has been dispensed with to allow the facing soffit material to bridge between them – in this case extra care must be taken to ensure that the facing materials are in line and the soffit width is uniform along its length.

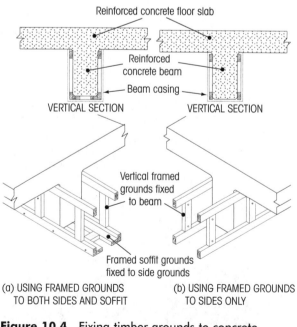

(a) USING FRAMED GROUNDS
TO BOTH SIDES AND SOFFIT

(b) USING FRAMED GROUNDS
TO SIDES ONLY

Figure 10.4   Fixing timber grounds to concrete beams

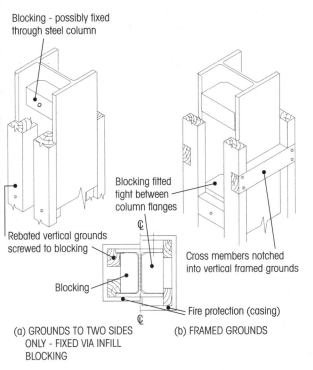

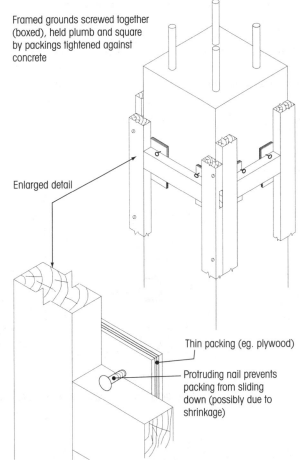

Figure 10.5 Fixing timber grounds around steel columns

10.1.3 Steel Columns

These can be dealt with in much the same way as steel beams except that in a free-standing column all faces of a steelwork will need to be encased. Figure 10.5 shows two alternative ways of fixing timber grounds to the column. For example, if provision has been made for fixing timber blocks between the flanges grounds may only be necessary on two sides (Figure 10.5a). Alternatively, the blocking can be trapped between flanges by fixing cross members to adjacent framed grounds (Figure 10.5b).

10.1.4 Concrete Columns

In this case it is usual to use framed grounds all round as shown in Figure 10.6. Before proceeding however checks must be made to ascertain the largest cross-section of the column to ensure that the framework will be slightly oversize at all points. Any differences between the backs of the framed grounds and the column will be taken up by inserting back packings behind the cross members, as shown in Figure 10.6. These packing must be prevented from falling down behind the grounds, if, or more likely when, any shrinkage takes place due to moisture movement. Therefore as a precaution a nail can be driven into the packing just above the cross member.

Figure 10.6 Fixing timber grounds around a concrete column

10.1.5 Fire Resistance

Steelwork, although non-combustible will lose structural strength and distort in temperatures at about 800°C and above. Steelwork must therefore be protected from heat generated by fire for a specific period of time to enable a compartment or building as a whole to be safely evacuated.

To achieve this integrity structural steelwork must be able to attain a prescribed period of fire resistance. This may be obtained either be spraying with a fire-resistant coating and/or being encased with a non-combustible material. Materials such gypsum-based plasterboard, and many other proprietary products manufactured specifically for this purpose could be used.

10.2 CASING WATER PIPES AND CABLES (Gas supply pipes should be left exposed)

Exposed pipes and cables are not normally acceptable to the building occupant. Therefore, they can be encased and hidden from view, however there are cases and situations where access may be required. For example, turning off a stoptap in an emergency or to enable repairs or maintenance to be carried out. In these cases some provision should be built into the casing design by way of a detachable facing (panel) or access trap.

Figure 10.7 shows several methods of forming a casing around otherwise exposed cables and pipe work. Where a backboard (pipe board) is not provided at the first fix stage the arrangements shown can be modified by fixing timber grounds directly to the walls as shown in Figure 10.8.

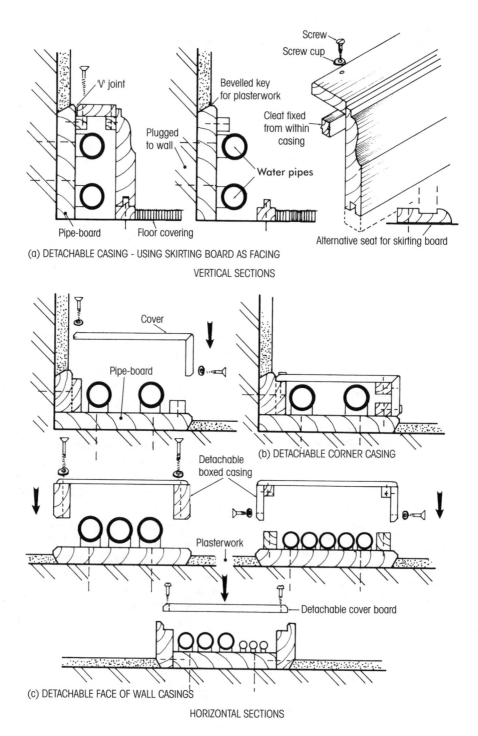

(a) DETACHABLE CASING - USING SKIRTING BOARD AS FACING

VERTICAL SECTIONS

(b) DETACHABLE CORNER CASING

(c) DETACHABLE FACE OF WALL CASINGS

HORIZONTAL SECTIONS

Figure 10.7   Casing around Water pipes and cables

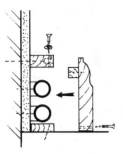

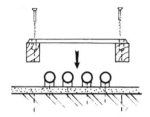

VERTICAL SECTION THROUGH
DETACHABLE SKIRTING BOARD

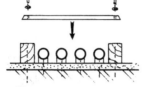

HORIZONTAL SECTION THROUGH
CORNER CASING WITH DETACHABLE
FACING

HORIZONTAL SECTION THROUGH
DETACHABLE WALL CASING

HORIZONTAL SECTION THROUGH
WALL CASING WITH DETACHABLE
FACING

NB. Stop or regulating valves should always have a means of access -
see Fig. 10.9

Figure 10.8 Casing in pipes without a pipe backing board

In the majority of cases the edge/edges of the box casings will require scribing to the wall [BENCH AND SITE SKILLS, SECTION 7.5.10].

Note. Where services pass through a ceiling or wall of a fire-resistant construction any gaps or openings around these pipes or cables, etc. must be properly fire stopped and sealed before they are encased or hidden.

10.2.1 Means of Access

Where for reasons of disturbing a decorative finish a detachable facing (panel) is not practicable, then some form of access traps may be provided. Figure 10.9 shows a few examples of how such a trap may be constructed over a regulating or stop valve.

10.3 BATH PANELLING

The type of panel will depend on where the bath is sited in the room for example:

(a) With only one long side exposed.
(b) With one long side and one end exposed.
(c) With two long sides and one end exposed.

We are going to consider two types of panelling.

1. Pre-formed panels made from various materials
2. Panels built-up in situ.

Pre-formed panels (Figure 10.10) – usually only require a small sectioned timber plate attached to the floor as a restraint and to provide means of fixing the bottom edge of the panel or toe piece. The top edge is held in place by a groove formed by a pre-fixed timber or metal sectioned stiffener to the underside of the rolled edge of the bath. The arrangement and fixing of the stiffener can vary between makes of bath.

Built-up bath panels (Figure 10.11) – require a supporting timber back framework constructed with simple framing joints secured to the under edge of the bath stiffener and the floor. It is important that

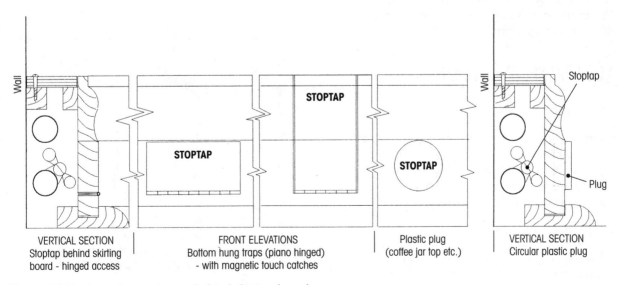

VERTICAL SECTION
Stoptap behind skirting
board - hinged access

FRONT ELEVATIONS
Bottom hung traps (piano hinged)
- with magnetic touch catches

Plastic plug
(coffee jar top etc.)

VERTICAL SECTION
Circular plastic plug

Figure 10.9 Accessing a stoptap behind skirting board

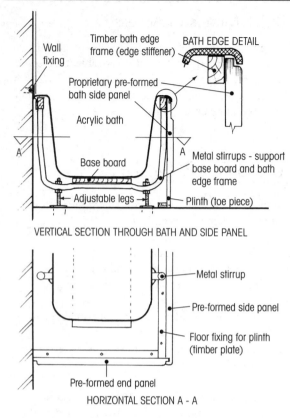

VERTICAL SECTION THROUGH BATH AND SIDE PANEL

HORIZONTAL SECTION A - A

Figure 10.10   Fixing a pre-formed bath panel

these frames are not over tightened between these points otherwise there will be a tendency to lift the front edge of the bath and possibly cause it to distort.

To this sub-framework can be attached the panelling material. The panel material may be in the form of a framed, strip, or sheet panel (see Section 10.4.4), with various decorative finishes.

Alternatively with metal baths, as shown in Figure 10.11 pre-made panels can be fitted under the rolled edge of the bath and adjusted for height by fixing to a set-back plinth (toe piece).

10.4 WALL PANELLING

Wall panelling may be used to:

- concealed unsightly services or uneven surfaces
- provide a decorative feature
- provide hygiene and easy clean surfaces
- provide special acoustic properties
- incorporate thermal installation.

The panelling may also be:

- full height of the room (Figure 10.12a)
- three-quarter room height – leaving the upper portion as a frieze (Figure 10.12b)

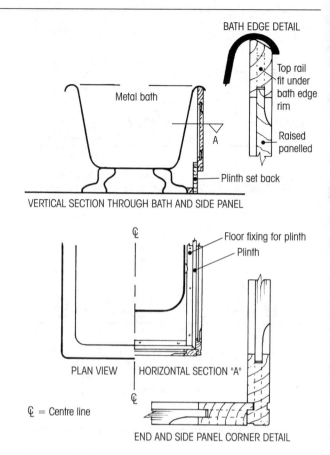

END AND SIDE PANEL CORNER DETAIL

Figure 10.11   In situ built-up bath panel

- waist height (about 1 m above the floor) better known as a dado panelling (Figure 7.2, Table 7.1 and Figure 10.12c).

Panelling of various heights can be further subdivided by type (Figure 10.13):

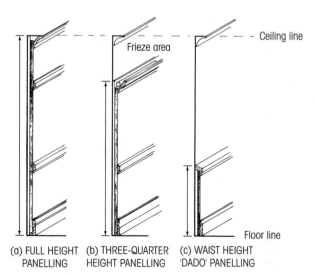

(a) FULL HEIGHT PANELLING   (b) THREE-QUARTER HEIGHT PANELLING   (c) WAIST HEIGHT 'DADO' PANELLING

Figure 10.12   Panelling by weight

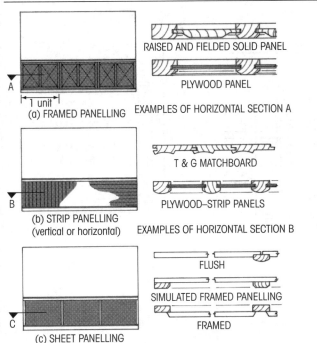

Figure 10.13 Panelling types

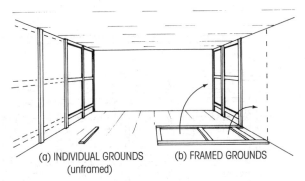

Figure 10.14 Unframed (individual) and framed grounds

- *Framed panelling* (Figure 10.13a) – a framework of timber grooved to house a series of panels.
- *Strip panelling* (Figure 10.13b) – made up of a series of grooved muntins to hold intermediate panels, or horizontal or vertically positioned tongued and grooved matchboard.
- *Sheet panelling* (Figure 10.13c) – large panel's face fixed to timber grounds.

10.4.1 Timber Grounds

There are two main types:

- unframed, and
- framed.

Unframed grounds (Figure 10.14a) – consist of a series of timber laths individually fixed to the wall at pre-determined centres, these may be positioned vertically all horizontally. Grounds must be fixed plumb and in line, great care must a being taken to ensure or that internal and external angles are always plumb.

Framed grounds – a pre-fabricated open framework of timber jointed and assembled as shown in Figure 10.14(b). Typical framing joints used for these frames is shown in Figure 10.15.

10.4.2 Traditional Framed Panel Construction (Figures 10.13a and 10.16)

A framework consisting of grooved or rebated timber sections, joined [BENCH AND SITE SKILLS, SECTION

7.5.7] to form a framework capable of housing or holding a series of small panels (see Figures 6.10–6.13). The panelling is framed up in the workshop ready for fixing on-site to the grounds. In some cases the panelling will be pre-stained and polished.

Fixing traditional panelling – Figure 10.17 shows examples of how the panelling is fixed to the grounds using drop-on fixings. Any fixings made through the face of the panelling should as shown in Figure 10.18 be unobtrusive or masked, by using cover laths, beads, fillets, pellets, skirting board and/or a capping.

10.4.3 Strip Panelling (Figure 10.13b)

A series of narrow boards or panels fixed vertically to accentuate height, or horizontally to give the appearance of lengthening the wall. Both will require a background of framed or unframed grounds.

Fixing strip panelling – as shown in Figure 10.19 fixings can be made by secret nailing or by using proprietary clips.

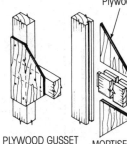

Figure 10.15 Framing joints for framed grounds

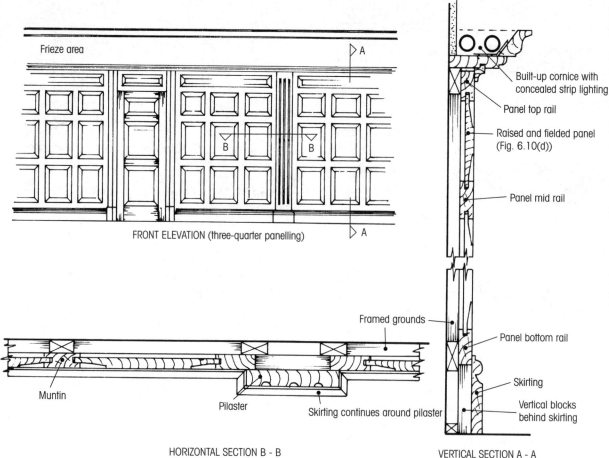

Figure 10.16 Traditional framed panelling

FRONT ELEVATION (three-quarter panelling)

Frieze area

HORIZONTAL SECTION B - B

Muntin

Pilaster

Skirting continues around pilaster

VERTICAL SECTION A - A

Built-up cornice with concealed strip lighting

Panel top rail

Raised and fielded panel (Fig. 6.10(d))

Panel mid rail

Framed grounds

Panel bottom rail

Skirting

Vertical blocks behind skirting

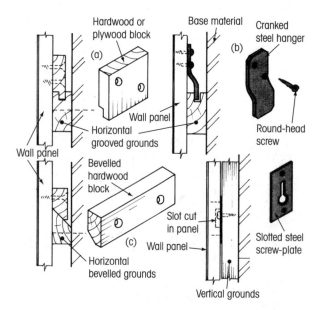

Figure 10.17 Fixing panelling by using drop-on fixings

Hardwood or plywood block

Base material

Cranked steel hanger

Wall panel

Horizontal grooved grounds

Round-head screw

Wall panel

Bevelled hardwood block

Slot cut in panel

Wall panel

Slotted steel screw-plate

Horizontal bevelled grounds

Vertical grounds

10.4.4 Sheet Panel Construction (Figure 10.13c)

This type of panelling often known as *flush panelling* has been made possible by the development of stable sheet materials [BENCH AND SITE SKILLS, CHAPTER 2], such as:

- veneer plywood
- core plywood, such as block board and laminboard
- fibre boards, such as medium board, hard board, and medium density fibre board (MDF).

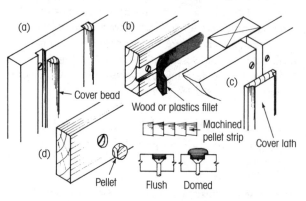

Cover bead

Wood or plastics fillet

Machined pellet strip

Cover lath

Pellet

Flush

Domed

Figure 10.18 Concealing face fixings

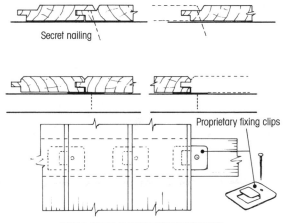

Secret nailing

Proprietary fixing clips

SECRET FIXING USING PROPRIETARY CLIPS

Figure 10.19   Fixing tongued and grooved strip panelling

These materials may be used with their own self finish, or used as a base for other finishes such as:

* polished wood veneer
* stainless steel
* brushed aluminium
* plastics and their laminates
* woven fabrics.

Fixing to grounds – grounds will normally be framed and fixed with centres to accommodate the sheet panels.

Panel connections can as shown in Figure 10.20 be made using the following methods:

* loose tongues (Figure 10.20a)
* tongued and grooved edges (Figure 10.20b)
* extruded metal or plastics channels (Figure 10.20c)
* extruded metal or plastics 'T' sections (Figure 10.20d).

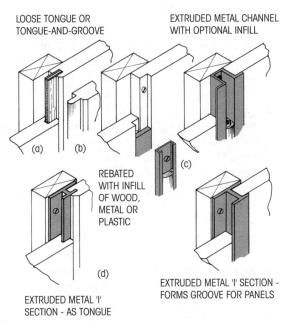

LOOSE TONGUE OR TONGUE-AND-GROOVE

EXTRUDED METAL CHANNEL WITH OPTIONAL INFILL

(a)   (b)

(c)

REBATED WITH INFILL OF WOOD, METAL OR PLASTIC

(d)

EXTRUDED METAL 'I' SECTION - AS TONGUE

EXTRUDED METAL 'I' SECTION - FORMS GROOVE FOR PANELS

Figure 10.20   Fixing sheet panelling

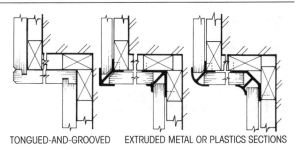

TONGUED-AND-GROOVED   EXTRUDED METAL OR PLASTICS SECTIONS

Figure 10.21   Sheet panelling internal and external corner details

Proprietary plastics and metal profiles as shown in Figure 10.21 are available to accommodate both internal and external angles.

Lightweight sheet panels may in certain circumstances be attached directly to the wall surface without the use of timber grounds. This is by the use of either proprietary purpose-made adhesive pads or a suitable panel adhesive as shown in Figure 10.22. However if the wall surfaces are uneven or out of plumb then timber grounds should be employed.

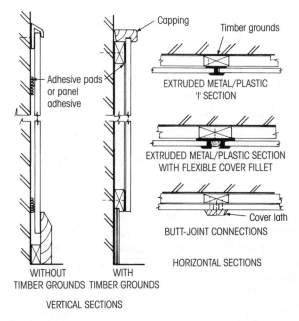

Capping

Timber grounds

Adhesive pads or panel adhesive

EXTRUDED METAL/PLASTIC 'I' SECTION

EXTRUDED METAL/PLASTIC SECTION WITH FLEXIBLE COVER FILLET

Cover lath

BUTT-JOINT CONNECTIONS

HORIZONTAL SECTIONS

WITHOUT TIMBER GROUNDS   WITH TIMBER GROUNDS

VERTICAL SECTIONS

Figure 10.22   Direct fixing sheet panels to wall

10.5 FIRE PERFORMANCE

Because most of the materials used in the construction of wall panelling are combustible the risk of surface spread of flame is greater than for non-combustible material. Therefore limitations are put on the use of panelling unless they are suitably treated with a fire retardant to satisfy current building and fire regulations relating to the rate of spread of flame and material classification.

11
Staircases

11.1 INTRODUCTION

The function of a stairway is to provide safe access by a series of steps from one level to another (Figure 11.1).

The Building Regulations 1996 and the Approved Documents B: Fire safety, K: Stairs and Ladders, M: Facilities for Disabled People, together with BS3595 Part 1 1977, BS3595 Part 2 1984 and BS585 Part 1 1989 lay down the requirements for the design, manufacture and fixing of staircases.

Staircases are divided into three categories by the Approved Document K.

1. Private – stairs for use in dwellings occupied by a single household.
2. Institutional and assembly – stairs in a place where large numbers of people will congregate, concert halls, colleges, hospitals and other such places.
3. Other – stairs in all other buildings not covered in (1) and (2).

These three categories relate to the steepness of the stair plus its associated landings and guarding. This chapter will deal mainly with private stairs as the others are not usually manufactured exclusively from timber.

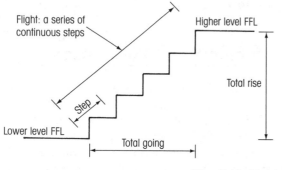

Figure 11.1   **Access between different floor levels**

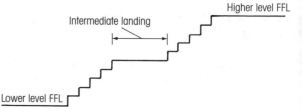

Figure 11.2   **Staircase**

11.2 TERMINOLOGY

(a) *Staircase*. The collective name for the total number of steps between one floor level and the next, but which may be broken up by one or more landings (Figure 11.2).
(b) *Flight of stairs*. A consecutive number of steps without a break, which may run from floor to floor, floor to landing, or landing to landing (Figure 11.3a). It is good practice to limit the flight to no more than 16 steps, and this number should not be exceeded if the stairs are used in a shop or for assembly buildings. Stairs having more than 36 risers in consecutive flights must make a least one change of direction between flights of at least 30° (Figure 11.3b).
(c) *Landing*. A floor or platform at the commencement and termination of a flight of steps (Figure 11.3).
(d) *Going*. The horizontal distance measured from the face of the front edge of the step to the front edge of the next step above or below (Figure 11.4). It can also refer to the total horizontal distance taken up by the staircase.

   Note. When setting out a staircase it is probably easier to consider the going as the horizontal distance between the face of one parallel riser and the next (Figure 11.5).
(e) *Nosing*. The moulded front edge of the tread which projects beyond the face of the riser.

(a)

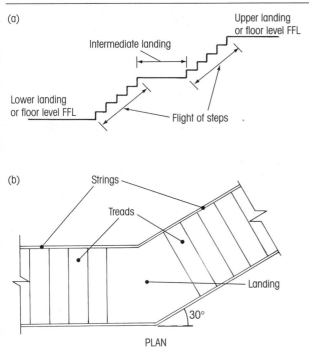

(b)

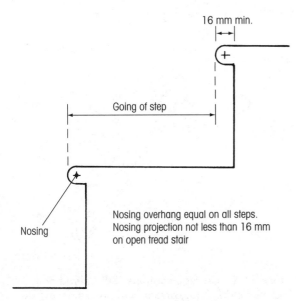

NB. If the staircase has more than 36 treads there must be a turn in the staircase of at least 30°

Figure 11.3 Flight of stairs and plan of staircase

Figure 11.4 Going of steps measured from face of nosings

(f) *Pitch.* The slope of a flight of stairs measured in degrees. For a private stair this should not exceed 42° (Figure 11.6).

(g) *Pitch line.* A line that touches the nosing of each step.

(h) *Rise.* The vertical distance measured from the top of one step to the top of the next step (Figure 11.6). Also the total vertical height the staircase rises between one floor and the next floor above (Figure 11.1).

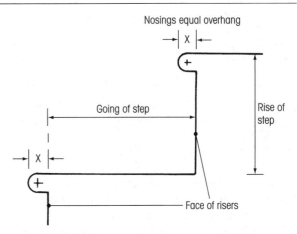

Figure 11.5 Going measured from face of risers for purposes of setting out

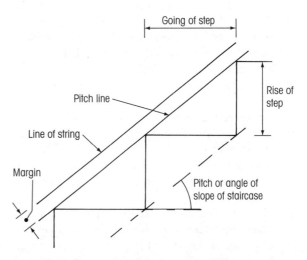

Figure 11.6 Slope or pitch of staircase

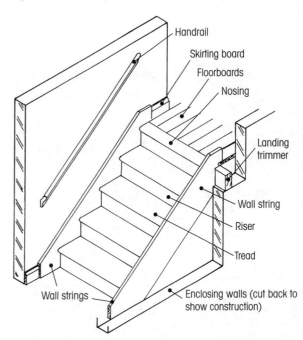

Figure 11.7 Enclosed staircase (straight flight)

11.3 COMPONENTS OF A STAIRCASE

These terms apply to closed and open string staircases.

(a) *Apron lining*. Finishing trim to the vertical face of a trimmer forming the edge of a stairwell (Figure 11.8).

(b) *Balusters (spindles)*. Vertical members forming part of an open frame work filling in the space between the string and handrail of a staircase.

(c) *Balustrade*. An open or panelled frame guarding the opening edge of a landing or staircase.

(d) *Bracket*. The cleats on a carriage piece used to support the treads on a wide staircase (Figure 11.9a) or the moulded bracket under the return nosing of the tread on a cut and mitred string (Figure 11.9b).

(e) *Capping*. A moulded timber member sitting on top of the outer string to carry the lower ends of the balusters, or the lower edge of a panelled balustrade (Figure 11.8).

(f) *Carriage*. An inclined member on the underside of a staircase to which is attached brackets to give extra support to the centre of the treads on wide staircases. Used when staircases are wider than 900 mm (Figure 11.9a).

(g) *Handrail*. A guide for the hand on a staircase. The top member on a level or inclined balustrade (Figure 11.8).

(h) *Newel*. Vertical posts usually square in section into which the open strings and handrails are morticed and tenoned (Figure 11.8).

(j) *Riser*. The vertical face of a step (Figure 11.8).

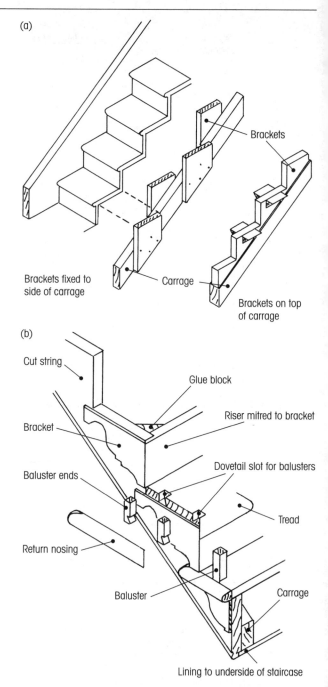

(a)

Brackets

Brackets fixed to side of carriage

Carriage

Brackets on top of carriage

(b)

Cut string

Glue block

Bracket

Riser mitred to bracket

Baluster ends

Dovetail slot for balusters

Return nosing

Tread

Baluster

Carriage

Lining to underside of staircase

Figure 11.9 Carriage supports. Cut, mitred and bracketed strings: (a) carriage and bracket step supports; and (b) cut mitred and bracketed string

(k) *Stair string cut*. A stair string which is cut to the profile of the steps (Figure 11.9b).

(l) *Stair string wall*. A stair string secured to an abutting wall (Figure 11.7).

(m) *Stair string open*. A stair string carried by newels and facing into the open space of a stairwell (Figure 11.8).

(n) *Step*. Consists of a tread and riser (Figure 10.9c).

(o) *Spandrel*. The triangular area between the inclined string and the horizontal floor. (Figure 11.8).

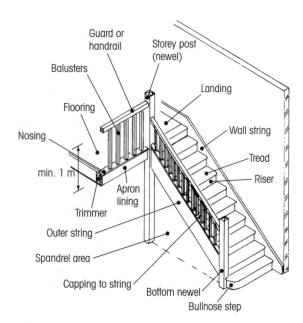

Guard or handrail

Storey post (newel)

Balusters

Landing

Flooring

Nosing

Wall string

min. 1 m

Tread

Apron lining

Riser

Trimmer

Outer string

Spandrel area

Capping to string

Bottom newel

Bullnose step

Figure 11.8   Open well staircase

Table 11.1 Requirements for the construction of staircases

Item	Figure	Description	Type of staircase		
			Private	Industrial. Assembly	Other
1		Stair widths	Stair width, no recommendation but notice should be taken of Approved Document "B" Escape from Fire and "M" Access for the Disabled	A staircase wider than 1.8 m. Should be divided by a central handrail so that the width between handrails is never greater than 1.8 m	
2	11.6	Pitch of stairs	Maximum pitch 42°	Not indicated but governed by the permitted maximum and minimum of the step rise and going	
3		Maximum number of risers in a flight of stairs	No requirement, but it is good practice not to exceed 16 risers	No more than 16 risers per flight in staircases serving shops and other buildings of assembly	As for private stairs
4			Staircases having more than 36 risers in consecutive flights should make at least one change of direction between flights of at least 30°		
5	11.25b	Rise of steps	Maximum rise 220 mm. Provided the pitch does not exceed 42°	Maximum rise 180 mm	Maximum rise 190 mm
6	11.25b	Going of steps	Minimum going 220 mm. Provided the pitch does not exceed 42°	Minimum going 280 mm	Minimum going 250 mm
7		To make walking up and down stairs as easy as possible	Twice the rise plus the going should not be less than 550 mm or greater than 700 mm	Twice the rise plus the going should not be less than 550 mm or greater than 700 mm	Twice the rise plus the going should not be less than 550 mm or greater than 700 mm
8	11.25c	Gap between open treads and step overhang	The gap between open treads should be divided by a bar or rail so that the gap will not allow a 100 mm diameter sphere to pass through. In open tread stairs the treads must overlap by at least 16 mm		
9		Minimum number of steps	This is not stated in Approved Document K, but a difference of levels less than 600 mm between floors can be a hazard for purposes of escape and at least three steps should be considered		
10		Width of landings	The width or going of a landing should be at least as wide as the staircase it serves		
11	11.1	Landings	A landing should be formed at the top and bottom of a flight of stairs. The landing may form part of the floor		
12	11.25h	Clearance on landings	Landings must be kept clear of permanent obstructions, but a door may open across a bottom landing provided it leaves a clear unobstructed space of 400 mm		

Table 11.1 Continued

Item	Figure	Description	Type of staircase		
			Private	Industrial. Assembly	Other
13	11.25i	Clearance on landings	Doors to cupboards on top landings are allowed to open onto the landing provided there remains an unobstructed space of 400 mm. Reference should be made to Approved Document "B" Means of Fire Escape		
14		Landings. Head room	Not less than 2 m	Not less than 2 m	Not less than 2 m
15	11.25d	Staircase head room	Head room measured vertically off the pitch line of the staircase should be at least 2 m. This height requirement also applies to any beams that may project below the line of the ceiling.		
16	11.25e	Stair to loft conversions	In the case of staircases used in loft conversions the head room may be reduced to 1.9 m measured on the centre line of the stair reducing to 1.8 m at one side of the staircase		
17	11.25a	Handrails	Stairs should have at least one handrail and be guarded at an open edge. If the staircase is wider than 1 m two handrails must be provided		
18	11.25g	Handrails		Handrails should be provided for the bottom two steps and project 300 mm along the landing. Approved Document "M"	
19	11.25f	Handrail heights	900 mm on the pitch of the staircase and on landings	900 mm on the pitch of the staircase and 1000 mm on the landings	
20		Guarding stairs	Stairs and landings with a drop of 600 mm or more must be provided with guard rails	Stairs and landings with a drop of two or more risers must be provided with guard rails	Stairs and landings with a drop of two or more risers must be provided with guard rails
21	11.25c	Balustrading	Balustrades should be so designed that they are not readily climbable by children under the age of 5 and that the gap between any member of the framing must not allow a 100 mm diameter sphere to pass through		
22	11.40	Tapered steps	Staircases with tapered steps should be designed and constructed to conform to Approved Document "K" and BSS 585 Wood Stairs Part 1 1989		

11.4 STAIRCASES MANUFACTURED TO STANDARD DIMENSIONS

For private housing most staircases can be bought from joinery suppliers as standard assembled units ready for fixing on site (see Figure 11.10). These standard staircases come assembled to site, and if there is an open string the newels will be attached. The balustrade and handrail can also be fitted, but it is probably more usual for the balustrade and handrail to come in kit form and be fitted during the second fixing stage when the plastering is completed. Manufacturers will also supply short flights of stairs to be used in conjunction with landings.

Figure 11.10 shows the plan of a variety of staircase configurations available from the manufacturers.

11.5 SITE FIXING OF ASSEMBLED STAIRCASES

As the staircases are delivered to the site almost completely assembled and finished ready to receive decoration it is necessary to treat them as items of finished joinery and to take care not to damage the exposed and finished surfaces.

11.5.1 Site Reception and Storage

The stairs may arrive enclosed in polythene sheeting. This gives some protection, but if the wrapping is torn it may indicate that some surface scuffing has occurred. Carefully unload and stack under cover in a clean storage area on bearers free from grit. Pieces of soft board on the bearers also give added protection against surface scuffing.

11.5.2 Checking the Stairwell for Size (Figure 11.11)

Before moving the staircase into the area where it is to be fixed check the following.

(a) That the finished floor levels are identified and that the total rise between floors is correct for the stairs that are to be fitted.
(b) The length of the stairwell is sufficient to give the necessary head room, that window openings in the stairwell are at the correct levels, and the total going of the staircase is not obstructed by an incorrectly placed door opening.
(c) Check the stairwell walls and trimmers for being square and parallel.
(d) Clean the area of builder's debris.

STRAIGHT FLIGHT ENCLOSED

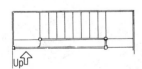
STRAIGHT FLIGHT OPEN STRING

FLIGHT WITH QUARTER SPACE LANDINGS

OPEN WELL STAIR QUARTER SPACE LANDING

HALF TURN STAIR TWO QUARTER SPACE LANDING

HALF TURN STAIR HALF SPACE LANDING

OPEN WELL STAIR QUARTER TURN OF WINDERS (Tapered steps)

Figure 11.10 Range of mass produced staircases plan layout

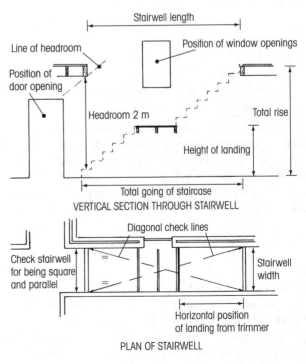
VERTICAL SECTION THROUGH STAIRWELL

PLAN OF STAIRWELL

Figure 11.11 Checks on stairwell dimensions

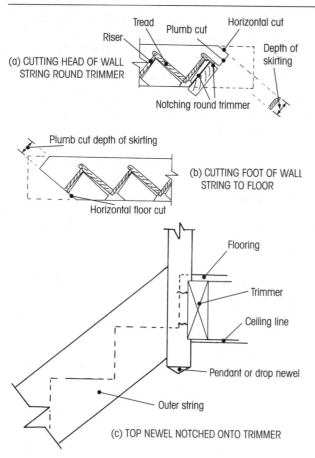

Figure 11.12 Fitting staircase to floor and landings

11.5.3 Cutting the End of the Strings to the Correct Profile

(a) Carry the staircase into the work area and place on batterns or folded dust sheets.
(b) Mark and cut the string where it fits over the landing and abuts the floor below, and cut to the correct vertical profile for the abuttment of the skirting board (Figure 11.12a, b).

Note. Use a fine-toothed saw, and saw from the exposed face so that splintering of the timber does not occur. Dress the cross-grain saw cuts with a block or smooth plane [BENCH AND SITE SKILLS, SECTION 5.4.1] If an open string is being used it may be necessary to notch the newel over the landing trimmer (Figure 11.12c).

11.5.4 Positioning and Fixing the Staircase

When the staircase is being raised into position, make sure the foot of the staircase is placed against a stop (kicker) (Figure 11.13a) which is securely held in position. This will stop the foot of the staircase from sliding out of place while being raised into position at the top landing (Figure 11.13b).

The wall string, after checking that the treads are level along and across their width (Figure 11.13c), is secured to the abutting wall by the use of:

(a) Nails driven into timber plugs, or pallets built into the mortar joints [BENCH AND SITE SKILLS, SECTION 11.6.1].
(b) Use of hardened steel pins driven through the wall string directly into the masonry using a ballistic tool [BENCH AND SITE SKILLS, SECTION 8.19].

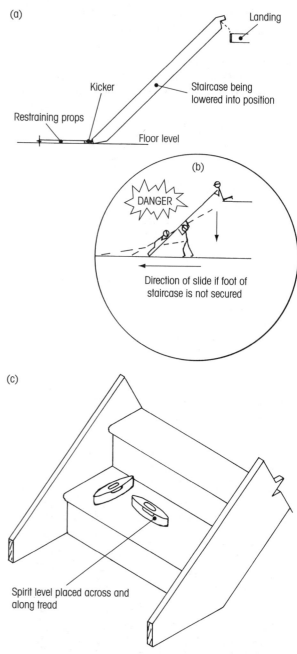

Figure 11.13 Offering upstairs and checking for level: (a, b) safe procedure in the offering up of a staircase; (c) checking tread for being level

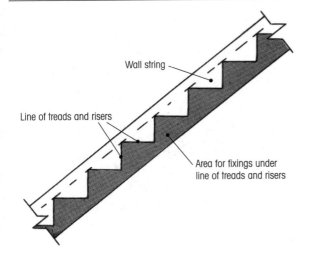

Figure 11.14 Fixing wall string to wall

Whichever fixing system is used, the fixings should be driven through that part of the string that is below the outline of the treads and risers, so that the fixings are hidden from the view of those who use the staircase (Figure 11.14).

When securing the newel to the trimmer the metal fixings should not be visible, and it is usual for the fixings to go through the trimmer into the newel (Figure 11.12c).

11.5.5 Finishing the Foot and Head of a Staircase

At the foot of the staircase the underside of the riser of the first step (which is usually supplied loose) should sit securely on the finished floor line (Figure 11.15a), and it is important when fixing the foot of the wall string that it is positioned accurately at the finished floor level.

Note. When fixing finished joinery the aim must be to work towards a good finish by accurately positioning all members as the work proceeds. It is difficult and costly to rectify mistakes at the end of a process. It is possible to packup or reduce the first riser at the floor

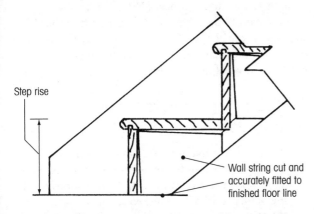

Figure 11.15 Accurate positioning of foot of wall string to finished floor level

level, but the rise of the first step is then altered, and saying any gap will be masked by a carpet is no solution to the poor positioning of a wall string.

As the bottom or first newel in the staircase is subject to a vertical and horizontal force (Figure 11.16a) it can be held in position by fixing a metal dowel into the end of the newel and bedding the projecting end of the dowel into the concrete sub floor or screed of a solid floor (Figure 11.16b).

In the case of a hollow suspended timber floor the newel can be:

- notched over and secured to a joist (Figure 11.16d)
- be dowelled or stub tenoned into the floor decking which is supported by a nogging (Figure 11.16c)
- be skew nailed or screwed through the floor decking into a joist or bearer (Figure 11.16e) making sure the fixings are unobtrusive.

At the top of the staircase the nosing which abuts the floor finish on the landing may be rebated rather than grooved (Figure 11.17) and if this is the case, then the space between the final riser on the staircase and the face of the trimmer will require a couple of packing strips to hold the top or last riser in place. The packings can be in the form of folding wedges.

If it is a neweled staircase with an open well, the exposed face of the trimmers will be masked by an apron lining, nosing and cover mould to form a finish at the edge of the stairwell for the structural framing, floor decking, and ceiling edge (Figure 11.18). The linings, nosings and cover moulds are fixed using oval nails, hidden as far as possible by placing in areas of shade, and punching the nail heads below the surface of the timber.

Careful measuring, accurate cutting and fitting and the absence of hammer marks will lead to a satisfactory finish.

11.5.6 Staircase Carrages

These are fixed down the centre line on the underside of staircases which are wider than 1 metre and subject to heavy traffic. In certain circumstances they are fixed at the lower inside edges of the strings.

The carrage is an inclined beam, which is there to support and stiffen with the use of brackets, the backs of treads and risers, and should be of sufficient cross-sectional size to carry the loads placed upon them.

They should be securely anchored at the top to the landing trimmer and at their foot to the floor structure (Figure 11.19).

11.5.7 Protection of Staircases When in Place

In an ideal world finished joinery, such as staircases, would not be fixed until after all the wet trades (plaster-

(a)

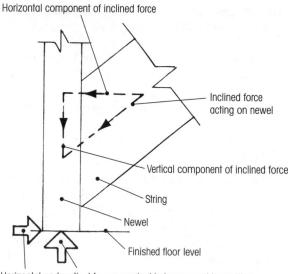

Horizontal and vertical forces required to keep newel in position

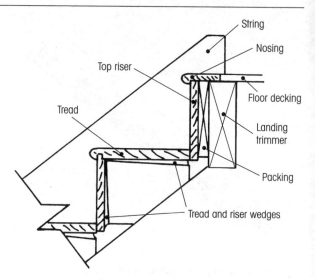

Figure 11.17 Fixing of top riser

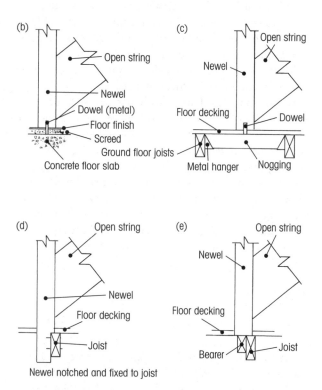

Figure 11.16 (a) Forces acting on a bottom newel; (b–e) principle and method of securing bottom newels to floor

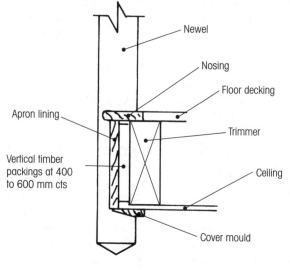

Figure 11.18 Apron lining to stairwell

lining them with cheap sacrificial materials such as chip and hardboard (Figure 11.20a).

Newels are wrapped in heavy duty polythene and encased in chip and hardboard (Figure 11.20b).

If the balustrade and handrail come fixed to the staircase they will also require protection by the use of heavy duty polythene taped in place and then boxed in with hardboard (Figure 11.20c).

11.5.8 Fixing Handrails and Balustrades

(a) When the wet trades have completed their work carefully remove all the protective material from the staircase and clean down, being careful not to tread any loose grit into the finished surfaces.

(b) Take the outer string cappings supplied by the staircase manufacturer, and if they are not precut to

ers) had completed their work, but it seems more customary in domestic house building to put the staircases in during the first fixing stage, and then to fit the handrails and balustrades, if they come as a separate pack, at the second fixing stage.

To protect the finished joinery from damage by heavy footwear, builder's detritus, and the moving about of heavy equipment, the treads and risers are protected by

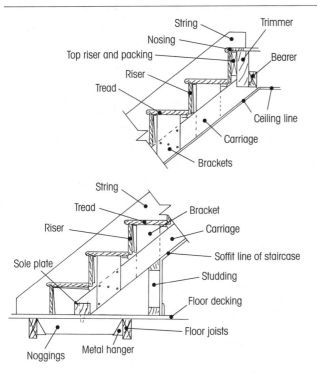

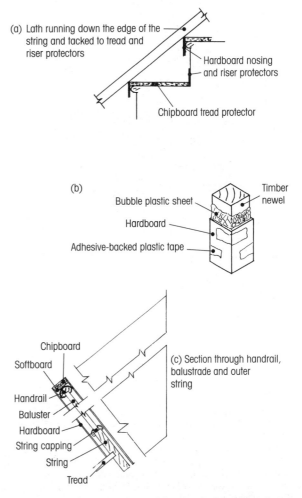

Figure 11.19 Carrage and brackets to a wide staircase

Figure 11.20 Protection of finished staircase joinery

length, measure the inclined distance between the adjacent newels and cut the capping to the correct length and bevel (Figure 11.21a).

Note. If the capping is moulded, making a bevelled mark over template would be useful (Figure 11.21b).

If the capping is morticed to take the ends of the balusters make sure the space between the newels and the first and last baluster are equal and all spaces are less than 100 mm (Figure 11.21c).

Some manufacturers supply a string capping piece with a continuous groove into which the ends of the balusters fit, with spacing blocks fitted to keep the balusters (spindles) in place (Figure 11.21d).

To determine the length of the spacing pieces between the balusters count the number of treads from the centre of one newel to the next and multiply by two, this will give the approximate number of balusters required.

Example: (11 steps × 2) = 22 balusters (see Figure 11.21e). The number of spaces will be one more than the number of balusters 22 + 1 = 23 spaces.

In Figure 11.21(e) subtract from the horizontal distance between the newels (2650 mm), the horizontal space taken up by 22 balusters 40 mm square. That is 2650 − (22 × 40) = 1770 mm. Divide 1770 mm by the number of spaces. 1770 ÷ 23 ≅ 77 mm.

Note. A 100 mm diameter sphere must not be able to pass through the space between

the balusters [*Building Regulations Approved Document "K"*]. This means that for a staircase pitched at 42° the inclined distance between adjacent balusters measured along the top edge of the string should be not more than 134 mm. In Figure 11.21(f) this inclined distance is only about 99 mm.

(c) *Fixing the handrail.* If there is a morticed underrail to carry the top ends of the balusters cut to fit (as described in Figure 11.21a) and secure to the newels by screwing or nailing in position (Figure 11.22a). The hand rail cap can be fixed to the underrail. The tenons on the ends of the handrail, by an adjustment to their profile can be with slight springing slid into the mortices cut into the newels to receive them (Figure 11.22b).

Some manufacturers of staircases do supply proprietary metal fixing brackets to secure the handrail to the newels (Figure 11.22c). If the handrail has a continuous groove on its under side to take the top ends of the balusters, spacing pieces are cut and fitted as described as for the string capping (Figure 11.21).

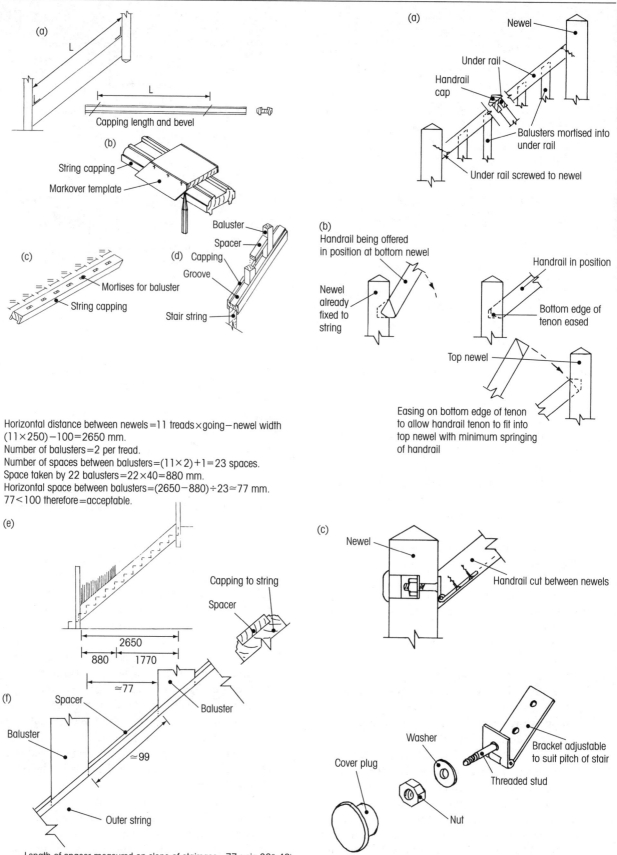

Horizontal distance between newels=11 treads×going−newel width
(11×250)−100=2650 mm.
Number of balusters=2 per tread.
Number of spaces between balusters=(11×2)+1=23 spaces.
Space taken by 22 balusters=22×40=880 mm.
Horizontal space between balusters=(2650−880)÷23≈77 mm.
77<100 therefore=acceptable.

Length of spacer measured on slope of staircase=77÷sin 38° 42'
77÷0·7804=98·66 or 99 mm

Figure 11.21   Spacing and fixing balusters

Figure 11.22   Methods of fixing handrails: (a, b)
fixing under rails and handrails to previously fixed
newels; (c) patent fixing bracket handrail to newel

11.6 SETTING OUT AND MAKING A STAIRCASE

The plans of a building should show the size and position of a stairwell into which the staircase is to be fitted, but before setting out the staircase the stairwell should be checked for size and position (see Section 11.5.2). Figure 11.23 shows the various arrangement of stairwells to accommodate the various types of staircase.

For setting out the stairs you will require:

(a) *A storey rod.* A lath which when held upright gives the vertical height between the finished floor levels and the vertical position of any landings can also be marked (Figure 11.24). This is done in preference to using a rule as these marked positions on the storey rod can be used in determining the number of risers in the staircase.

(b) Checking the stairwell should be carried out as described in Section 11.5.2 and any other dimensions or information you might require to help with the setting out of the stairs recorded in a notebook while taking the site measurements.

11.6.1 Regulations Governing the Setting Out of a Private Staircase

The functions of the Building Regulations Approved Documents B, K, and M, and the British Standards listed in Section 11.1 are to assist in constructing staircases which are safe to use and free from any unnecessary hazards. See Table 11.1 and Figure 11.25(a–i) for the regulations governing the design and construction of staircases.

11.6.2 Materials Used in Private Stair Construction

Stairs constructed of timber should comply with BS1186 Part 1 1986. This standard gives the minimum finished size of members where the treads and risers are fully housed into the strings. Open tread stairs where the tread is not supported by a riser should be of sufficient thickness to resist undue deflection of the tread under load.

Newels attached to outer strings and forming part of the guarding at the edges of landings should have a square cross-section of at least 5626 mm² (75 mm × 75 mm).

11.6.3 Setting Out the Staircase

Figure 11.26 shows the site measurements taken for the staircase including:

• The size of the stairwell opening: 2400 × 970 mm.
• The total rise of the staircase: 2700 mm.
• The total available going: 3150 mm.

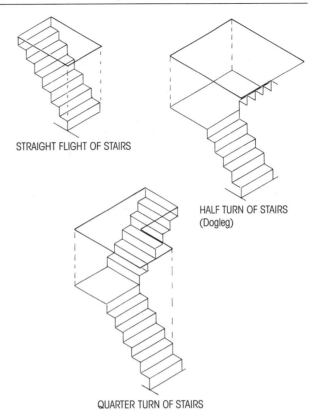

STRAIGHT FLIGHT OF STAIRS

HALF TURN OF STAIRS (Dogleg)

QUARTER TURN OF STAIRS

Figure 11.23   Stairwell configurations

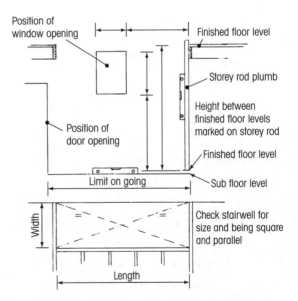

Figure 11.24   Site dimensions to be checked before setting out a staircase

(a) To estimate the number of risers and their height.
    Total rise divided by the maximum step rise for a private stair. 2700 ÷ 220 = 12.27 say 13 risers.
    Step rise 2700 ÷ 13 = 207.692 mm. This value cannot be measured on a joiners rule but using a pair of dividers [BENCH AND SITE SKILLS, SECTION

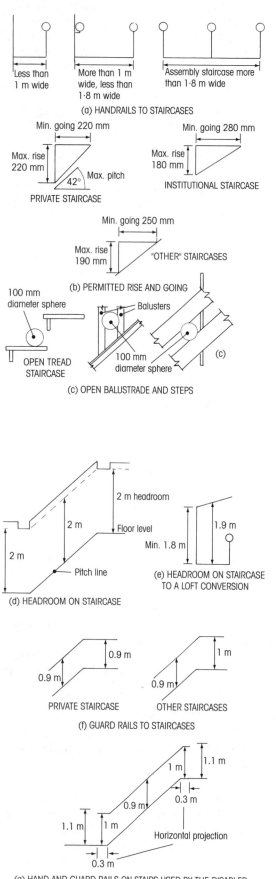

(a) HANDRAILS TO STAIRCASES

Min. going 220 mm
Max. rise 220 mm
42° Max. pitch
PRIVATE STAIRCASE

Min. going 280 mm
Max. rise 180 mm
INSTITUTIONAL STAIRCASE

Min. going 250 mm
Max. rise 190 mm
"OTHER" STAIRCASES

(b) PERMITTED RISE AND GOING

100 mm diameter sphere
Balusters
100 mm diameter sphere
(c)
OPEN TREAD STAIRCASE

(c) OPEN BALUSTRADE AND STEPS

2 m headroom
2 m
Floor level
2 m
Pitch line
(d) HEADROOM ON STAIRCASE

1.9 m
Min. 1.8 m
(e) HEADROOM ON STAIRCASE TO A LOFT CONVERSION

0.9 m
0.9 m
PRIVATE STAIRCASE

1 m
0.9 m
OTHER STAIRCASES

(f) GUARD RAILS TO STAIRCASES

1.1 m
1 m
0.9 m
0.3 m
1.1 m
1 m
Horizontal projection
0.3 m

(g) HAND AND GUARD RAILS ON STAIRS USED BY THE DISABLED

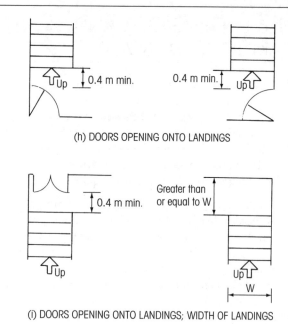

(h) DOORS OPENING ONTO LANDINGS

(i) DOORS OPENING ONTO LANDINGS; WIDTH OF LANDINGS

Figure 11.25 (a–i) Staircase construction regulations

5.2] set to the approximate rise step off along the storey rod, and adjust the dividers until 13 equal divisions are achieved.

(b) To find the step going that will give the staircase a pitch (slope) of not more than 42°.

Step going.   Rise × cotangent 42° (note from tables the cotangent of 42° is 1.1106).
207.692 × 1.1106 = 230.66,
say 231 mm.

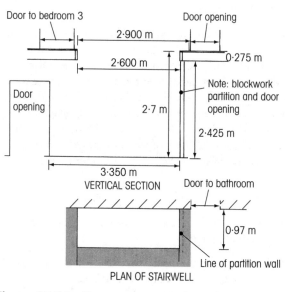

Figure 11.26 Site measurement for staircase

(c) Does twice the rise plus the going give a figure of between 550 and 700?

   $2 \times 207.692 + 231 = 646.384$. Yes.

(d) Will the staircase fit into the horizontal space available? 3150 mm.

   $12 \times 231 = 2772$ mm allow a further 50 mm at the top of the stair giving 2282 mm.

(e) Is there sufficient head room, i.e. 2 m?

   As the head room between the floor and the underside of the ceiling above is 2425 mm, the staircase can rise 425 mm, that is just over two steps. Will the remaining ten steps of the flight come within the length of the stairwell? Stairwell 2400 mm. Going of ten steps ($10 \times 231 = 2310$ mm), as this figure is less than 2400 mm there will be sufficient head room. As a check on the calculation draw a single line elevation of the staircase and the stairwell opening (Figure 11.27).

11.6.4 Templates Required for Setting Out the Staircase

For setting out a staircase four templates are required.

(a) *Margin template*: this gives the parallel distance (margin) between the top edge of the string and the line of the steps (Figure 11.28a).
(b) *Pitch board*: a triangular template giving the rise and going of the step (Figure 11.28b).
(c) *Tread template*: an end profile of the tread and fixing wedge (Figure 11.28d).
(e) *Riser template*: an end profile of the riser and fixing wedge (Figure 11.28e).

Note. The margin template and pitch board can be combined as a single template (Figure 11.28c). If the slope for the wedge on the tread and riser template are made the same it means that only one type of wedge needs to be cut.

Time and care must be taken to make these templates as accurate as possible as they will be used in setting out the staircase.

11.6.5 Setting Out the Stair Strings

The strings are set out in pairs as left and right, or wall and open string, pairing or handing the strings is very important.

(a) Length of strings required, scaled off the drawing or using pythagoras.

   The square root of the total going squared plus total rise squared [BENCH AND SITE SKILLS, SECTION 13.10].

   $\sqrt{(R^2 + G^2)} = \sqrt{(2700^2 + 2800^2)} = 3889$ mm, say 4 m.

(b) The strings are planed straight and true, and the face side and edge of each string is marked and then paired (Figure 11.29a) [BENCH AND SITE SKILLS, SECTION 7.5.4].
(c) Using the margin template set against the face edge draw a parallel line to the top edge of both strings. This is the pitch line (Figure 11.29b).

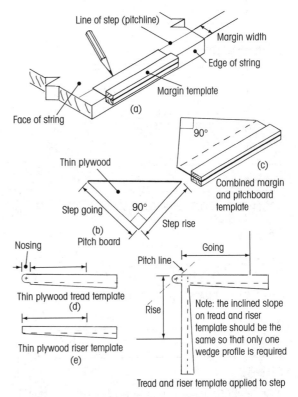

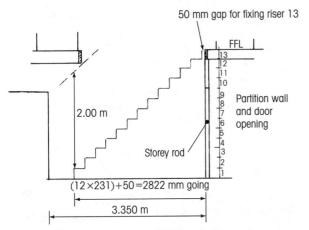

Figure 11.27 Checking staircase dimensions are to regulations

Figure 11.28 Templates required for setting out a staircase

(d) Take a pair of dividers and carefully set to the hypotenuse of the pitch board [BENCH AND SITE SKILLS, SECTION 13.10] (Figure 11.29c) and step off along the pitch line on both strings (Figure 11.29d).

Note. If you use the pitch board alone for marking the distances along the pitch line, inaccuracies will occur.

(e) Using the margin template and pitch board, mark out on both strings the rise and going of each step against the marks made by the dividers (Figure 11.29e).

(f) To cut the bottom of the wall string for the skirting board to butt against, measure up at right-angles to the floor line the height of the skirting board plus 5 mm using the riser edge of the pitch board (Figure 11.29f).

(g) To cut the top of the wall string for the skirting board to butt against, measure up at right-angles from the last tread line the height of the skirting board plus 5 mm using the going edge of the pitch board (Figure 11.29g).

(h) Setting out the tenon at the foot of the outer string to fit into the newel. A bullnose step is being used

at the bottom of the staircase and it is assumed that the centre of the face of the newel will line up with the face of the riser. From the face of the second riser measure back along the line of the second tread half the thickness of the newel, and draw a plumb line using the riser edge of the pitch board. From this line measure forward along the tread line 5 mm, this will give the depth the string is housed into the newel and the shoulder line for the tenon. From this shoulder line again measure forward along the tread line three quarters the thickness of the newel to give the depth of the tenon. Then divide the tenon into two parts with a central haunch. All marking out can be done using the pitch board (Figure 11.29h).

(j) Joint to the newel at the top of the stair on the outer string. From the face line of the last riser measure along the tread line towards riser 12 (Figure 11.29h) half the thickness of the newel, then 5 mm back towards riser 13 to give the depth of housing for the string and the shoulder line for the tenon. From the shoulder line of the tenon mark three-quarters the thickness of the newel to give the depth of tenon. Then divide the tenon into two parts with a central haunch. All marking out lines can be done using the pitch board (Figure 11.29h).

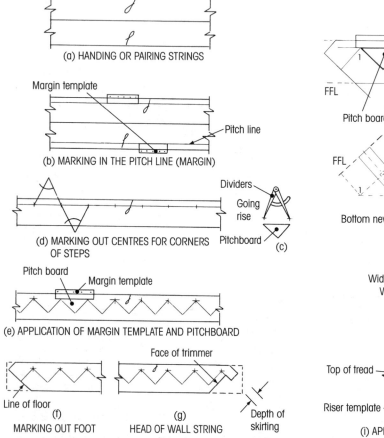

Figure 11.29 showing the various setting out stages:

(a) HANDING OR PAIRING STRINGS

(b) MARKING IN THE PITCH LINE (MARGIN) — Margin template, Pitch line

(c) Dividers, Going rise, Pitchboard

(d) MARKING OUT CENTRES FOR CORNERS OF STEPS

(e) APPLICATION OF MARGIN TEMPLATE AND PITCHBOARD — Pitch board, Margin template

(f) MARKING OUT FOOT — Line of floor

(g) HEAD OF WALL STRING — Face of trimmer, Depth of skirting

(h) WALL STRING — Margin template, Pitch board, FFL, Top newel, Newel thickness, String housed 6 mm into newel, Outer string, Bottom newel, Pitch line, Margin; OUTER STRING — Width of housing in newel for string, Width of mortise in newel for string, Width of newel

(i) APPLICATION OF TREAD AND RISER TEMPLATES — Pitch line, Top of tread, Face of riser, Riser template, Tread template

Figure 11.29 (a–g) Setting out stair strings; (h–i) setting out stair strings

(k) The housings on the strings for the treads and risers can be marked out as shown (Figure 11.29i).

11.6.6 Housing the Strings

The housings can be formed by hand using the techniques illustrated in BENCH AND SITE SKILLS, SECTION 7.7.5. However little or no housing is done by hand, even for single flights of stairs, as the use of electrically powered routers make for a much quicker operation [BENCH AND SITE SKILLS, SECTION 8.15]. Figure 11.30 shows a template for cutting tread and riser housings using a plunge router. Specialist stair trenching machines are used in the batch production of stair strings.

11.6.7 Treads and Risers

Figure 11.31(a) shows a number of ways of jointing the treads to the risers.

In open riser stairs it is necessary to ensure that the gap between the treads will not allow a 100 mm diameter sphere to pass through. To achieve a gap of less than 100 mm a tie bar can be used as it serves the purpose of not only reducing the gap, but also tying the strings together (Figure 11.31b). If tie bars are not used a proportion of the treads would tenoned into the strings and drop or upstand risers used (Figure 11.31c).

116.8 Assembly of the Outer String to the Newels

Figure 11.32 shows the step housings and the general assembly of the string to the newels. It is traditional practice to house the strings into the newels to mask any shrinkage, but a barefaced tenon can be used as an alternative to housing (Figure 11.32).

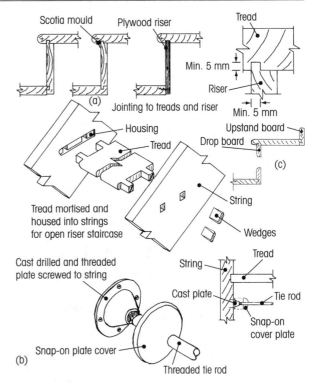

Figure 11.31 Tying strings together in open riser stairs and fixing treads to risers

11.6.9 Marking Out the Newels

When marking out a newel:

(a) the face of the riser is normally on the centre line of the newel, except when winders (triangular steps) are used
(b) the rise of the step is measured vertically up the newel.

These key lines for the marking out of the newel are shown in Figure 11.33.

The four faces of the bottom newel are stretched out to show the marking out for the housings and mortice to receive the treads risers and string in two situations (Figure 11.33a) and Figure 11.33(b) shows the worked newels in isometric projection.

11.6.10 Assembly of Staircases

As much as possible of the staircase is assembled in the workshop, but the amount of assembly will depend upon:

(a) the shape of the staircase
(b) available size of access into a building and the space for manoeuvring the assembled components. It is usual to almost completely assemble straight flights, and for staircases with turns to assemble them in short flights between landings and turns.

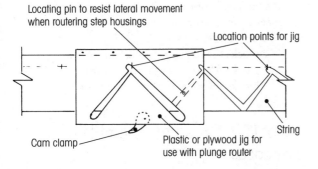

Figure 11.30 Router jig for cutting tread and riser housings

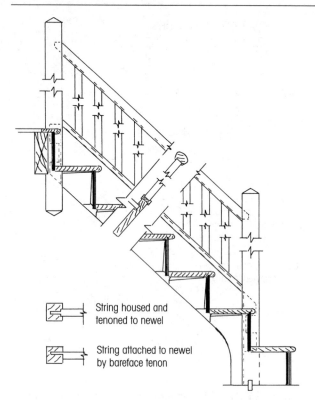

String housed and tenoned to newel

String attached to newel by bareface tenon

Figure 11.32   General construction details of a newel staircase

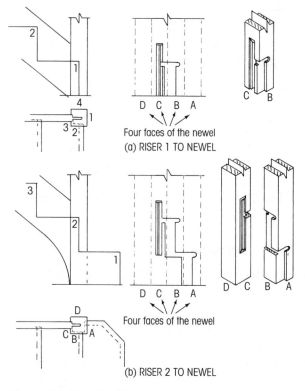

D C B A

Four faces of the newel

(a) RISER 1 TO NEWEL

D C B A

Four faces of the newel

(b) RISER 2 TO NEWEL

Figure 11.33   Marking out and working mortices and housing to newels

Stairs can be assembled flat or on edge to conform to BS585 Part II 1989. In factories specialist stair cramping jigs are used to cramp the components together and hold the staircase square.

11.6.10a Flat assembly of the staircase

(a) Pairs of strings are placed on the cramping frame with the top edge of the strings resting on the bed of the frame (Figure 11.34a).

(b) Apply adhesive to the string housings and tread ends. Slide the treads into position in the string housings and tighten the cramping frame.

(c) Two people working in unison apply adhesive to the wedges, push them into the wedge housings and with a mallet tap the treads into their correct position in the housings so that:
  (i) the tread nosings are correctly fitting into the housings
  (ii) the face of the groove on the underside of the tread to receive the riser is in line with the riser housing in the tread (Figure 11.34b)

(d) then drive the wedges home and cut them back with a chisel clear of the riser housing (Figure 11.34c).
  Note. Treads and risers partially housed into the newels are not usually fixed at this stage.

(e) Apply adhesive to the risers and slide into position, and tap down to make sure they are fully home in the groove cut for them on the underside of the tread (Figure 11.34d).

(f) Drive home the wedges and with a chisel cut off the surplus length as necessary.

(g) Screw the bottom edge of the risers to the back edge of the treads using counter sunk 10 gauge (5 mm diameter) screws of sufficient length to penetrate into the back of the tread at least 23 mm and spaced at a maximum distance apart of 230 mm (Figure 11.34e).

(h) Triangular sectioned glue blocks 75 mm long, cut from 50 mm square timber, are glued and tacked on the underside of the treads and risers where they meet at an internal angle (Figure 11.34f) are placed 150 mm apart.
  2 blocks for steps up to 900 mm wide.
  3 blocks for steps up to 990 mm wide.
  4 blocks for steps over 990 mm wide.

11.6.10b On edge assembly of staircases

One string is placed on the bed of the cramping up frame (jig) (Figure 11.35a) and the pre-assembled steps

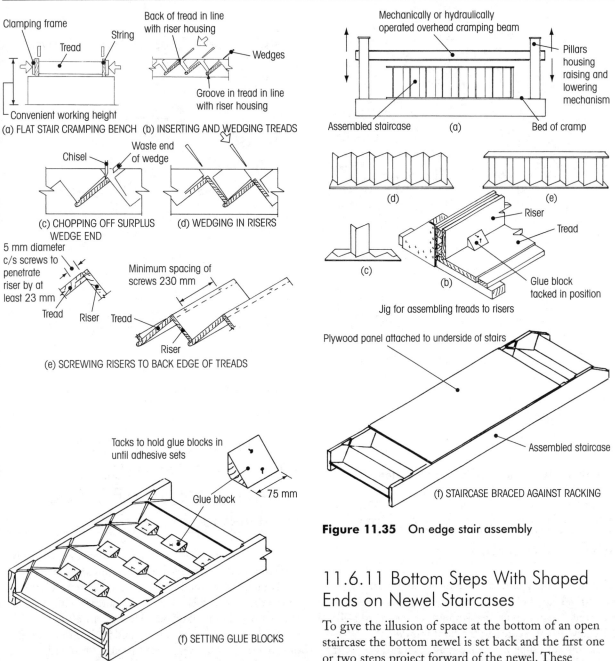

Figure 11.34 Staircase assembly

Figure 11.35 On edge stair assembly

(Figure 11.33b) are glued and placed in the string housings (Figure 11.35c, d). The other string is then positioned on top of the steps so that they fit into the housings cut into the string to receive them (Figure 11.35e), and the cramping frame closed. The treads and risers are then wedged in position as previously described.

When stairs have been assembled and to ensure they are not racked out of square while being moved about, they are diagonally braced, or a panel of chip board or plywood is temporarily attached to the back of the strings (Figure 11.35f).

11.6.11 Bottom Steps With Shaped Ends on Newel Staircases

To give the illusion of space at the bottom of an open staircase the bottom newel is set back and the first one or two steps project forward of the newel. These steps usually have shaped ends, are made separately and built in during the on-site fixing of the staircase, with the shaped end of the step being screwed to the newel.

Splayed end step: the tread and riser are slayed, usually at an angle of 45°. With the riser being mitred, or mitred and tongued and fixed to a backing block. The riser is then pocket screwed to the riser (Figure 11.36a).

Bullnose step (quadrant end): that portion of the riser forming the curve is reduced to veneer thickness of 3 mm, then bent round and glued to a built up block (Figure 11.36b).

'D' or round ended step: the rounded end is formed in the same way as for a bullnose step (Figure 11.36c).

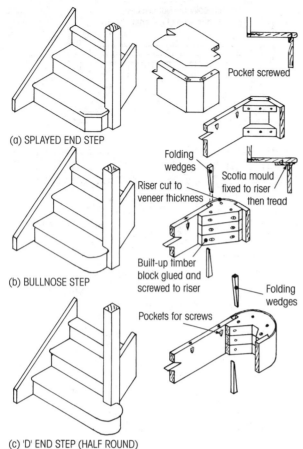

(a) SPLAYED END STEP

Pocket screwed

Folding wedges

Riser cut to veneer thickness

Scotia mould fixed to riser then tread

(b) BULLNOSE STEP

Built-up timber block glued and screwed to riser

Folding wedges

Pockets for screws

(c) 'D' END STEP (HALF ROUND)

Figure 11.36   Shaped ends to the first step

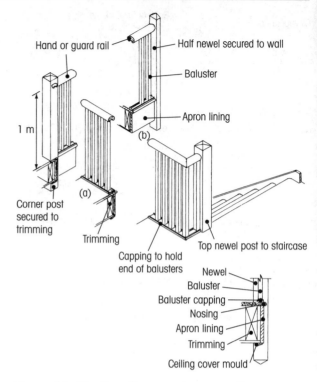

Hand or guard rail

Half newel secured to wall

Baluster

Apron lining

1 m

(b)

Corner post secured to trimming

(a)

Trimming

Top newel post to staircase

Capping to hold end of balusters

Newel
Baluster
Baluster capping
Nosing
Apron lining
Trimming

Ceiling cover mould

Figure 11.37   Guarding round a stairwell

11.7 GUARD RAIL, TO STAIRWELL AND LANDINGS

The balustrade (guard rail) is returned along the unguarded edges of the landing opening onto the stairwell. To give additional strength to the guard rail false newels are secured to the floor framing (Trimming) to support the handrail (Figure 11.37a). If the balustrade finishes against a wall, a half newel will be used and attached to the wall (Figure 11.37b).

11.8 FORMING BULKHEADS IN STAIRWELLS

The length of a stairwell may be reduced while still giving a head room of 2 m by the use of a bulk head to form a sloping ceiling pitched at the same rake as the staircase. Figure 11.38(a) shows the general arrangement for a bulk head. Their use is to provide additional floor space if the stairwell limits the size of a room. The isometric sketch in Figure 11.38(b) shows the arrangement of the framing in a bulk head.

11.9 POSITIONING AND FRAMING INTERMEDIATE QUARTER AND HALF SPACE LANDINGS

Once the story rod has the height of the risers correctly marked out, it can be used to determine the height of the landing above the lower floor level. Figure 11.39(a) shows the plan and elevation of a staircase with the landing position marked out. The structural part of the landing can then be framed and built into the wall or supported on hangers [BENCH AND SITE SKILLS, SECTION 11.5.4]. One edge of a quarter space landing can be supported by a dwarf support wall (Figure 11.39b), or one corner can be attached to a storey post (a newel rising up the height of the staircase (Figure 1.39c).

11.10 A STAIRCASE CONTAINING TAPERED STEPS (WINDERS)

Before setting out this type of staircase reference must be made to the Building Regulations Approved Documents A, D and K, the sections dealing with tapered steps (winders) for staircases less than 1 m wide. This must conform to the following:

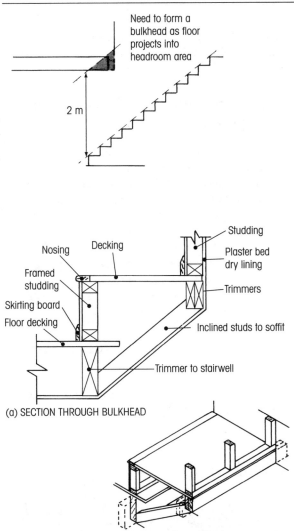

(a) SECTION THROUGH BULKHEAD

(b) ISOMETRIC SKETCH OF BULKHEAD FRAMING

Figure 11.38 Bulkhead construction

(a) The rise of the parallel and tapered steps must be the same.

(b) The narrow end of the tapered step must not be less than 50 mm wide.

(c) The going of the tapered step measured on its centre line must not be less than the going on the parallel steps.

(d) Twice the rise of the step plus the going measured on the centre line should not be less than 550 or more than 700 mm.

(e) Meet the general requirements of Approved Document K paragraphs 1.1–1.5.

Figure 11.40(a) shows the setting out of a quarter turn staircase with three winders. Figure 11.40(b) shows how the minimum width of the tapered steps are determined. Figure 11.40(c, d) shows how the treads and risers on the lower and upper wall strings are developed from the plan and how the strings are built up to allow the easings on the strings (see Figure 11.44) to be formed. Figure 11.40(e) shows the single line setting out for the treads and risers which are housed into the newel. The isometric sketch (Figure 11.40f) shows how the wall strings are jointed at their intersection.

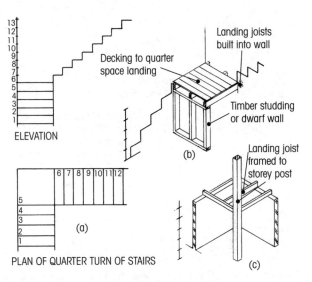

Figure 11.39 Staircase quarter space landings

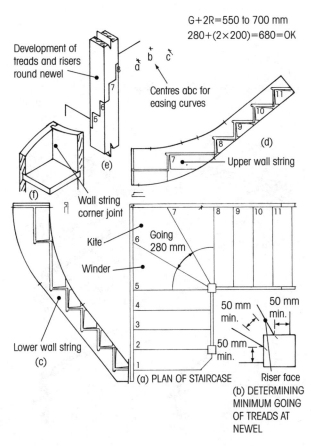

Figure 11.40 Setting out quarter turn of winders

11.11 GEOMETRICAL STAIRCASES

This term is given to open well staircases where the outer string, cut or enclosed runs as a continuous ribbon from the bottom to the top of a staircase having one or more quarter or half turns (Figure 11.41).

Note. Traditional practice was to use cut strings (Figure 11.9b) in geometrical staircases.

A geometrical string is made up of a number of sections with the curved parts of the string being formed, then jointed to the straight strings (Figure 11.41).

11.11.1 Jointing Continuous Stair Strings

The jointing method will depend on

(a) How the curved portion of the string is formed.
(b) The thickness of the string.
(c) The decorative finish: paint or clear varnish.

For thick closed strings handrail bolts can be used [BENCH AND SITE SKILLS, SECTION 11.4].

Use of counter clamps [BENCH AND SITE SKILLS, SECTION 7.2.8] (Figure 11.41b).

If the curved portion of the string is formed by veneering, a halving or bridle joint can be formed

[BENCH AND SITE SKILLS, SECTIONS 7.51 AND 7.54] (Figure 11.41c, d). As an alternative the bridle joint can be modified to give the appearance of a butt joint on the top edge of the string (Figure 11.41e).

11.11.2 Forming the Curved Portion of the String

(a) *Staving*: the string thickness is reduced on the curved portion to a veneer thickness of 3 mm onto which is glued a series of staves tapered in their cross-section to form a core which is then covered, if exposed to view with a second veneer (Figure 11.42a).

(b) *Veneering*: the veneers are coated with adhesive and built up to the required thickness of the string. A vacuum is formed round a former to the curve required, by placing in a flexible bag before extracting the air (Figure 11.42b).

11.11.3 Finding the Curved Shape of the String

This is probably the most difficult part and requires some knowledge of solid geometry (orthographic

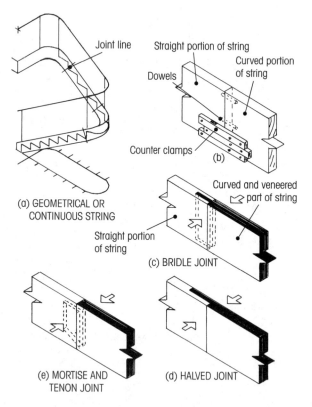

(a) GEOMETRICAL OR CONTINUOUS STRING

(b)

(c) BRIDLE JOINT

(e) MORTISE AND TENON JOINT

(d) HALVED JOINT

Figure 11.41   Continuous string jointing methods

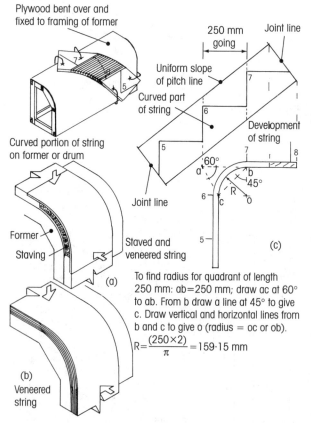

To find radius for quadrant of length 250 mm: ab=250 mm; draw ac at 60° to ab. From b draw a line at 45° to give c. Draw vertical and horizontal lines from b and c to give o (radius = oc or ob).

$$R = \frac{(250 \times 2)}{\pi} = 159 \cdot 15 \text{ mm}$$

Figure 11.42   Developing the shape and forming a curved string

projection) [BENCH AND SITE SKILLS, SECTION 12.4.1] as the nosing (falling) line of the staircase falls or rises as it goes round the curve.

Probably the most pleasing nosing line is formed when the pitch of the stair remains constant as it turns round the curve (Figure 11.42). It may not always be possible for this to occur, as the going may change if tapered steps are used at the turn of the stair rather than a landing.

In the following example see (Figure 11.42) a continuous string turns through 90° at a quarter space landing with a step going of 250 mm and a rise of 200 mm. To maintain a uniform nosing line the length of the quadrant (quarter circle) must be equal to the step going.

Note. If the quadrant length of the curve has to be 250 mm then the radius for that quadrant will be 250 mm = $2\pi R \div 4$. Then radius $R = 250 \times 2 \div \pi = 159.15$ mm. For all practical purposes the value of the radius would be considered to be 159 mm.

Figure 11.42c shows how the radius may be found by the use of geometry.

(a) Draw a line *ab* equal to the length of the going 250 mm.
(b) From *a* draw a line downward at 60° to line *ab*.
(c) From *b* draw a vertical line downwards and an inclined line at 45° to *ab*.
(d) Where the 60° and 45° line meet at *c* draw a horizontal line to *o*.
(e) Distance *ob* will be the radius for a quadrant of 250 mm.

In a further example (Figure 11.43) using the same rise and going of 200 and 250 mm for the steps a 40 mm thick closed string rises up to a half space landing turns through 180° to rise to a higher landing. The width of the stairwell is 360 mm and the centre line of the string has a radius of 200 mm with the radius on the step side

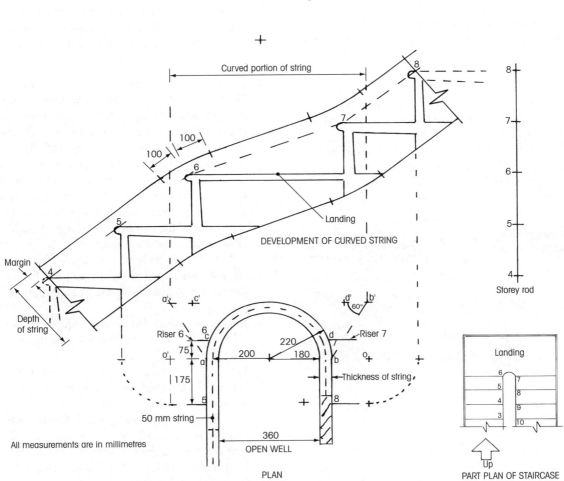

Figure 11.43   Development of a continuous string. A half turn landing

of the string of 220 mm (Figure 11.43). The distance *ac* and *bd* in the plan is the distance risers 6 and 7 are set back from the springing line of the semicircle. This distance is 75 mm. To develop the shape of the curved portion of the string.

1. Draw the plan of the staircase as shown in Figure 11.43.
2. To the right of and above the plan in elevation set up a storey rod showing four risers.
3. Stretch out the plan semicircle *ab* in elevation using the method shown in Figure 11.42(c).
4. To find the position of the face of riser 6 on line *a'b'* step off *ac* round the curve in plan and mark off this distance from *a'* at *c'*. For riser 7 measure in plan round the curve *bd* and mark off this distance as *d'b'* in elevation.
5. To find the position of riser 5 rotate the distance *o'5* about *o'* and project into elevation and the same for riser 8.
6. Having established the outline of the steps in elevation draw the pitch (nosing) line through the points where the tread and riser lines intersect, i.e. points 4, 5, 6, 7 and 8.
7. Draw in the top edge of the string by drawing lines 40 mm above but parallel to the pitch line.
8. The depth of the string 250 mm is measured square off the line that represents the top edge of the string.
   Note. The line of the string looks angular at the differing pitch line intersect, and it is necessary to curve (ease) one line into the other to form what is termed an easing.
9. At *a* in Figure 11.44 on the intersection of the changes of slope on the top edge of the string mark out point *b* and *c* equal distance from *a* (in this case 100 mm, Figure 11.43) from point *b* draw a line at

right-angles to the pitch line and from *c* do the same, extend these two line until they intersect (meet) at *d*. Which will be the centre from which a curve tangential to both pitch lines can be drawn to form the easing [BENCH AND SITE SKILLS, SECTION 13.5.3].

10. The housings for the treads and risers can now be set out.

Note. The housing for tread 6 which is the landing is a parallel housing as no wedges are required. In setting out the slope for the wedges they should be the same for both treads and risers, or you would have to cut two different types of wedge.

11.11.4 Handrail to a Geometrical Stair

Determining the shape of the handrail as it falls around a quadrant or semicircle as the handrail changes direction needs some understanding of solid geometry, as a geometrical staircase cannot be considered without reference being made to the handrail. The simplest form of curved and twisted handrail will only be considered. This is when the curved and twisted part (wreath) of a handrail rakes upwards or downwards and turning throughout 90° to a level rail, known as '*a rake to level wreath*'.

1. Draw the plan of the staircase shown in Figure 11.46. The stairwell is 360 mm wide between the strings which turn through 180°. The steps have a rise of 200 mm and a going of 250 mm. A handrail with a width of 80 mm and a depth of 60 mm with its centre line following the centre line of the string as it turns round the semicircle, gives the centre line of the handrail a radius of 200 mm. Figure 11.45(a) shows the inclined plane (surface) on which the centre line of the handrail lies as it turns and falls, or rises round the quadrant.
   Note. The centre line of the handrail in setting out is always shown laying on the pitch line (nosing line) of the steps forming the straight flight of stairs.
2. In plan (Figure 11.46) draw the square *abcd* with sides equal to the centre line radius of the curve (200 mm).
3. In elevation set up a storey rod for steps 4, 5, 6, 7 and 8. Step 6 is the landing and should be drawn in elevation above the plan, 100 mm above the landing draw line *ef*. This height is the difference in height between the handrail 900 mm on the pitch of the stair and the guardrail (handrail) 1000 mm on the level landing. 1000 − 900 = 100 mm (the difference in height). Project vertical lines from *d* and *c* in plan to meet *e* and *f* in elevation.
4. From point *e* in elevation draw the pitch line to

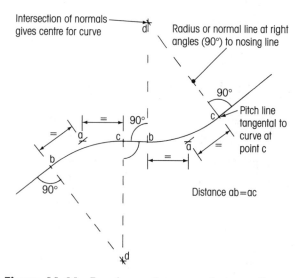

Figure 11.44 Forming easing curves between lines at different pitches

intersect with the landing line at 6″. From this point draw a vertical line downwards to intersect with extended line *dc* in plan at 6′. Set the compass point in *d* and with distance *d6′* scribe a quadrant finishing at point 6 on plan line *ad*. This is the position of the landing riser 6 in plan. In plan the positions of risers 5, 7 and 8 can then be positioned and the elevation of steps 5, 6, 7, and 8 drawn in.

5. In plan, with compass point at *d* and radius *da* strike an arc to intersect on extended line *dc* at *a″* and project vertically into elevation to give point *a‴* on the pitch line. From points *e* and *a‴* draw two lines at right-angles to the pitch line. With compass point at *e* and radius *ef* draw an arc to terminate at *f′*. From *a‴* mark off the same distance *ef′* to give *b″* and complete the rectangle *a‴ b″ f′ e* which is the true shape of the inclined surface of the square prism and on which the falling line of the wreathed part of the handrail sits (Figure 11.45a).

6. Project the plan width of the handrail vertically into elevation until the pitch line is intersected at *g* and *h* from points *g* and *h* project lines at right angles to

the pitch line to intersect line *b″ f′* at *i* and *j*. On line *b″a‴* mark either side of *a‴* half the plan width of the handrail to give points *k* and *l*. Points *ki* and *lj* have to be joined by elliptical curves.

7. To draw the elliptical curve joining *l* to *i* (see Figure 11.45b for general principles for drawing an elliptical curve), *b″* is the centre of the ellipse and *b″j* is half the major axis while *b″l* is half the minor axis. Mark on a straight edge (trammel) the distance of half the major axis *b″j*, turn the straight edge through 90° and place point *b″* on the straight edge at point *b″* and mark on the minor axis at *l* on the straight edge (Figure 11.46). Keeping *j* and *l* on the minor and major axes respectively while rotating the straight edge, mark the line of the elliptical curve from *b″* (Figure 11.46). Repeat for the elliptical curve joining points *k* to *i*. When the two elliptical curves are drawn we have a pattern (**face mould**) for cutting the timber blank from which the handrail wreath is formed.

8. For the purposes of joining the wreath to the straight handrail a shank length of between 75 and 100 mm is added. See the face mould (Figure 11.46).

9. As the wreathed part of the handrail twists as it turns the bevel, the risers make with the pitch line, in this case, gives the angle of twist or (**twist bevel**) Figure 11.47(a).

10. To find the required thickness of the blank from which the wreath is cut (Figure 11.47a), draw a cross-section through the handrail and at *a* draw the pitch line. From *c* draw another line parallel to the pitch line. This if measured between points *d* and *e* will give the required thickness.

11. Figure 11.47(b) shows the cut out handrail blank with a face moulds placed on the top and bottom of the blank. The depth of the blank can also be seen with the twist bevel marked in and the handrail sections drawn in their relative positions on the ends of the blank.

Note. To make it easy to set out the positions of the handrail sections on the blank, a template of the cross-section of the handrail is cut-out of thin plywood or tinplate with a small hole at the centre. Using a pin the template can be centred on the blank and rotated until centre lines on the blank and template line up, and the outline of the section can be drawn in by tracing round the template with a pencil.

12. To determine the position of the curves giving the required twist to the wrath, draw the pitch line (twist bevel) on the edge of the blank at *df* passing through the centre point *o* (Figure 11.47b). Moving the top face mould forward to *d* and the bottom face mould backwards to *f* will give the curves on the blank for cutting the required twist.

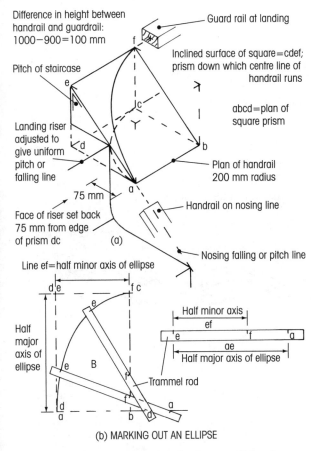

Figure 11.45 Isometric sketch of centre line of a handrail running down the inclined surface of a square prism

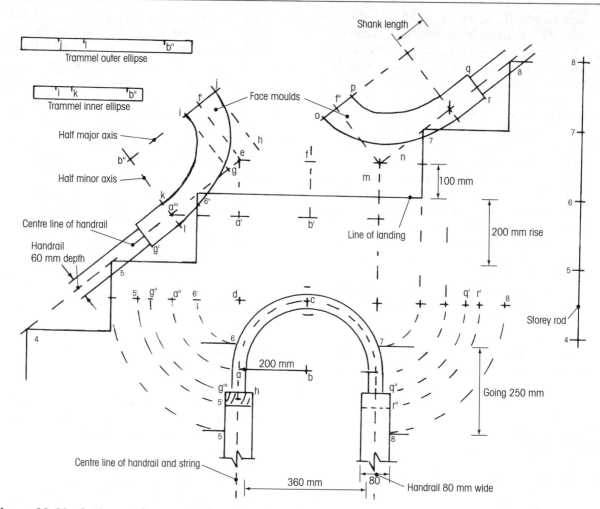

Figure 11.46 Setting out face moulds for rake to level handrail wreath

13. Once the two twisted surfaces on the edge of the wreath have been worked the top and bottom faces of the wreath can be made square with the edges.

To give a pleasing look to the downward curving and twisting around the quadrant turn, and if you do not trust your eye, an inner and outer falling mould can be developed.

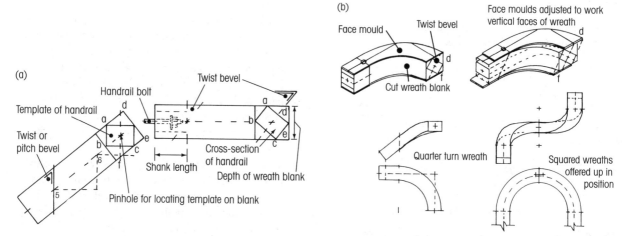

Figure 11.47 Application of face mould to the blank: (a) handrail wreath blank thickness and twist bevel and (b) squaring the wreath

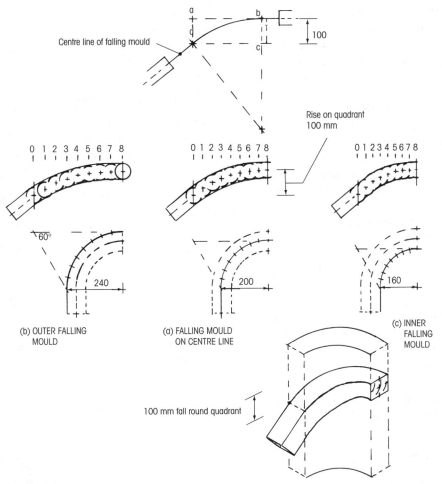

Figure 11.48 Falling mould development

11.11.6 Inner and Outer Falling Moulds for the Handrail Wreath

1. With reference to Figure 11.48 draw the plan of the handrail as for developing the wreath face mould and stretch out the centre line quadrant *0.8* to *0.8'* in elevation as shown in Figure 10.48(a). Draw the rectangle *abcd* with *bc* being 100 mm (the difference in height between the handrail on the slope of the staircase and on the level landing). Through point *d* draw the pitch line of the staircase. Scribe an arc to

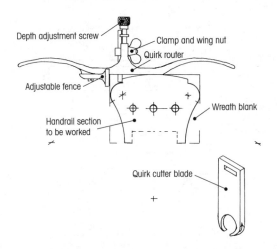

Figure 11.49 Quirk router for working quirks on wreath

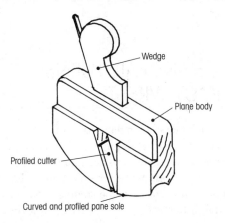

Figure 11.50 Thumb plane for working moulding on wreaths

be tangential to the pitch line at *d* and tangential to line *ab*. Divide the rectangle *abcd* into eight vertical and equal strips, and using as centres the points where the vertical lines touch the curved line scribe circles equal in diameter to the handrail thickness. Draw a curved line tangentially to the top and bottom of the circles to give the falling mould for the centre of the handrail.

2. Having established the falling mould for the centre line of the handrail the falling mould for the inner and outer curves of the handrail are required. Stretch out the quadrant lengths as shown and lay out the rectangles as the elevations in Figure 11.48(b, c), and divide the two developed rectangles into eight equal parts. From the determined centres on the central falling mould project the points horizontally across to the rectangles *b* and *c* containing the outer and inner falling moulds to intersect with the vertical lines dividing the rectangles, and at these points of intersection scribe circles of diameter equal to the thickness of the hand rail. Draw a curved line tangentially to the top and bottom of the circles to give the inner and outer falling mould for the handrail wreath.

3. Once the vertical faces of the wreath have been worked the two falling moulds can be applied and eyed in to see if they give a pleasing falling line.

11.11.7 Working the Moulds on the Squared Wreath

Once the wreath has been squared up the actual moulding of the wreath can be worked. To give guidance and limit to the curved portions of the mould on the handrail a series of small quirks (grooves) are worked using a quirk router (Figure 11.49) which follow the twist and curve of the wreath. When the quirks are cut the mouldings can be worked using thumb planes (Figure 11.50). As an alternative small hand held routers can be used for cutting the quirks and mouldings.

11.12 HANDRAIL TURNS WITHOUT WREATHS

Provided the handrail turn remains in the horizontal plane and the sloping handrail is eased into the horizontal there is no need for a wreath. However, some adjustment of the position of the risers as they come onto and leave the landing is necessary if the handrail is to remain at the correct height above the landing and the nosing line of the steps. Figures 11.51 and 11.52 shows how the riser of the step is adjusted in relationship to the edge of the landing.

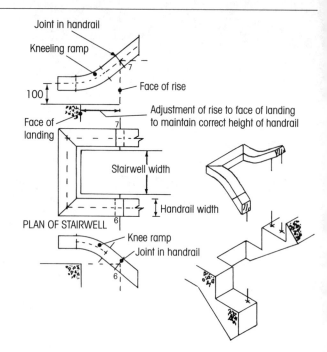

Figure 11.51 Continuous handrail square turn with ramps and easings

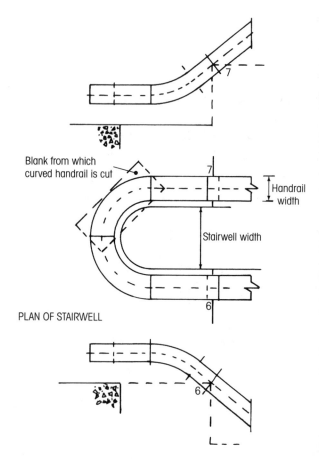

Figure 11.52 Continuous handrail quadrant turn with ramps and easings

12
Joinery fitments

12.1 INTRODUCTION

As opposed to loose furniture bought by the customer, joinery fitments can be regarded as those items of furniture which require fixing in place to a floor, wall or ceiling. They are commonly associated in the home as:

- kitchen fitments
- bedroom fitments
- bathroom fitments.

Joiners are also required to fit-out offices, workshops, and warehouses with a variety of fitments. Commercial premises such as retail shops and banks, etc. usually employ specialist shopfitters or joinery firms who specialise in purpose-made items.

For the purpose of this chapter we can divide fitments into the following:

(a) *Unit fitments* – these are modular units of uniform depth and varying widths, usually described as:
 (i) Kitchen fitment units – floor units (base units, sink units, corner units, appliance housings).
 (ii) Wall units (corner units, overhob bridging units).
 (iii) Bedroom fitment units – wardrobe units, cabinets (corner units), shelving.
 (iv) Bathroom fitment units – floor units (vanity units).
 (v) Wall units – (mirrored cabinets, shelved cabinets).

 They can be individually produced or manufactured on a mass production basis, then sold on to the installer as a flat-packed kit or partly assembled unit. Or partly pre-built in the workshop by the joiner to be fitted later in situ.

(b) *Built-in fitments* – these often consist of a pre-made front frame and possibly side frames or panels, they rely fully or partly upon their attachment to a floor, wall, or ceiling for stability of the fitment. A good example would be the use of a house wall as a backing when attaching a series of units and/or shelves to it.

12.2 DESIGN CONSIDERATIONS

The main advantage of built-in fitments and purpose-built units over mass-produced manufactured units, is that end-user requirements, especially for people with special needs or disabilities, can more easily be implemented at the design stage.

Before starting work the following considerations should be taken into account:

- Fitment height should suit the user.
- Work surfaces, cupboards, and shelving in regular use should be easily accessible.
- Wall cupboard doors – where there is a risk of walking into an opened door, these should, where practicable, be made to slide.
- Tall units (cupboards, housing units, wardrobes, etc.) should be securely fastened back to the wall and/or ceiling to prevent them toppling over – such units are particularly at risk when all the hinged doors are open. (There has always been this danger with some tall free-standing wardrobes, particularly if they have heavy mirrored doors.)
- Unit width and length to accommodate a worktop housed sink.
- Unit width and length to accommodate a worktop housed vanity basin.
- Provision for housing electrical and/or gas appliances.
- Provision behind the base unit backing or plinth so as not to interfere with the existing skirting board.
- Provision behind the base unit backing to accommodate any services (water, gas, waste pipes, and cables, etc.).
- Type and method of construction.

Figures 12.1–12.3 can be used as a guide to sizing the various styles of domestic fitments. These will however vary slightly between manufacturers.

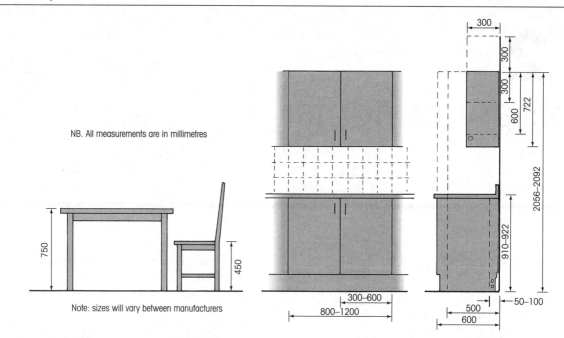

Figure 12.1 Guide to the dimensions of kitchen fitments

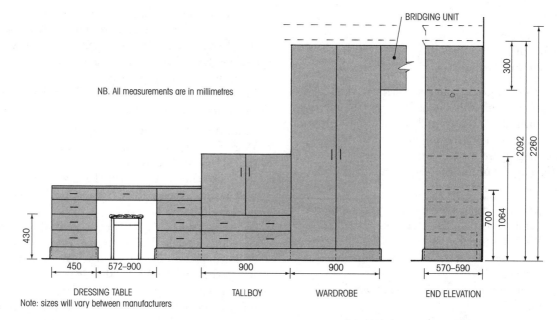

Figure 12.2 Guide to the dimensions of bedroom fitments

12.3 FLOOR STANDING BASE UNITS

These units either provide cupboard space, house drawers or a combination of both. Provision can also be made to house sinks, and basins and appliances such as cooker hobs, ovens, and fridges or hanging clothes (wardrobe). As with most proprietary products, the method of construction and materials used will almost certainly reflect the cost and quality of the end product.

12.3.1 Carcass construction

Carcass work usually refers to the main framework of the unit as a whole, excluding doors and drawers. The cornice and plinth may be built separately or attached to the carcass at a later date as an add-on.

Figure 12.4 shows how construction methods can vary, and these can be named accordingly as:

- boxed or rail and slab construction
- front frame and slab construction
- front and back framed skeleton construction
- cross-framed skeleton construction.

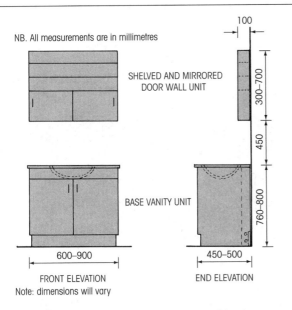

NB. All measurements are in millimetres

100

SHELVED AND MIRRORED
DOOR WALL UNIT

300–700

450

BASE VANITY UNIT

760–800

600–900

450–500

FRONT ELEVATION
Note: dimensions will vary

END ELEVATION

Figure 12.3   Guide to the dimensions of bathroom fitments

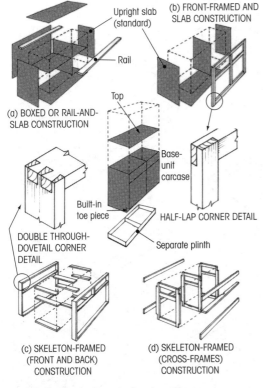

Upright slab
(standard)

(b) FRONT-FRAMED AND
SLAB CONSTRUCTION

Rail

Top

(a) BOXED OR RAIL-AND-
SLAB CONSTRUCTION

Base-
unit
carcase

Built-in
toe piece

HALF-LAP CORNER DETAIL

DOUBLE THROUGH-
DOVETAIL CORNER
DETAIL

Separate plinth

(c) SKELETON-FRAMED
(FRONT AND BACK)
CONSTRUCTION

(d) SKELETON-FRAMED
(CROSS-FRAMES)
CONSTRUCTION

Figure 12.4   Variations in base unit carcass construction

Boxed or rail and slab construction (Figure 12.4a): most unit manufacturers favour this method of construction. Base, ends, and intermediate upright panels and shelving are all cut from manufactured boards – usually 15–18 mm thick plastics faced chipboard. Modular sizes often reflect standard metric sheet sizes. Unit backs, which hold the unit square and either made from the same material, or pre-finished hardboard or MDF.

Interconnecting joints can in some cases be made by traditional means, but some manufacturers prefer to use knock-down fittings that enable the whole unit to be packed flat for on-site assembly. All that is then required is an ability to understand simple working drawings and the use of a screwdriver, hammer, and soft mallet. Originally these knock-down fittings were only available to the manufacturers of assembly-line fitments, but alternatives are now freely available from most hardware suppliers.

The plinth (or toe piece) may form part of the carcass, or a separately built unit, but it will more likely be an add-on board, which would be attached to the front of a series of height adjustable legs.

Front frame and slab construction (Figure 12.4b): again, ends and intermediate upright panels are made from manufactured boards [BENCH AND SITE SKILLS, SECTIONS 2.3–2.5.6]. A back piece is optional. The front frame is framed-up from small timber sections (say 25×50 mm) jointed by mortice and tennons and/or halving joints [BENCH AND SITE SKILLS, SECTIONS 7.5.4–7.5.7].

Front and back framed skeleton construction (Figure 12.4c): front frame (face edge to the front) and back frames (face side or edge facing forward) are joined to cross-rails which serve as shelf bearers and/or drawer runners. A back piece is optional.

The plinth is usually built separately.

Cross-framed skeleton construction (Figure 12.4d): framed uprights serve as shelf bearers or supports for drawer runners. They are positioned to suit door widths or drawer lengths and are notched to receive front and back runners.

The plinth may be separate, but is more likely to be formed by notching the cross-frames back to receive a toe piece.

Note. End panels and divisions (standards/uprights) as shown in Figure 12.5 may consist of edged or lipped manufactured boards, open or panelled frames, or framed flush panels.

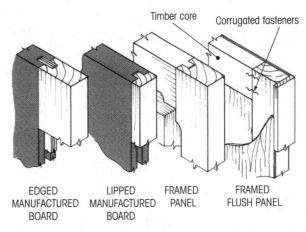

Timber core   Corrugated fasteners

EDGED
MANUFACTURED
BOARD

LIPPED
MANUFACTURED
BOARD

FRAMED
PANEL

FRAMED
FLUSH PANEL

Figure 12.5   Fabrication and treatment of end panels and divisions

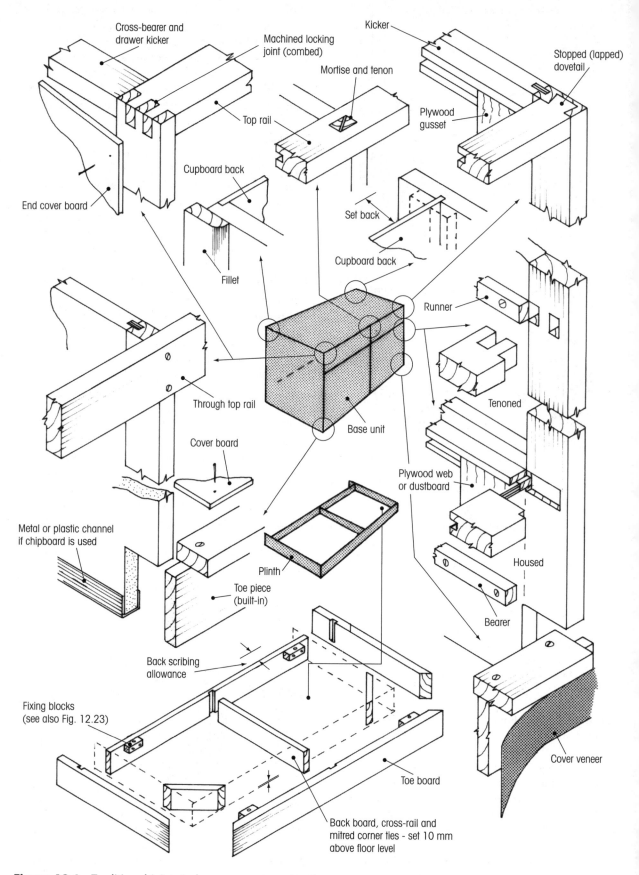

Figure 12.6 Traditional joints in base carcass construction

12.3.2 Joints Between Members in Carcass Construction

[BENCH AND SITE SKILLS, SECTION 7.5.]

Figure 12.6 shows a few examples of how traditional joints between rails, slabs and frames can be made. An alternative method would be to use proprietary knock-down connectors. Unfortunately, as seen in Figure 12.7 when viewed from inside the unit, some of these connectors can be unsightly.

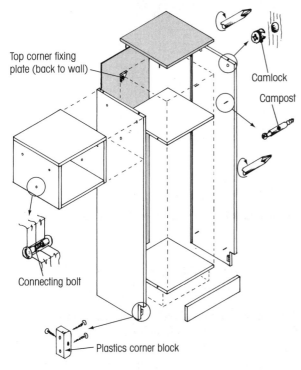

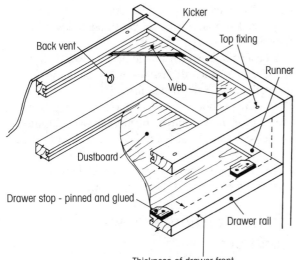

Figure 12.7   Tall base unit of knock-down assembly and hardware

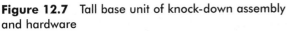

Figure 12.8   Traditional drawer opening

12.4 WOOD DRAWERS

Drawer construction is a skill in its own right. In the joinery trade it has always been regarded as a major achievement when an apprentice manages to make and fit a drawer successfully by using traditional materials and techniques.

12.4.1 Drawer Openings

The opening should match the type of drawer. For example, a traditional drawer opening, like the one shown in Figure 12.8 would house the drawer front on all of its four edges.

If the drawer recess is to be enclosed by a solid top, sides, and a dust tray, then the action of closing a well-fitted drawer and thereby compressing the air behind it, will, unless the drawer back is lower than its sides or a vent hole is cut into the back of the recess, be difficult to close.

In all cases, the run of the drawer is checked by glueing and tacking two hardwood or plywood stops. They are positioned near to the ends of the drawer rail and set in from its front edge by as much as the thickness of a drawer front.

12.4.2 Drawer Fronts

The arrangement of various types of drawer front is shown in Figure 12.9 – as can be seen the drawer edges

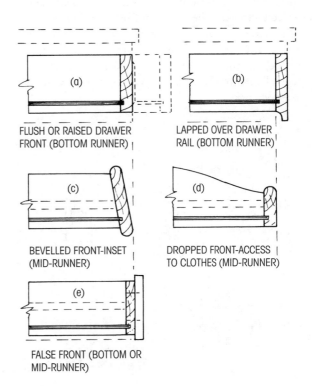

Figure 12.9   Types of drawer front

of types (a) and (b) are either fully or partly housed within the drawer recess. With single drawers, the overhang below the drawer sides can be cut away to form a drawer pull. The drawer front at (c) is sloping and drops below the sides; therefore the drawer rail can be omitted or concealed – the runners in this case are inset mid-way into the drawer sides. The drawer at (d) has a drop front to allow easy access to clothing, etc., and is often found within a wardrobe.

However the drawer front shown in Figure 12.9(e) unlike the others, is not an integral part of the drawer construction but merely a false front, and when fixed in place it laps the uprights of the main carcass. These, like the doors, can therefore be interchangeable, giving the customer a choice of finish whilst the main carcass remains a standardised item.

12.4.3 Drawer Runners

Traditionally, as shown in Figure 12.8 and 12.10(a) draw runners were built into mainframe carcass, and the bottom of the drawer sides slid over them. This meant that wood was being constantly run over wood, to reduce wear and to help movement candle wax was, and is still, used as a lubricant. Even so, with drawers which are in constant use it is inevitable that wear will take place not only to the runner but also where the drawer runs over its front rail.

Drawers these days are more often as not suspended from a runner as opposed to running on it. But many of these are not very satisfactory, for example a hardwood mid-runner as shown in Figure 12.10(b) can also be subjected to wear, the plastic mid-runner (Figure 12.10c, d) can have restrictive movement particularly with heavier or laden drawers. Metal runners (Figure 12.10e) running on a ball race or roller system are generally the modern answer to achieve an easy trouble-free operation.

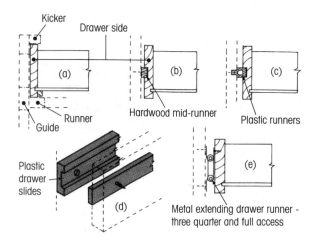

Figure 12.10 Drawer runners

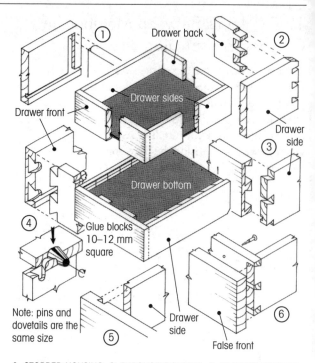

Note: pins and dovetails are the same size

1: STOPPED HOUSING. 2: THROUGH DOVETAIL. 3: STOPPED DOVETAIL (lapped). 4: MACHINED DOVETAILS. 5: DOVETAILED HOUSING. 6: THROUGH DOVETAILS AND FALSE FRONT.

Figure 12.11 Joints used in drawer construction

12.4.4 Joint Used in Drawer Construction (Figure 12.11)

Traditionally drawer sides to a non-over hanging front will be dovetailed (either cut by hand [BENCH AND SITE SKILLS, SECTION 7.5.2] or by machine). If the face of the drawer front is seen, then stopped dovetails will be used, otherwise through dovetails can be used.

Overhanging drawer fronts can be trenched to receive dovetailed housed drawer sides. Alternatively they can be planted as a false front onto a through dovetailed framework.

Drawer backs, which should be set lower than the sides, can be joined to them by either using through dovetail joints or stopped housing joints.

The drawer bottom, responsible for holding and keeping the drawer square, fits into grooves cut into the drawer sides and front (or add-on bead (Figure 12.12, 5a)). It is then nailed onto the bottom edge of the drawer back. Glue blocks are then stuck between the drawer sides, front and the drawer bottom.

12.4.5 Traditional Drawer Construction

The following stages should be read in conjunction with Figure 12.12.

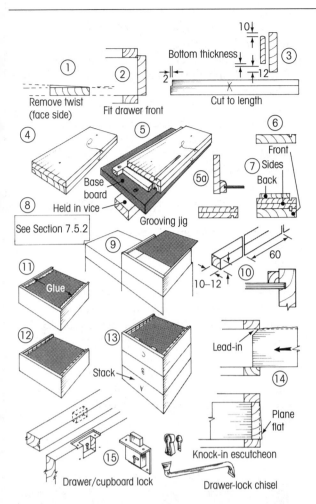

Figure 12.12 Traditional drawer construction – sequence of assembly

1. Check the inside face of each drawer front for twist – remove any twist by planing.
2. Fit and label each drawer front one at a time into its appropriate opening – for example 'A', 'B', 'C' and so on. Stack to one side.
3. Cut to the draw backs shorter than the fronts by about 1.5–2 mm. Label to suit each front. Mark the face side and edge. Round the opposite edge. Stack to one side.
4. Cut the drawer sides to length. Mark the face side and face edges. Tack them together in marching pairs with panel pins – face sides outer most. Label, then stack them to one side.
5. With a hand or powered router groove all the sides (while tacked together) to receive the drawer bottoms – use a grooving jig. Alternatively use lengths of grooved bead (5a).
6. Using the same jig, groove the drawer fronts to the same depths as the sides.
7. Match the entire drawer sets together – one front, two sides (still tacked together), and one back.

8. Mark, cut, and fit all the dovetailed joints [BENCH AND SITE SKILLS, SECTION 7.5.2]. Stopped housings can be used between the backs and sides (Figure 12.11).

   Note. If drawer locks are to be fitted, their housings into the drawer fronts should be cut at this stage.
9. Using a PVAc adhesive [BENCH AND SITE SKILLS, SECTION 10.1.6] make all the joints – use cramps as necessary. Square the drawer framework over the drawer housing of the base carcass: insert the drawer bottom from the back and nail down to the drawer back.
10. For each drawer, cut up two drawer depth lengths of 10–12 mm square section into glue blocks each approximately 60 mm long using a mitre block. By cutting them up in this way, it ensures that no matter what shape they and the bottom are, their fit into the corners is assured.
11. For each drawer, applying PVAc adhesive along the outside joint between the drawer sides and the bottom. Push the glue blocks into the corner and along the length of each drawer side.
12. Two square-ended blocks are then added between the underside of the drawer bottom and the drawer front.

    Note. Glue blocks have the effect of keeping the drawer square, and stiffing the drawer bottom against 'drumming' (rattling).
13. Stack the drawers one on top of the other until the adhesive has set.
14. Dress all the joints with a sharp smoothing plane. Plane a bevelled 'lead-in' off the back top edge of each drawer side. Reduce the sides to fit under the kickers (Figure 12.8). If required, dress the faces of the drawer fronts flush with a drawer opening one at a time.
15. Fixed drawer handles or pulls. If drawer locks are fitted, cut a mortice hole (used a drawer lock chisel) into each lock rail to receive the bolts.

Finally, wax (with candle wax) the drawer runners, kickers, and sides to reduce friction and wear.

12.5 WALL UNITS

Wall units consist of an arrangement of closed (doors) and open shelves – available in a variety of sizes as shown in Figures 12.1 and 12.3.

12.5.1 Carcass Construction

The carcass is usually constructed by the method termed 'Boxed or rail and slab construction' similar to that previously described for base units. The cornice is either built-on or attached to the carcass at a later time as an add-on [BENCH AND SITE SKILLS, SECTION 12.9.1].

12.6 CUPBOARD DOORS

Depending on the design of the fitment cupboard doors are either inset, flush, or protrude in front of the carcass framework. Inset doors·will slide on a track [BENCH AND SITE SKILLS, SECTION 12.6], whereas flush and protruding will be hung on hinges.

Hung doors can give full and easy access to the cupboard content, but the swing of an open door in some situations can be a hazard. In this case as shown in Figure 12.15 a slid-away door could be a consideration – particularly at head height. Sliding doors on the other hand can restrict access to the cupboard, since only half, or in some cases a third of the opening is open at any one time.

Doors can be made-up in several ways (Figure 12.13), for example:

- *Panelled doors* (Figure 12.13a) – framework of timber with an inset panel of wood, manufactured board, or safety glass (Figure 12.13b) which must comply with current regulations and standards relating to its safety in its application and end use. [*Workplace (Health,*

Safety & Welfare) Regulations 1992: BS6206: 1981. BS6262: Part 4: 1994.]
- *Flush wood doors* (Figure 12.13c) – sheet material of single or double veneered blockboard or laminboard – edges lipped all round.
- *Flush particle board doors* (Figure 12.13d) – chipboard faced and edged with melamine, or laminated plastics.
- *Rebated doors* (Figure 12.13e) – timber framework faced with plywood or MDF.

12.6.1 Hanging Cupboard Doors

Figure 12.14 shows some alternative methods of hanging a cupboard door and the appropriate hinge. For example:

- *Butt hinge (solid drawn brass) uncranked* (Figure 12.14a) – set back in line with the carcass. For neatness only half of the hinge knuckle projects. Unless there is a vertical mid division, a doorstop will be required.
- *Singe cranked (steel) two leaf hinge* (Figure 12.14b) – allows the door to fit partly or fully (lay-on fixing) over the door frame. Open doors could interfere with adjacent fitments.

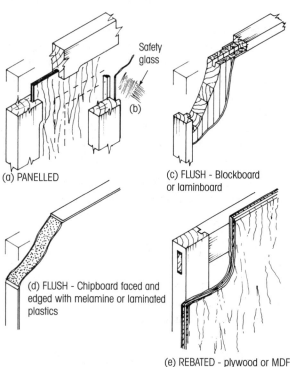

(a) PANELLED

(b) Safety glass

(c) FLUSH - Blockboard or laminboard

(d) FLUSH - Chipboard faced and edged with melamine or laminated plastics

(e) REBATED - plywood or MDF on framed backing

Figure 12.13   Cupboard doors

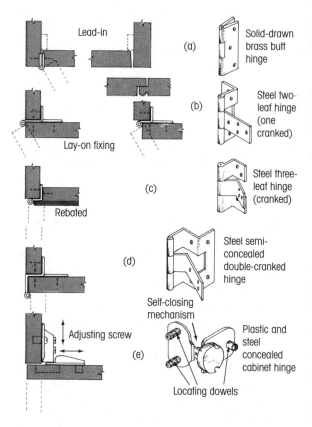

Lead-in

Lay-on fixing

(a) Solid-drawn brass butt hinge

(b) Steel two-leaf hinge (one cranked)

Rebated

(c) Steel three-leaf hinge (cranked)

(d) Steel semi-concealed double-cranked hinge

Self-closing mechanism

Adjusting screw

(e) Plastic and steel concealed cabinet hinge

Locating dowels

Figure 12.14   Alternative methods of hanging cupboard doors

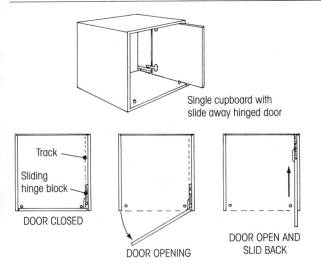

Single cupboard with slide away hinged door

Track

Sliding hinge block

DOOR CLOSED

DOOR OPENING

DOOR OPEN AND SLID BACK

Figure 12.15 Slide-away hinged cupboard door

- *Double cranked (steel) three-leaf hinge* (Figure 12.14c) – used with rebated doors.
- *Double cranked (steel) two-leaf semi-concealed hinge* (Figure 12.14d) – allows similar fitments to abut the hanging side without door opening interference.
- *Concealed (steel and/or plastics) cabinet hinge* (Figure 12.14e) – they allow the door to be adjusted for height, sideways movement and tilt. The last portion of their closing arc is self-closing, and they do not interfere with adjacent doors or fitments when open.
- *Slide-away hinged door* (Figure 12.15) *more usually associated with wall units* – where cupboard doors are in constant use, or when their opening presents a hazardous situation, 'Ziel' disappearing hardware could be the solution.

12.6.2 Catches and Stops (Figure 12.16)

Catches are available to suit the varying sizes and weights of door, collectively this would mean that the catch should be capable of holding the door closed whilst resisting varying degrees of force to open it. Probably the most popular types are:

- *Ball catches (Figure 12.16a)* – steel ball set within a spring-loaded barrel kept in place when the door is shut by a stopped metal striking plate. Only suitable for small light doors
- *Roller catch (Figure 12.16b)* – plastics roller or wheel operating as a latch, which when the door is shut latches over a plastics striking plate and is held in place by the raised keep.
- *Plunger catch (Figure 12.16c)* – plastics spring loaded plunger catch, which when the door is shut catches onto a plastic raised keep as shown at Fig. 12.16.

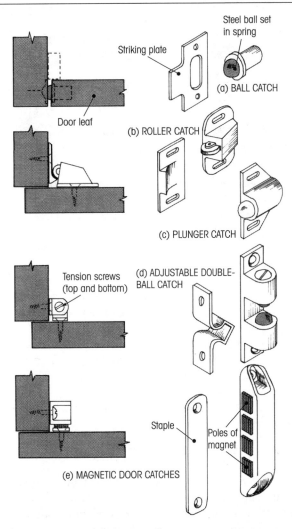

Steel ball set in spring

Striking plate

(a) BALL CATCH

Door leaf

(b) ROLLER CATCH

(c) PLUNGER CATCH

Tension screws (top and bottom)

(d) ADJUSTABLE DOUBLE-BALL CATCH

Staple

Poles of magnet

(e) MAGNETIC DOOR CATCHES

Figure 12.16 Cupboard door catches and stops

- *Double ball catch (Figure 12.16d)* – available in various sizes, some models have adjustable spring tension which can be adjusted to meet the required force requirement.
- *Magnetic door catches (Figure 12.16e)* – catches are available for face fixing (as shown) or mortice fixing (not illustrated). The partly encased magnets are mobile to the degree that they will take up small amounts of tolerance and follow the staple should the door become slightly distorted. Depending on type they can resist forces from 3.5 kg up to 8.0 kg.

12.7 SLIDING CUPBOARD DOORS

As shown in Figure 12.17 the arrangement will depend on the opening width and number of doors. For example, a single door using a single track would have to slide over and beyond the fitments opening. Two

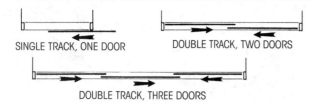

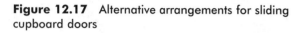

Figure 12.17 Alternative arrangements for sliding cupboard doors

doors on the other hand using two tracks could bypass one another, but the total opening width at any one time would be reduced by less than half. The same principal would apply using three doors on two tracks, by reducing the total opening width by less than a third.

Thin sliding door panels (Figure 12.18) – whether safety glass, plywood, or hardboard – use either the grooves cut into their surrounding framework (plastics inserts provide a smooth action) or purpose made channels of plastics. Which ever method is used, the top grooves will need to be deeper than the bottom grooves, this enables doors to be positioned and lifted out for cleaning purposes, etc.

Thicker and heavier doors can be run off a bottom – or top track system.

12.7.1 Bottom Track Sliding Doors

The system shown in Figure 12.19 is suitable for small lightweight wood based doors. Arrangements at the top are such that the door is kept from deviating from its straight path – this is achieved by pinning guide laths to the door head – one to each side of the door, or by using a channel to house the top edge of the doors. Another alternative is to use retractable or non-retracted guide-pins.

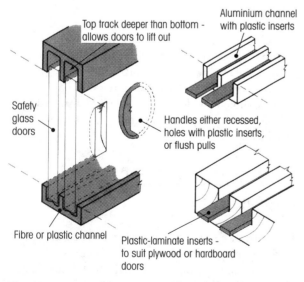

Figure 12.18 Sliding thin cupboard doors

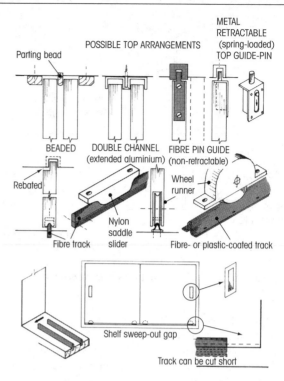

Figure 12.19 Bottom track system for sliding lightweight doors

The bottom arrangement can include the use of a nylon saddle slide (two per door) to carry the door over the top of a fibre track, which is either housed into, or surfaced fixed onto the door frame. Alternatively, wheel runners can be used in place of the nylon saddle. One problem associated with this type of track system is that it stands proud – making cleaning difficult. This can in part be overcome by leaving a gap (by cutting the track short) at each end.

12.7.2 Top Hung Sliding Doors

Figure 12.20 shows a two-track two-door by-pass system. Screws attach two nylon hangers (slides) to the top edge of each door, which slide within aluminium top fixed channel. Nylon guides screwed to the bottom shelf (no grooves or track makes for an easy-clean shelf) to keep both doors apart and vertical. This arrangement is suitable for doors not exceeding 9 kg in weight and between 16–30 mm thick.

12.8 FIXING BASE UNITS

Firstly a horizontal datum line at about a height of 1 m above the floor is marked onto the wall (Figure 12.21a). [BENCH AND SITE SKILLS, SECTION 6.6]. This line should extend along the wall or walls as far as the units go. Because the floor may not be level, take several

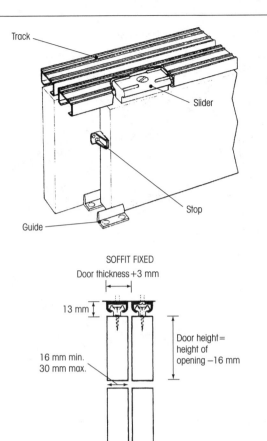

Figure 12.20 Henderson 'Slipper' top hung sliding door gear

measurements down from this datum line to establish the highest point of the floor (shortest distance). Then, if appliances are to be fitted under or into the units, measure from this point on the floor the height of the unit – bearing in mind that any difference at floor level can be taken up when fitting the plinth. However, if no appliances are involved, or if the plinth is of a fixed height (not on adjustable legs or feet) then the shortest distance can be used and the plinth or pieces scribed to the floor.

12.8.1 Worktop Base Units

The height of these units, less the thickness of the worktop is marked onto and around the wall parallel to the datum line (Figure 12.21b). A back rail (say 25×50 mm) is fixed to the wall up to this new line as a back support for the worktop. This rail is used as a datum for levelling across the units – adjustments being made at plinth level either by packing off the floor or by the use of proprietary adjustable legs.

Note. At this stage, if wall units are to be fixed, it is worth considering fixing them before the base units. However, if tall base units are involved these can determine wall unit height.

As shown in Figure 12.22 the best place to start positioning the units is a corner. Once the units are set in place and in line with one another away from the wall to allow for the run of any services, they can be attached to the floor with screws and to one another with screws or connecting bolts (Figure 12.7).

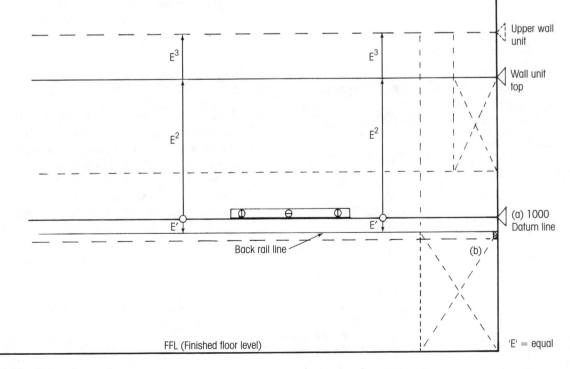

Figure 12.21 Fixing datum lines

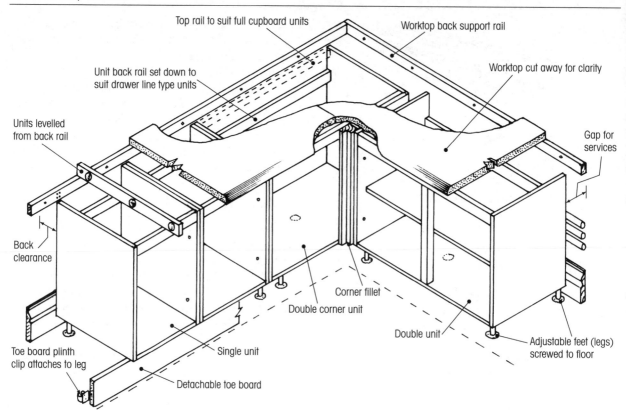

Figure 12.22 Fixing worktop base units

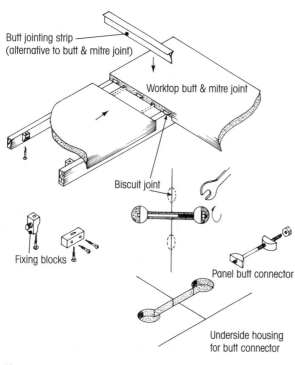

Figure 12.23 Workshops – their jointing and fixing

12.8.2 Tall Base Units (Figure 12.7) (larder or appliance housing unit)

Their height will determine the top datum line (Figure 12.21) for fixing the wall units. Because of the risk of these being pulled over it is very important that they are securely fixed back to the wall.

12.9 WORKTOPS

Worktops usually consist of 40 mm thick chipboard 600 mm wide in lengths of 1500, 3000 and 3600 mm lengths. Veneered with laminated plastics with a square, or post-formed [BENCH AND SITE SKILLS, SECTION 2.6] self edge, or edge lipped with wood.

Where one worktop abuts another at internal corners, those with pre-formed edges will, as shown in Figure 12.23 require either a butt and mitre joint formed with the aid of a power router and special jig (the joint can be reinforced with 'biscuits' [BENCH AND SITE SKILLS, SECTION 7.2.8] or 'panel butt connectors'), or a worktop jointing strip.

Worktops are attached to the undercarcass with plastics fixing blocks.

As shown in Figure 12.24 openings cut out of the worktop for a sink top or cooker hob, etc., are usually undertaken with a jigsaw after first drilling 12 mm holes

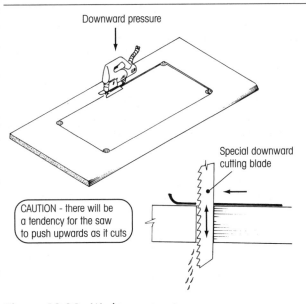

Figure 12.24   Worktop cut-outs

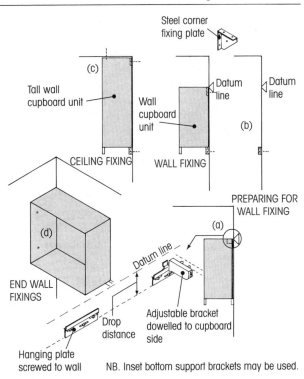

NB. Inset bottom support brackets may be used.

Figure 12.25   Fixing wall units

at each corner as starting points. Special jigsaw blades are available which cut on the downward stroke and thereby reduce chipping of the laminates face – because of this action the saw will have a tendency to lift as it cuts, a firm grip and downward pressure will therefore be required.

Note. The edges of all cuts made into a worktop must be sealed with a waterproof sealant.

12.10 FIXING WALL UNITS

A new datum line is now required – its position as shown in Figure 12.21 can depend on:

- The top height of the tall base unit(s)
- Height of the wall unit
- Margin between the top of the worktop and underside of the wall unit
- Head clearance height of unit doors
- Room ceiling height
- This line (new datum) should extend along the wall or walls horizontally as far as the units go.

Work should start from the same corner as the base units. If a proprietary adjustable hanger bracket is to be used, like those shown in Figure 12.25(a) these should be securely attached to the sides of the wall unit. Then using a paper or preferably a card template, accurately mark the positions of the screw fixings on the wall for the steel wall fixing hanger plates. When these are all securely fixed [BENCH AND SITE SKILLS, SECTION 11.6.4], hook the tongues of the brackets over the plates. Unit height adjustments can now be made as and when they are fixed to meet the level datum line, on completion the bracket must be tightened over the steel plate. Alternatively as shown in Figure 12.25(b) a back

support rail can be fixed to the wall at a distance equal to the depth of the unit below the datum. Then the units are screwed back to the wall via steel fixing plates, and in the case of ceiling height units to the ceiling (Figure 12.25c) as shown in Figure 12.25(d) units abutting an end wall should be fixed to it.

Adjacent units are then joined one to another either with screws or connecting bolts (Figure 12.7), ready for any cornice and/or pelmet.

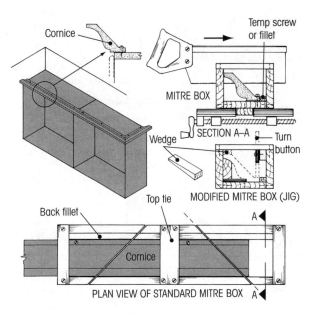

Figure 12.26   Mitred cornice

12.10.1 Cornices (Figure 12.26)

These have two functions; they physically bind the units together one to another, and provide a decorative finish across the head of the units.

Generally available in 3 m lengths and various profiles to match the units. Many are now manufactured from MDF then covered with a thin decorative film. Alternatively these can be purpose made from solid wood to any particular profile [BENCH AND SITE SKILLS, SECTION 1.9.4].

Mitres can be cut by hand using a purpose made mitre box and tenon saw, or proprietary adjustable mitre box with fixed saw [BENCH AND SITE SKILLS, FIGURE 5.114, PAGE 148], or an electric powered mitre saw [BENCH AND SITE SKILLS, SECTION 8.21].

Before any cuts are made the moulding must be held firm. Either by the use of a jig, or as shown temporarily screwing it down to a base board. All cuts should be made towards the decorative face to minimise the risk of damage to its surface.

Internal corners should be scribed [BENCH·AND SITE SKILLS, SECTION 7.5.9], particularly if a solid timber moulding is used to minimise any shrinkage gaps.

12.10.2 Pelmets

Their function is similar to the cornice in that they help to bind the units together and provide a decorative finish, plus they can conceal underunit strip lights.

Corners are mitred or scribed in the same way as the cornice. As shown in Figure 12.27 they can be fixed to the underside of the shelf by one, or the combination of three ways, for example:

- Screwing through the bottom shelf into the top edge of the pelmet.
- Using small steel angle brackets.
- Using plastic corner fixing blocks.
- If the pelmet is used to conceal an under-shelf light fitting it must be a tight fit to the shelf – use the first method. Using a self-adhesive opaque sealant along the joint can seal small amounts of light leakage.

12.11 BUILT-IN FITMENTS

As previously stated and shown in Figure 12.28, these fitments usually use a wall, or walls, floor and/or ceiling as part of the fitment component.

12.11.1 Shelving

The simplest of all the built-in fitments but, they must be of adequate strength to support their intended load and their support (bearers) must be strong enough to carry both the shelf and its load. Figure 12.29(a) shows the possible result of using a shelf of inadequate thickness. But, by providing a mid-support (Figure 12.29b) the span can be suitably halved. Alternatively (Figure 12.29c) the shelf thickness can be increased.

Shelves – alternative materials and methods of construction are shown in Figure 12.30 as:

(a) solid timber single board
(b) double board with loose tongue
(c) tongued and grooved boards with front stiffener
(d) open slatted shelf with front stiffener
(e) particle board – veneered or non-veneered
(f) particle board with front and back stiffeners
(g) plywood with front and back stiffeners
(h) lipped blockboard or laminboard.

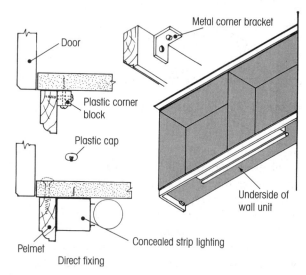

Figure 12.27 Pelmet fixing

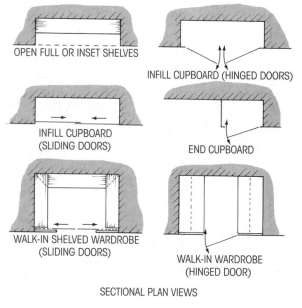

Figure 12.28 Built-in fitments

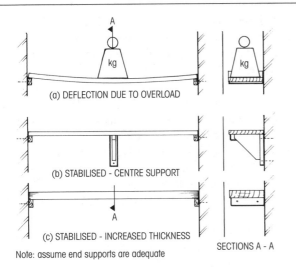

(a) DEFLECTION DUE TO OVERLOAD

(b) STABILISED - CENTRE SUPPORT

(c) STABILISED - INCREASED THICKNESS

SECTIONS A - A

Note: assume end supports are adequate

Figure 12.29 Shelves in relation to loading

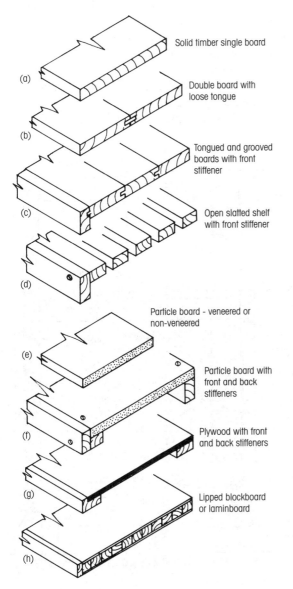

(a) Solid timber single board

(b) Double board with loose tongue

(c) Tongued and grooved boards with front stiffener

(d) Open slatted shelf with front stiffener

(e) Particle board - veneered or non-veneered

(f) Particle board with front and back stiffeners

(g) Plywood with front and back stiffeners

(h) Lipped blockboard or laminboard

Figure 12.30 Shelf material and fabrication

Shelf bearers – as can be seen from Figure 12.31 there are several ways of providing support – from using bearers fixed directly onto the wall, or indirectly via brackets. Independent support can be provided via a system of ladder bearers – usually associated with shelf racking.

Figure 12.32 shows a proprietary system for shelving and its support.

12.11.2 Floor Standing Cupboards

Built-in cupboards may simply consist of a front frame to cover a series of shelves, or a front and one side, or front and two sides, but hardly ever a back.

Figure 12.33 shows a tall corner cupboard, which could be fitted out with shelves, or used as a top cupboard and single wardrobe. Using the same method of construction it could be extended to a double tall cupboard/wardrobe. The front frame (with pre-hung doors) is prefabricated together with a handed (right- or left-hand) side panel – existing walls serve as the back and other side.

Options – two alternative door types and front frame are shown. The doors may be either in line with the door frame face or protecting. The bottom rail of the frame

NB. Bearers will be concealed below eye line

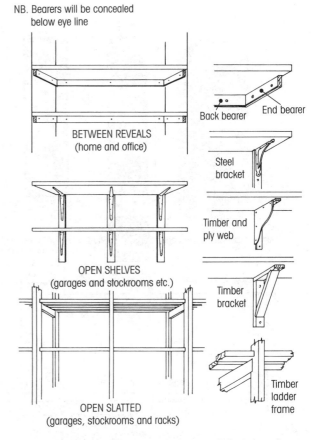

Back bearer   End bearer

BETWEEN REVEALS (home and office)

Steel bracket

OPEN SHELVES (garages and stockrooms etc.)

Timber and ply web

Timber bracket

OPEN SLATTED (garages, stockrooms and racks)

Timber ladder frame

Figure 12.31 Traditional shelves and their bearings (supports)

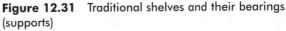

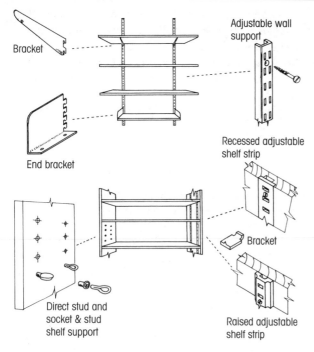

Figure 12.32 Proprietory shelves and their bearings (supports)

may depending on the door hanging have its face edge or face side to the front – on the other hand it may be set back (not shown) to allow an inset door to lap over it.

Fitting and fixing the fitment (Figure 12.33)

- From the corner measure and mark the width and depth of the fitment onto the wall.
- Mark the inside dimensions square onto the floor.
- Mark a plumb line from both these points to extend to the full height of the fitment.
- Fix the front frame back bearer to the wall with plugs.
- Fit and scribe the front frame to the floor and wall.
- Mark the position of the shelf bearers – level these around the wall.
- Fix the shelf bearers to the wall with plugs.
- Fit and scribe the end panel to the floor and wall.
- Fit shelves, floor piece, top (if used) into the wall corner – cut to length and width.
- Fix the front frame to wall bearer.
- Fix side panel to a wall bearer or shelf bearers, and front frame.
- Shelves – if the shelves can be removed these can be refitted at this stage.
- Fix shelves, floor piece, top, and hanging rails, etc.

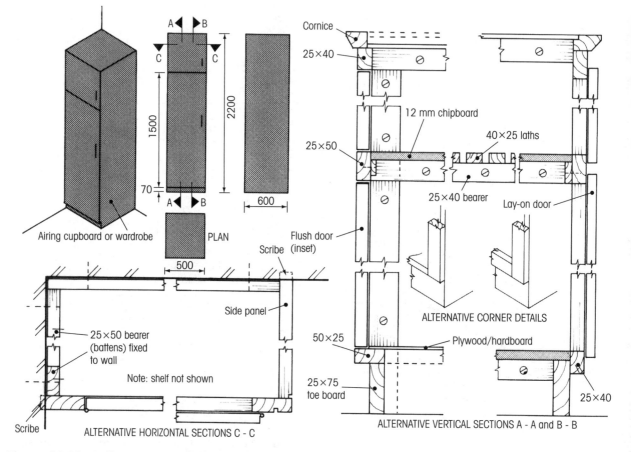

Figure 12.33 Built-in corner cupboard

- Fit and fix cornice or ceiling trim.
- Re-hang doors.
- Fix handles and catches.
- Dress and fill joints, etc. in preparation for finish.

12.11.3 Sliding Doors to Walk-in Cupboards and Wardrobes

Sliding doors are a useful alternative to hinged doors – particularly with walk-in wardrobes.

Figure 12.34 shows an arrangement for a single-track door with two alternative methods of fixing. The track may either be face or soffit fixed.

Where doors can pass one another a double hung track system like the one shown in Figure 12.35 can be used – in this case the track is soffit fixed.

Opening widths: 600–900 mm
Max. leaf size: 2400×900×40 mm
Max. leaf weight: 45 kg

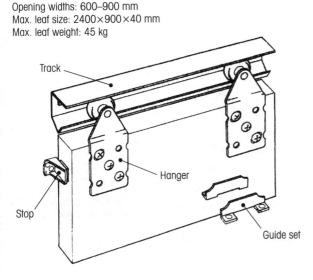

Opening widths: 1200 - 1800 mm (2 leaves),
2400 mm (3 leaves)
Max. leaf size: 2400×900×40 mm
Max. leaf weight: 45 kg

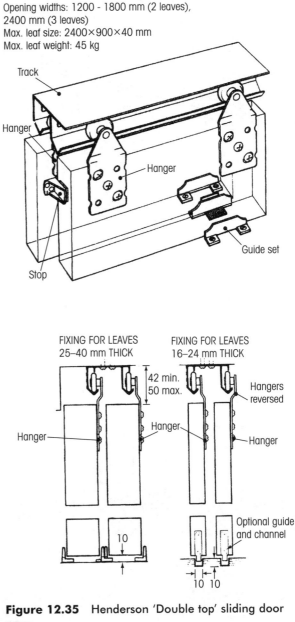

Figure 12.35 Henderson 'Double top' sliding door gear

Where there is no room or wall space to accommodate the door when open or complete open access is required a bi-fold (two folding doors) system like the one shown in Figure 12.36 could be used.

12.12 FITMENT FURNITURE (HANDLES, KNOBS AND PULLS) (FIGURE 12.37)

Probably the most important aspect here is not the pattern or style, but their location and position so that the door or drawer opens or slides with ease and that they are sited at a convenient position to avoid overreaching.

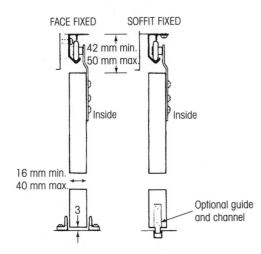

Figure 12.34 Henderson 'Single top' sliding door gear

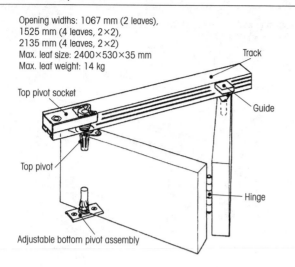

Opening widths: 1067 mm (2 leaves),
1525 mm (4 leaves, 2×2),
2135 mm (4 leaves, 2×2)
Max. leaf size: 2400×530×35 mm
Max. leaf weight: 14 kg

Track

Top pivot socket

Guide

Top pivot

Hinge

Adjustable bottom pivot assembly

Bar handles, knobs, and pulls are attached with screws or bolts. Flush pulls – associated with sliding doors – are recessed into the face of the door.

Door and drawer edge grips are extruded from aluminium or plastics. They are usually attached via a barbed tongue, which is driven into the grooved edge of the door or drawer. Tongues and grooves are cut and stopped short of the ends so that they do not show on the ends of door or drawer.

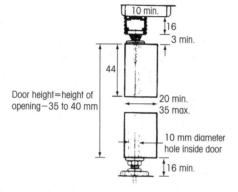

Figure 12.36 Henderson 'Bifold' sliding door gear

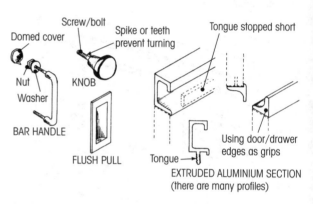

Figure 12.37 Cupboard door and drawer handles

13 Repairs and maintenance

13.1 INTRODUCTION

The assumption is that most wood-based components and their fittings will, over a period of time while being subjected to wear and tear eventually require some form of maintenance. This may simply involve routine checks on external protection with a periodic repainting programme, and internal decoration, etc., also lubricating any moving parts of associated hardware.

Failure to carry out a general building maintenance programme can effect not only the building fabric, but the components contained within it.

Firms who specialise in repair and maintenance tend to employ operatives who are flexible in the sense that they may be expected to carry out basic skills outside their specialism. For example, the joiner may be expected to glaze or re-glazed a window, use a pointing trowel, or plasterer's float to do some patchwork on a plaster wall or concrete floor, or even make good defective brickwork or, damaged wall or floor ceramic tiles. The ability to lay or renew broken roof tiles or slates may also be a requirement. It follows therefore that a sound knowledge of building construction is essential for all skilled operatives engaged on this type of work.

In these days of short-term training schemes and or apprenticeships, multi-skilled training to a high standard is not practicable. This, among other factors, has led to the setting up of firms which specialise in specific areas of work – for example, timber preservation to include remedial work which has been affected by fungal or insect attack, damp proofing, thermal insulation, and replacement windows, to mention just a few.

We will start by carrying out a general survey of the property as a whole then deal with some of the areas of work and individual items separately. For example:

1. Repair/replacement survey.
2. Customer relations.
3. Tools and equipment
4. Remedial treatment
   (i) Repairing defective and damaged components of:
      – doors
      – door frames
      – window casements and sashes
      – window frames
      – door and window hardware
      – floorboards
      – staircases
      – eaves
      – gates.
   (ii) Removal renewal or replacement of:
      – doors
      – door frames
      – window casements and sashes
      – window frames
      – door and window hardware (making good holes and recesses)
      – renewing sash cords
      – reglazing a window
      – eaves gutters.
   (iii) Making good small areas of brickwork.
   (iv) Making good small areas of plasterwork.
   (v) Making good ceramic wall and floor tiles.

13.2 GENERAL INSPECTIONS AND REPAIR SURVEYS

It is always desirable when making provisional inspections to take notes and sketches of all your observations, and in many cases measurement [BENCH AND SITE SKILLS, SECTIONS 6.1, 6.6, 13.1, 13.3, 13.9]. Not only is this a record of your visit, but will also form a basis to estimate the amount of work involved.

If a client requests a full survey this could involve gaining access under the floor and into the roof space, and possibly taking test samples. In this case a more detailed report would be produced enabling a more realistic analysis to be made.

A true record of the existing conditions of the property including items outside your specialism should always be made. In some cases external items surrounding the property should also be included in your survey.

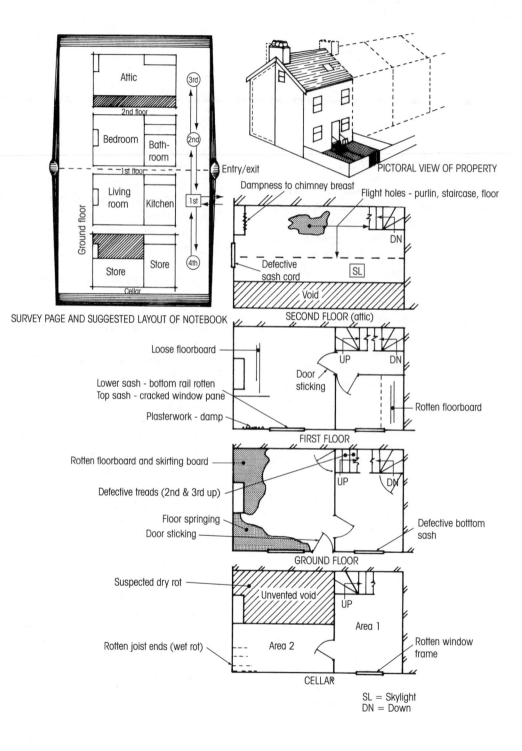

Figure 13.1 Typical floor plan layout to show possible defects to a pre-war back-to-back terrace house

There are three ways a provisional record can be made while surveying the property.

1. A detailed list of all the defects together with their location (schedule).
2. Annotated sketches.
3. The combination of (1) and (2).

13.2.1 Internal Survey

The way in which we view the property can vary between individuals, but more often than not you will be directed to a major trouble spot by the occupants and take it from there. You would however be advised to adopt a set way or procedure, for example, Figure 13.1 shows a suggested route to take around a pre-war back-to-back end terrace house. By starting at ground floor level you can meet the client and draw up an outlined sketch of the property before moving around the property as indicated.

Figure 13.1 shows how by using a pencil and straight edge (rule) the outline of each room floor plan can be quickly drawn onto a page or pages of a notebook.

Door openings, windows, and stairs, etc., can be added as you enter each room, together with any defects you find. The accuracy of your observations are very important at this stage – you should concentrate on the job in hand and take your time to add as much detail as possible without cluttering the page. If a lot of detail is required, consider using a single page for each floor plan, and then make out an accompanying list of defects – don't forget to include an exact defect location (measurements may be a requirement).

Table 13.1 shows how such a list can be tabulated, but in this case possible causes and remedial action have been included.

Figure 13.2 shows a similar survey, but in this case we are looking at a post-war 1950–1960 semi-detached house, again a schedule of defects together with possible causes and remedial action is shown in Table 13.2.

13.2.2 External Surveys

External repairs are usually listed separately – preferably after the internal inspection. This is because defects found within the dwelling could be as a result of an exterior defect which, without close scrutiny, might otherwise have been missed. For example, a damp patch on the inner face of an external wall could be the result of surface compensation. On the other hand it could have been caused by an excessive amount of rainwater running down the outside wall. Possibly due to a blocked or broken fall pipe, or even a defect gutter joint. Provisional observation from the ground can be enhanced by using binoculars, more detailed reports may require the use of some form of scaffold.

A survey of this nature should cover the following:

1. *Face walls*
   - Stability (no structural movement – settlement cracks) – note close proximity of any trees.
   - Height of the damp proof course (DPC) above ground level (should not be less than 150 mm).
   - Air grate (bricks) clear and away from overgrown vegetation.
   - Cement pointing – no open joints.
   - Cement rendering – no cracks or loose parts (no hollow sound when gently tapped).
   - Staining (no undue discolouration) – particularly just above the DPC line or under the roof eaves, and behind rain water fall pipes.

2. *Wall openings*
   - Doors – general condition of the woodwork, and particular attention to weather seals around the frame, threshold and weatherboard.
   - Window – general condition of the woodwork, glass and putty pointing, and particular attention to weather seals around the window frame.

3. *Fall pipes*
   - Soil pipes – joints fully caulked (sealed).
   - Vent pipes – joints fully caulked (sealed), and extend above and beyond any window or house vent openings (as per Building Regulations).
   - Particular attention should be paid to the backs of cast iron pipes, as this is usually the first place to corrode (rust) due to lack of paint protection.

4. *Gully grates*
   - Check for signs of water being discharged over the edges into the soil, onto a path, or up against the house wall.
   - Check the water level in the 'U' bend by pouring down a bucket of water.

5. *Eaves guttering (spouting)*
   - Falls – no standing water. On a dry day test by pouring a bucket of water into the gutter at its highest point.
   - Joints – check for water staining on the under-side of the gutter, and signs of moss growth.
   - Vegetation growth within a wood gutter – good sign of poor falls and possible decay.

6. *Eaves (pitched and flat roofs)*
   - Fascia boards (open, closed, or flush eaves) – for signs of decay and/or distortion.
   - Soffit boards (closed eaves) – for, the presence of any ventilation grills or gaps.

7. *Flat roofs*
   - Surfaces – visible falls, signs of ponding (retained grit residue on surfaces without solar chippings), check for signs of vegetation growth.

Table 13.1 An example of how an interior repairs survey of a pre-war back-to-back terrace house could be tabulated – based on the defects shown in Figure 13.1

Location	Defect	Possible cause	Remedial action/Remarks
1. Living room	Rotten floor boarding	Dry rot (unvented under floor space)	Investigate under floor space Check surrounding moisture content of and skirting board room structure [BENCH AND SITE SKILLS, SECTION 1.7.3] Check for dry rot [BENCH AND SITE SKILLS SECTION 3.3]
	Floor area under window springing	Cellar rot (wet rot) – confirmation in cellar	Check for wet rot [BENCH AND SITE SKILLS SECTION 3.3]
	Exterior door sticking	Over-painted; moisture movement; door sagging; defective hinge; etc.	See Table 13.4 Reassemble door – see Figure 13.5 Replace hinges
2. Stairs to first floor (second and third treads)	Tread nosing split	Heavy wear and tear; timber defect; riser groove cut too deep	See Section 13.5.1g and Figure 13.13
	Creaking tread	Shrinkage due to moisture movement; treads and risers unhoused	Insert glued wood veneer shims Screw treads to risers – see Figure 13.13
3. Bathroom	Rotten floorboard under bath	Wet rot, due to water spillage or leaking waste water pipe	Make good seal around bath edge Check for leaks Check for wet rot [BENCH AND SITE SKILLS SECTION 3.3]
4. Bedroom	Defective door latch Door sticking Loose floorboard. Cracked window pane (glass) – top sash Rotten sash rail (bottom sash) Damp plasterwork	Worn and overpainted Defective hinge Access to services Unknown Window condensation – Figure 13.10 Surface condensation – penetrating damp	Replace Replace both hinges Repair or renew – refix with screws Replace – see Table 13.5 Renew casement – see Table 13.5 See Table 13.3
5. Stairs to second floor (attic)	Flight holes in treads, risers, and strings	Woodworm (Common furniture beetle?)	Investigate, recheck remainder of house, for sign of insect infestation [BENCH AND SITE SKILLS, SECTION 3.4]
6. Second floor (attic)	Flight holes in purlin and floor boards	Woodworm (Common furniture beetle?)	Investigate, recheck remainder of house, for sign of insect infestation [BENCH AND SITE SKILLS, SECTION 3.4]
	Gable-end window unable to open Dampness to chimney breast	Defective sash cords Defective cement pointing, interstitial condensation, flashings or soakers around the chimneystack	Replace – see Section 13.5.2f Investigate and make good Make good as necessary Thermal insulate face wall

Table 13.1 *Continued*

Location	Defect	Possible cause	Remedial action/Remarks
Note: on returning down stairs, recheck rooms and staircases.			
7. Kitchen	Window (bottom sash) – joints between stile and bottom rail	Defective haunch or wet rot	Repair or replace – see Table 13.5
8. Cellar			
Area 1	Rotten window frame	Cellar rot (Wet rot)	Repair or replace – see Section 13.5.1d
Area 2	Rotten joist ends (under living room floor)	Cellar rot (Wet rot)	Check surrounding moisture content of structure [BENCH AND SITE SKILLS, SECTION 1.7.3]
			Check for wet rot [BENCH AND SITE SKILLS, SECTION 3.3]
7. Cellar void	Dampness (see living room – dry rot)	No ventilation	Provide through under floor ventilation (via airbricks to external wall)
			Check surrounding moisture content of structure [BENCH AND SITE SKILLS, SECTION 1.7.3]

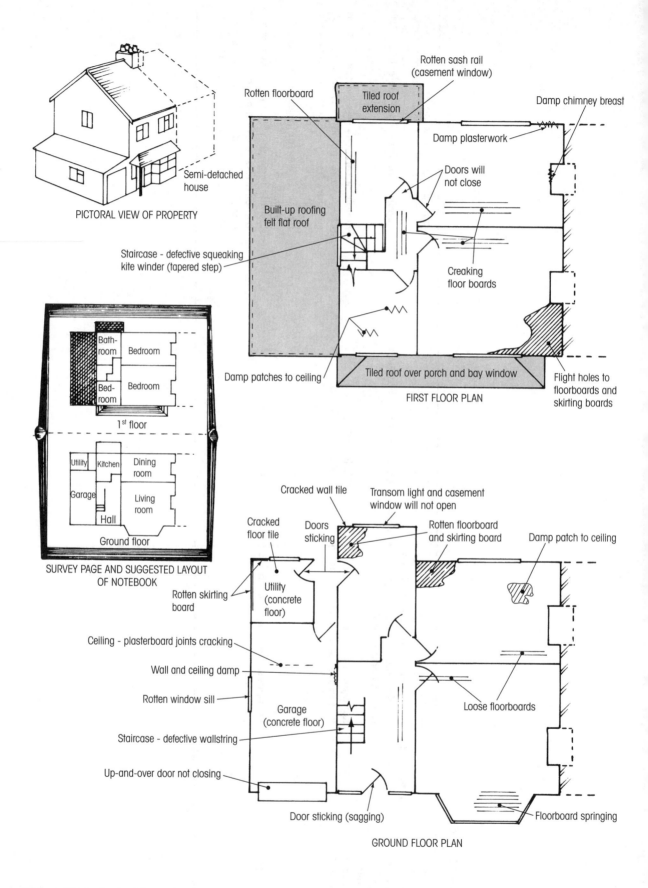

PICTORAL VIEW OF PROPERTY

Semi-detached house

SURVEY PAGE AND SUGGESTED LAYOUT OF NOTEBOOK

1st floor

Bathroom

Bedroom

Bedroom

Bedroom

Ground floor

Utility | Kitchen | Dining room

Garage | Living room

Hall

FIRST FLOOR PLAN

Rotten floorboard

Tiled roof extension

Rotten sash rail (casement window)

Damp chimney breast

Damp plasterwork

Doors will not close

Built-up roofing felt flat roof

Creaking floor boards

Staircase - defective squeaking kite winder (tapered step)

Damp patches to ceiling

Tiled roof over porch and bay window

Flight holes to floorboards and skirting boards

GROUND FLOOR PLAN

Cracked wall tile

Transom light and casement window will not open

Doors sticking

Rotten floorboard and skirting board

Damp patch to ceiling

Cracked floor tile

Rotten skirting board

Utility (concrete floor)

Ceiling - plasterboard joints cracking

Wall and ceiling damp

Rotten window sill

Garage (concrete floor)

Staircase - defective wallstring

Loose floorboards

Up-and-over door not closing

Door sticking (sagging)

Floorboard springing

Figure 13.2 Typical floor plan layout to show possible defects in a post-war semi-detached house

Table 13.2 An examples of how an interior repairs survey of a post-war semi-detached house could be tabulated – based on the defects shown in Figure 13.2

Location	Defect	Possible cause	Remedial action/Remarks
1. Lounge	Loose floor board	Used as means of access to services	Repair or renew – refix with screws
	Springing floorboards within the bay window area	Undersized joists Fungal decay	Investigate under floor space [BENCH AND SITE SKILLS, SECTION 3.3]
2. Dining room	Loose floor board	Used as means of access to services	Repair or renew – refix with screws
	Rotten floor boards and skirting board	Fungal decay – dry rot (unvented under floor space)	Investigate under floor space Check surrounding moisture content of structure [BENCH AND SITE SKILLS, SECTION 1.7.3] Check for dry rot [BENCH AND SITE SKILLS, SECTION 3.3]
3. Hall	Loose floor board	Used as means of access to services	Repair or renew – refix with screws
	Front door sticking	Door sagging (defective joints, and/or worn hinges)	Replace door or reassemble door – see Table 13.4 & Figure 13.5 Replace hinges
4. Staircase	Wall string coming away from the wall.	Defective fixings Packings detached Structural movement	Provide new fixings Check for alignment with adjacent wall Replace and make good
	Creaking kite winder	Movement within wall string housing due to moisture movement shrinkage Moisture movement shrinkage between tread and riser	Providing tread is fully housed, insert glued wood veneer shims – see Section 13.5.1g Screw treads to risers – see Figure 13.13
5. First floor landing	Floor boards creaking	Shrinkage due to moisture movement – boards cupping	Refix with screws – always check for underfloor services before fixing
6. Bedroom (small – front)	Damp patch to ceiling.	Roof space condensation running down rafters onto the ceiling below Defective roof covering Leaking water services within the roof space	Check for, or provide eaves ventilation – Table 4.1 Renew or replace defective tiles/slates (see exterior schedule) Investigate water storage tanks (cisterns); service pipes etc. for leaks
7. Master bedroom (front)	Floor boards creaking	Shrinkage due to moisture movement – boards cupping	Refix with screws – always check for under floor services before fixing
	Flight holes in floor-boarding and skirting board.	Woodworm (Common furniture beetle)	Investigate underfloor space and recheck remainder of house, including the roof space for sign of insect infestation [BENCH AND SITE SKILLS, SECTION 3.4]
8. Bedroom (rear)	Door – unable to close	Door twisted Hinge bound (build-up of paint) Defective latch	Ease rebates or replace door Ease door and rehang Adjust or replace
	Damp patch on chimney breast	Defective chimney flashings (back gutter, apron, step flashing, and soakers) Defective chimney stack flaunching (collar around chimney pots), and/or cement pointing Interstitial condensation	Investigate and make good (see exterior schedule) Make good as necessary Thermal insulate face wall
	Damp patch to outer wall	Defective eaves gutter Defective eaves slates/tiles Defective face brick/blockwork	Make good joints as necessary Make good as necessary Make good by repointing as necessary (see exterior schedule)

Table 13.2 Continued

Location	Defect	Possible cause	Remedial action/Remarks
9. Bathroom	Door – unable to close	Door twisted	Ease rebates or replace door – see Table 13.4
		Hinge bound (build-up of paint)	Ease door and rehang.
		Defective latch	Adjust or replace.
	Rotten floor boards (wet rot)	High moisture content	Take readings with moisture meter [BENCH AND SITE SKILLS, SECTION 1.7.32]
		Defective seal around bath edge (water spillage)	Make good seal around bath edge
		Defective waste pipe	Check for leaks [BENCH AND SITE SKILLS, SECTION 3.3]
	Rotten sash rail to opening casement.	Window condensation	Renew casement – see Table 13.5 and Figure 13.10
	Note: on returning downstairs, recheck landing, staircase, and hallway.		
10. Kitchen	Damp patch to ceiling	Roof space condensation running down rafters, dripping onto the ceiling below	Provide roof space ventilation.
		Defective roof covering	Renew or replace as necessary (see exterior schedule)
	Cracked ceramic wall tile to sink splash back	Defective apron	Renew/replace – see Section 13.5.2 and Figure 13.21
	Rotten floorboards and skirting board.	Accidental	Investigate under floor space (see exterior schedule). Check for dry rot [BENCH AND SITE SKILLS, SECTION 3.3, PAGE 95]
		Fungal decay – dry rot (unvented under floor space).	Ease by planing – prime over bare wood – see Table 13.5
	Transom light and casement will not open.	Ceased-up with over-painting.	Confirm by taking moisture content readings – see Table 13.4
	Outer exterior door sticking	Expansion due to moisture intake.	If joints are intact and there is no distortion – ease by planing back to bare wood, then reseal with a breather paint/stain, with particular attention to end grain. Otherwise replace
11. Utility room	Outer door sticking	As above	As above
	Skirting board rotten (wet rot)	Constant water spillage.	Check for leaks
	Replace as necessary – check for wet rot [BENCH AND SITE SKILLS, SECTION 3.3]	Accidental defective waste pipe	Renew/replace – see Section 13.4.5 and Figure 13.21
	Cracked ceramic floor tile		
12. Garage	Ceiling plasterboard joints cracking	Deflection of under sized joists	Excessive moisture movement
		Joints not taped	Make structural calculations – amend accordingly
			Measure moisture content
			Consult plasterer
	Ceiling and inner wall – damp patch	Roof space condensation dripping onto the ceiling below	Provide roof space ventilation
		Defective roof covering	Renew or replace as necessary (see exterior schedule)
	Window sill rotten (wet rot)	Defective apron	Interior condensation onto direct glazing
		Moisture retained under sill	Provide permanent ventilation
		Lack of maintenance	Provide or reinstate a drip groove (see exterior schedule) [BENCH AND SITE SKILLS, SECTION 3.3]
	Up and over door not closing	Door frame out of square	See exterior schedule
		Defective operating gear	

- Solar chipping – in place on built-up felt roofs.
- Flashings and aprons (metal (lead) or felt) – intact and secure with their abutment (wall).

8. *Pitched roofs*
   - Surfaces – check for missing, displaced, or defective, roof tiles and slates, hip and ridge tiles.
   - Verges – cement pointing to verge tiles/slates. Check the finial, bargeboards and soffits for signs of decay and/or distortion, etc.
   - Roof lights – glazing and water seals.
   - Dormer windows – as above (roof surfaces, eaves, glazing and water seals).

9. *Chimney stacks*
   - Brickwork, blockwork, etc. – stability (no signs of structural movement), cement pointing or rendering intact check for sooty deposits between joints, particularly at perpends (vertical joints).
   - Flaunching (cement collar around the chimney pots) – intact, check for sooty deposits at joints with brickwork and chimney pots.
   - Chimney flashings – check for missing or displaced step flashings and soakers ('L' shaped pieces of metal – usually lead), which lap one another from the chimney back gutter to the front apron to form a weather seal between the under side of the stepped flashings and roof tiles/slates. The front apron can often be viewed from the ground, but the back gutter would require access to the roof.

10. *House steps*
    - Check for level and surface wear.
    - Handrails – stability, fixings, or fully intact.
    - Building regulations – does the pitch, rise and going, handrails and balustrade comply with current regulations.

11. *Garden paths and driveways*
    - Surface condition.
    - Drain inspection chamber – metal covers intact and secure.

12. *Boundaries*
    - Walls – make observations with regards to settlement and distortion (The local authority must be notified of any walls adjacent to a public thoroughfare or highway that appear to be in a dangerous condition.)
    - Wooden fences and gate posts – check for signs of decay particularly at around ground level and where water can be retained, such as where rails crossover posts, etc.
    - Wooden gates – check for sag, distortion (twist), and decay, particularly at joints, junctions and where members overlap, also behind metal fittings.

13. *Out-buildings (attached and/or detached).*
    - Garages – observations as above.
    - Porches and conservatories – observations as above.

If you are required to pass on this information to a colleague via a report, or write up a repair schedule, you could consider making simple annotated orthographic line diagrams of each elevation, like the ones shown in Figure 13.3. Then include these with a written schedule like the one shown in Table 13.3.

13.3 CUSTOMER RELATIONS

Whether you are self-employed or working for an employer, the property repair joiner will more often as not have to carry out his or her work while the occupants of the property are still in residence. The occupants must be inconvenienced as little as possible.

Prior notice of your visit should have been given well in advance, giving guidance to the nature of the work, and an estimate as to the length of its duration. In this way the occupant can prepare the site as required. You will usually be expected to provide dustsheets, and perhaps help with moving heavy items of furniture – the lifting of carpets can also be a common occurrence.

It goes without saying that when working inside the property great care must be taken to minimise any damage to decorations, and to always protect the floor and/or any floor covering. Trafficking through a person's home can very easily damage the surface of a polished floor and cause undue wear to carpets. Providing access for building materials into a furnished house can require a great deal of skill just as can working around expensive furniture – routes should be planned in advance which may mean accessing via a window opening rather than a doorway.

External work may involve working from or over a garden border and/or lawn, all of which can present problems. Many cultivated gardens can house collections of plants, which can cost many hundreds of pounds, and a good lawn can take years to recover.

One cannot emphasise enough that from the outset of the job that a little care and consideration can go a long way towards good customer relations, which will not only promote your company, but also add credits to your competence as a skilled, capable operative.

13.4 TOOLS AND EQUIPMENT

Once items have been installed and fixed (possibly with other trade involvement), any removal or repairs, and eventual reinstatement will require the use of tools and equipment not usually found as part of a bench or site joiners tool kit. In fact some of the hand tools found in a maintenance or a property repair joiners tool box, bass (basin-shaped canvas hold all), or bag, could be mistaken for those used by a bricklayer or engineer.

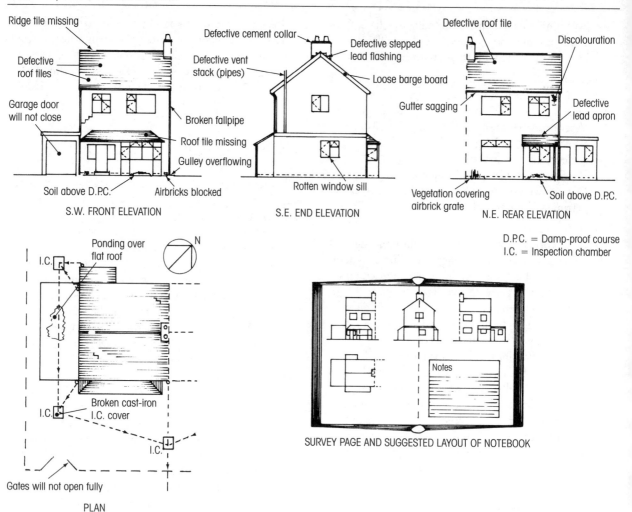

Figure 13.3 Use of line diagrams to identify external defects

13.4.1 Hand Tools and Ancillary Equipment

The list of tools and equipment listed below is not extensive, but it has over the years been shown to be necessary for this type of work. You should also make reference to the book [BENCH AND SITE SKILLS, SECTION 5 'HAND TOOLS AND WORKSHOP/SITE PROCEDURES']. Traditionally these hand tools would have been carried around in a joiner's bass – although still available the bass has more or less been superseded by the plastic box or chest. This, like the 'Porterbox' not only offers better tool protection against damage, but also provides greater security – a good reminder at this point is to always security mark your tools. Security marking should be such that if the markings were to be removed it would render the tool unsaleable, for example, by engraving your name and postcode onto a main metal component where removal by grinding would render the tool useless or not worth stealing.

1. *Measuring, marking, and checking tools.*
   - 1 m folding rule*.
   - 3 m steel tape measure*.
   - 1 m spirit level.
   - Short spirit level*.
   - Builder's line*.
   - Plumb bob and the line*.
   - Marking gauge*.
   - Sliding bevel*.

2. *Saws.*
   - Panel saw (hardpoint pattern)*.
   - Tenon saw.
   - Bowsaw (Bushman type).
   - Hacksaws – large and junior*.
   - Coping saw*.

All saw teeth should be protected with a sheath.

3. *Planes.*
   - Jack plane (wooden types are often preferred to metal, because they are lighter, and less likely to break).

Table 13.3 An examples of how an exterior repairs survey to a post-war semi-detached house could be tabulated – based on the defects shown in Figure 13.3

Location	Defect	Possible cause	Remedial action/Remarks
Front (SW) Elevation	*House roof*		
	Ridge tiles missing	Wind damage	Renew and make good
	Roof tiles	Wind damage	Renew/replace and make good
	Chimneystack stepped flashing	Unknown	Reset make good pointing
	Face wall		
	Cracking to face wall	Structural settlement	Consult structural engineer
	Rain water fallpipe (cast iron)	Lack of maintenance	Renew or replace whole stack with PVCu fallpipe
	Gully grate – signs of overflowing	Grate covered with leaves	Remove and clean grate
		Gully trap 'U' bend blocked	Loosen with bent rod – remove waste with drain scoop – plunge free with water and rubber plunger
		Drain blocked	Consult with Local Authority before engaging contractor, as it may be their responsibility
	Bay roof		
	Tiles missing	Unknown	Replace tiles
	Ground level		
	Damp proof course (DPC) covered and air grates (bricks) blocked	Build-up of soil and ground vegetation	Remove ground vegetation away from house wall, and reduce soil level to not less than 150 mm below the DPC
	Garage		
	Up and over garage door not closing	Door frame out of square – possible structural movement of building	Check face walls for structural movement – (advice from structural engineer may be required). Modify doorframe accordingly
		Defective operating gear.	Adjust and modify and/or replace components as per manufacturer instructions
End (SE) Elevation	*Gable end*		
	Bargeboard loose	Unknown	Refix and make good verge tiles.
	Soil vent stack coming away from the wall	Unknown	Dismantle as necessary – refix back to wall
	Garage		
	Water standing (ponding) on flat roof	Falls wrong	Check for falls.
		Roof sagging due to undersized joists, or their spacing in relation to span of decking	Check ceiling for cracks or distortion (see interior schedule)
		Defective decking	Renew components, and repair, as necessary Re-felt as necessary

Table 13.3 Continued

Location	Defect	Possible cause	Remedial action/Remarks
	Window cill rotten (wet rot)	Moisture retained under cill Lack of maintenance	Provide or reinstate a drip groove (see interior schedule [Table 13.2] and Section 13.5.1d)
Rear (NE) Elevation			
	House roof Ridge tiles Roof tiles Eaves gutter sagging	Wind damage Wind damage Distortion Snow damage. Defective brackets.	As front (SW) elevation Renew/replace and make good Refix to falls with provision for thermal movement Renew brackets
	Face wall Water stains (Discolouration) under eaves gutter.	Defective eaves tiles or slates – sarking felt behind gutter Defective gutter joint Blocked funnel outlet Eaves gutter falling wrong way Discolouration (dampness) just above ground level Defective DPC Build-up of soil and ground vegetation	Investigate Remake, or seal as necessary Test flow – pour water into gutter Investigate Remove ground vegetation away from house wall, and reduce soil level to not less than 150 mm below the DPC
	Air grate (brick) blocked	Build-up of soil and ground vegetation	Unblock
	Extension roof Defective lead apron Broken cast iron inspection chamber cover (IC) in driveway	Unknown Vehicle driven over the cover	Make good or replace – reset into face wall Replace with heavier duty IC cover
Outside	Drive gates – unable to close	Gates sagging Catching the drive when opened Rotten gate posts	Repair or renew gates Re-hang with clearance Renew gateposts

- Smoothing plane* (metal).
- Rebate plane.
- Block plane*.
- Surform types (planes and files).

4. *Boring tools.*
   - Carpenter's brace and an assortment of bits, including a counter sink, and a turn screw. * Drill bits with tapered square shank's (morse drills) and an expansive bit set should be included. **All loose bits should be kept within a stout purpose-made bit-roll.**
   - Hand drill (wheel brace), with a boxed set of twist drills*.
   - Set of bradawl*.

5. *Chisels* – an assortment of the following:*
   - Bevelled edge chisels.
   - Firmer chisels.
   - Gouges.

All loose wood cutting chisels must be kept within a stout purpose made chisel-roll.
   - Engineer's cold chisel – for cutting steel and the masonry*.
   - Masonry chisels*.
   - Seaming chisel -for plugging (cutting out mortar joints).
   - Bolster – for cutting brick and block work.
   - Flooring chisel (bolster)*.

Axe* – the blade must be protected at all times with a thick, strong leather or similar sheath.

6. *Hammers.*
   - Engineer's hammer*.
   - Club (lump) hammer.
   - Warrington hammer (Number 2).
   - Claw hammer*.
   - Tacker – staple gun.

7. *Screw drivers.*
   - Rigid-blade types* – large and small (electrician's) slotted and pozidriv.
   - Ratchet*.
   - Spiral ratchet, with an assortment of bits and drill points.
   - Turnscrew* (for use with the carpenter's brace).

8. *Pincers, pliers and shears.*
   - Carpenter pincers*.
   - Combination pliers*.
   - Waterpump pliers.
   - Self-grip (mole) pliers*.
   - Tin snips.

9. *Spanners*
   - Adjustable spanner*.
   - Adjustable wrench.

10. *Punches*
    - Nail punches*.

- Centre punches.
- Drifts*. [BENCH AND SITE SKILLS, SECTION 5.8.2.]

11. *Metal files*
    - Mill saw file (with a wood or plastics handle) or a farmer's friend file* (all-in-one metal handle).
    - Saw file.
    - Warding file.
    - Needle files (set)*.

12. *Knives*
    - Pocket knife*.
    - Trimming knife*.
    - Cobbler's hacking knife* – for the removal of hard dried putty pointing (fronting) and bedding from glazed window rebates.
    - Putty pointing knife.

13. *Trowels*
    - Bricklayer's pointing trowel*.
    - Mastic pointing trowel.
    - Plasterer's float.
    - Jointer towel.

14. *Ancillary tools and equipment*
    - Dust brush* (painters).
    - Oilstone* (combination stone within a box).
    - Cork block and an assortment of abrasive papers*.
    - Wrecking bar (tommy bar or jemmy).
    - 'G' cramps, or similar (minimum of two).
    - Nail pouch.
    - Tool belt and holsters.
    - Saw stool or 'Workmate' (or similar).
    - Plastics plugs (box of assorted sizes).
    - Nail and screw boxes.
    - Nail and screw bags* (circle of stout fabric with edge draw-string) for use with a 'bass' – each filled with assorted types and sizes.
    - Panel pin container* – filled with assorted and sizes.
    - Window cord accessory container – containing a 'mouse' (see Figure 13.16), pulley pegs (pointed dowel), sash nails, and 38 mm oval nails.
    - Plasterer's handboard (hawk).
    - Safety glasses (goggles)*.
    - Dust mask* (type must suit the risk factor).
    - Industrial gloves*.
    - Knee pads*.
    - Mastic gun with an assortment of cartridges.
    - Tool box/bag/bass*.
    - Dust sheets (fabric and polythene).
    - Polythene sheeting (weather protection to openings, etc.).

15. *Testing equipment:*
    - Moisture meter – [BENCH AND SITE SKILLS, SECTION 1.7.3].
    - Voltage, metal, and stud detector – see Section 2.1.7.

Note. Those items marked with an asterisk (*) can be regarded as standard for the immediate tool kit. The remaining items should however be easily accessible.

13.4.2 Portable Powered Hand Tools and Ancillary Equipment

More and more operations are becoming mechanised, particularly with advancement in battery technology giving cordless hand tools more power over a longer period. However, heavy duty operations will still require the greater power that is provided by the 110 V power tools. A few examples of both of these tools are listed below. You should also make reference to BENCH AND SITE SKILLS, SECTION 8.

Table 13.4 Common defects to door and doorframe and remedial treatment

Defect and effect	Possible cause	Remedial treatment
Sticking – stile to jamb/top rail to head/bottom rail to floor	Build-up of paint Moisture intake – increase in moisture content (check with moisture meter)	Ease* the offending closing edges (allow for clearance) – reseal edges as necessary
Wide gap – between stile and jamb	Shrinkage – reduced moisture content	Add a lipping (not less than 10 mm thick) to the hanging and/or closing edges – see Figure 13.4
Sagging (dropped) – stile/bottom rail – touching jamb/catching the floor. Binding (springing) – door springs open as it is being closed	Bad design or construction fault – see Section 6.1.1 Defective hinges Build-up of paint over hinges or within the hanging side rebate (Figure 13.6 hinge-bound) Oversized screw heads to hinge (Figure 13.6 hinge-bound) Hanging stile bowed (Figure 13.6)	Remove door – reassemble or reshape (see Figure 13.5) – refit and hang. Renew Remove build-up of paint* – reseal edges as necessary Rehang with longer smaller headed screws Add an extra middle hinge Slightly bevel hanging stile face Depending on degree of bow – renew the door
Twist (distortion) – gap at the top or bottom of the door when closed	Irregular grain – subjected to variable atmospheric conditions; reduction in moisture content after fixing Construction fault – components twisted Doorframe twisted – one jamb fixed out of line with the other	Usually with small amount of twist jamb rebates can be adjusted (Figure 13.7) Badly twisted doors should be renewed
Unable to stay closed (latched)	Twisted Door dropped Defective latch bolt	See above, then adjust striking plate. Adjust striking plate Renew or replace
Rotten or damaged stile and/or rails	Wet rot – end grain taking up moisture; unsealed Accidental damage Criminal damage	Cut away damage or defect (Figure 13.8a) – used keyed splices and false tenons as necessary (Figure 13.8b) – reseal and/or treat new and exposed timber
Defective door panel	Accidental damage Criminal damage	Cut out door panel – bore corner holes to help start saw cut (Figure 13.8c) Form rebates to three sides – cut back bead Fit new panel – fit and pin new beads in place (Figure 13.8d)

Note. *Old paintwork can be removed with a scraper (shave-hook) or 'Surform', etc. (always wear safety glasses or goggles during this scraping operation) – plane as necessary. *Caution:* Older house paintwork may contain lead.

If microporous (breather) paint is to be used only use it on the outer edges of outward opening doors.

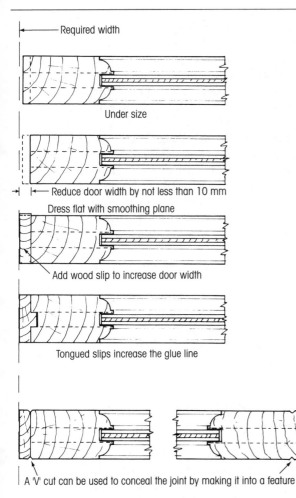

Required width

Under size

Reduce door width by not less than 10 mm

Dress flat with smoothing plane

Add wood slip to increase door width

Tongued slips increase the glue line

A 'V' cut can be used to conceal the joint by making it into a feature

Figure 13.4 Increasing a door's width

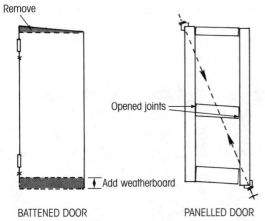

Remove

Opened joints

Add weatherboard

BATTENED DOOR

Remove top bevel, lift door, add weatherboard to cover gap.

PANELLED DOOR

Dismantle and reassemble, or corner-to-corner cramp then face one or both sides with plywood or hardboard.

NB. Refitting top rail of panelled door would ruin panelled effect

Figure 13.5 Sagging doors

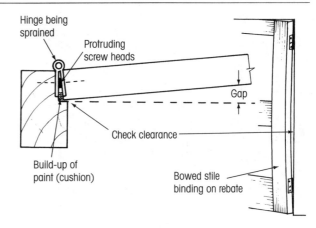

Hinge being sprained

Protruding screw heads

Gap

Check clearance

Build-up of paint (cushion)

Bowed stile binding on rebate

Figure 13.6 Doors which bind

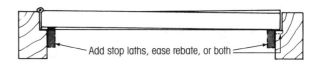

Add stop laths, ease rebate, or both

Figure 13.7 Possible remedy for a door which is slightly twisted

1. **110 V Power tools**
   - Heavy-duty rotary drill – with an assortment of twist drills and, auger and flat bits.
   - Hammer drill – with an assortment of masonry drills.
   - Circular saw.
   - Jigsaw.
   - Planer.

   Ancillary equipment.
   - 110 V transformer with a residual current device (RCD) mains connection.
   - Dust extraction device.
   - Face mask.
   - Goggles.
   - Ear defenders.

2. **Battery operated (cordless) hand tools.**
   - Drill (Rotary and percussion).
   - Screwdriver.
   - Circular saw.
   - Jig saw.
   - Sander
   - Tacker (staples and the nails).

   Ancillary equipment.
   - Spare battery.
   - Battery charger
   - Dust extraction.
   - Face mask.
   - Goggles.
   - Ear defenders.

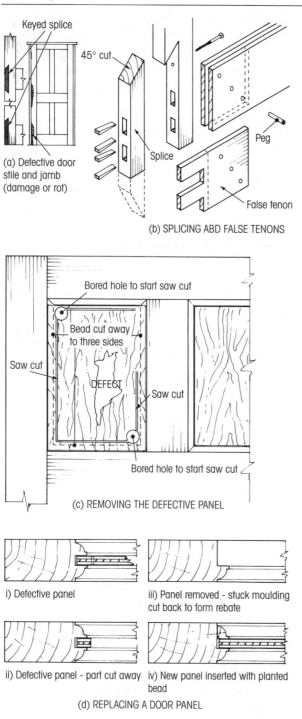

(a) Defective door stile and jamb (damage or rot)

Keyed splice

45° cut

Splice

False tenon

Peg

(b) SPLICING ABD FALSE TENONS

Bored hole to start saw cut

Bead cut away to three sides

Saw cut

DEFECT

Saw cut

Bored hole to start saw cut

(c) REMOVING THE DEFECTIVE PANEL

i) Defective panel

ii) Defective panel - part cut away

iii) Panel removed - stuck moulding cut back to form rebate

iv) New panel inserted with planted bead

(d) REPLACING A DOOR PANEL

Figure 13.8 Repairing damaged or defective doors

13.5 REMEDIAL TREATMENT

Any item of woodwork together with any accompanying hardware could at sometime become non-functional or liable to damage or deterioration.

In many cases such items can be modified or repaired but in others replacement can be the most practical alternative.

Probably the simplest and most common job is the easing of sticking doors or window casements. Door faults usually come to light after re-decoration (build-up of paint) or in the autumn, when unsealed or unprotected timber is exposed to a prolonged increase of moisture intake, causing it to swell. Sticking window casements are more likely to be noticed in the spring, when most of the casements are opened for the first time after being closed during the winter months.

Items with a high risk of deterioration – usually through lack of maintenance, are: external doors and windows, together with their frames (garages and outbuildings are particularly at risk), wood eaves guttering, driveway and a garden gates, their posts and boundary fencing.

Internal items are usually the subject of either damage or wear and tear, such as damaged or creaking floorboards and staircase treads. Sliding sash windows that won't open may be due to over-painting or defective sash cording.

Unfortunately glass also has the habit of breaking when least expected.

13.5.1 Repairing Defective and Damaged Components

The first consideration must be will it mend – can it be repaired? Judgement is very important as it is usually your advice to the client which will determine the outcome. Each case must be judged on its own merits – cost and longevity are usually the foremost criteria in the client's mind.

(a) *Doors.* Defects associated with doors are many and varied – treatment will largely depend on this type and location. Table 13.4 shows a few defect examples together with a possible cause and suitable treatment.

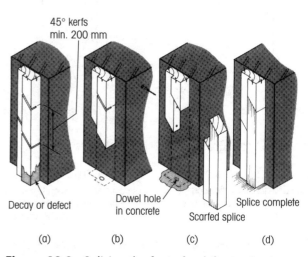

45° kerfs min. 200 mm

Decay or defect

Dowel hole in concrete

Scarfed splice

Splice complete

(a)   (b)   (c)   (d)

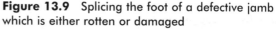

Figure 13.9 Splicing the foot of a defective jamb which is either rotten or damaged

(b) *Doorframes*. Defects of this nature are usually as a result of either damage by trying to force the door open (possibly criminal damage), or more commonly with older properties fungal decay.

If the foot of the doorjamb is found to the rotten (usually caused by wet rot) then one possibility, and the one causing least disturbance to interior plasterwork and decorations, is to make a repair by splicing. The following method of splicing a piece into the foot of a defective doorjamb should be read in conjunction with Figure 13.9.

1. Make a 45° cuts sloping down towards the exterior – in this way, surface water will be shed away from the joint (Figure 13.9a).
2. Remove the defective portion. Exposed and clean out the anchor – peg hole (Figure 13.9b). Use a sharp chisel to chop away the waste wood when forming the scarf.
3. Cut a replacement piece out of material with the same or similar profile and section (Figure 13.9c), or use a squared unrebated section – the profile of the jamb can be marked off the existing one after the joint has been made, then formed prior to fixing.
4. Bore holes for the anchor peg and fixing screws. Treat all end grain and the back with wood preservative and – when dry – seal as necessary.
5. Place cement grout (sand and cement 1:3 mix) around the peg hole. Coat the joint with waterproof mastic before tightening by leverage from the floor, then screw the joint together.
6. Dress the joint with a smoothing and rebate plane and make good cement work around the base (Figure 13.9d).

(c) *Window casements and sashes*. Small defects like the ones listed in Table 13.5 can usually be dealt with on site. Rotten or badly damaged members are not usually worth repairing.

(d) *Window frames*. The most common defect is as a result of wet rot at the junction or abutment of members where moisture can permanently be retained as a result of capillary attraction.

Always check for soft spots in the regions of:

- Junctions between jambs and mullions to sill.
- Underside of sill in contact with the face wall.
- Transom rail ends to jambs and mullions.
- Gaps between the bottoms of non-opening casements and the sill and transom rails (see Figure 13.10).

Jambs and mullion can be dealt with in a similar manner to door jambs shown in Figure 13.9. However a word of warning – if they are members of a bay window, it is likely that they may (although wrongly unless designed as such), have been used as load bearing members to support the structure above – possibly the roof. In which case, before any work is undertaken ensure that all dead and imposed (live) loads are suitably and safely supported. Defective load bearing members should not be spliced but replaced in their entirety.

Figure 13.11 shows a technique for repairing the front edge of a defective sill by splicing. If the defect extends back beyond the wall face it would be advisable to replace the whole window.

(e) *Door and window hardware*. Working mechanisms and/or moving parts of most hinges, locks, latches, fasteners, bolts and many of the security devices should be lubricated to ensure trouble-free use. Also keep free from paint and protected from corrosive conditions.

Table 13.5 Possible remedial treatment for common problems associated with window casements and sashes

Defect	Possible cause	Remedial treatment
Sticking	As doors (Table 13.4)	As doors (Table 13.4)
Sagging (dropped)	Defective or worn hinge(s) Inadequate or wrong positioning of glass setting and/or location blocks Defective corner joint(s)	Renew hinge(s) Reglaze and reposition blocks Repair, or remake casement/sash
Binding (springing)	As doors (Table 13.4)	As doors (Table 13.4)
Twisted	Construction fault Twisted (out of plumb) window frame	Remake. Realign and refix window frame
Rotten or damaged members	Wet rot Criminal or accidental damage	Opening lights should be remade Fixed lights may possibly be repaired by splicing – see Figure 13.8b
Broken glass	Forcing open a sticking opening light. Accidental or criminal damage	See Section 13.5.2g re-glazing a window)

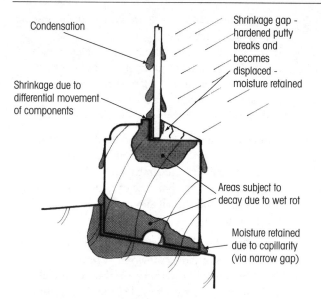

Figure 13.10   Decay as a result of permanent moisture presence

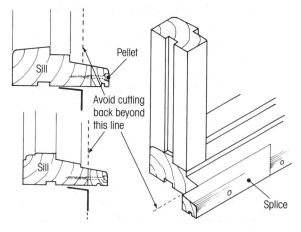

Figure 13.11   Repairing a defective window sill

Items found to be defective should be replaced with their nearest alternative. However in many cases the existing hardware may not be adequate to meet current standards required for home or business security, in which case these can be upgraded. An example of door hardware suitable to meet these requirements is shown in Table 13.6.

See also Section 13.5.2e and Chapter 6 (Door hardware).

(f) *Floorboards.* If the defect is due to rot, it should be dealt as you would any fungal decay [BENCH AND SITE SKILLS, SECTION 3.3]. If it is due to mechanical wear or damage as a result of carelessness in forming an access trap, such as shown in Figure 13.12(a), then as shown in Figures 13.12(b, c) two solutions are possible. Simply replace the damaged board or boards and provide an end bearing and joint cover as necessary – if the damage extends beyond one board, then extend the area of repair or trap.

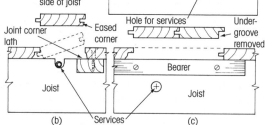

Table 13.6   Suitable door hardware for upgrading security

	Front door	Back door	Internal doors
Basic items			
High security cylinder lock (BS3621)	Yes	Yes	–
5 Lever mortice lock (BS3621)	Yes	Yes	–
3 Lever mortice lock	–	–	Yes
Added security items			
Hinge bolts	Yes	Yes	–
Security door bolt	Yes	Yes	Yes
Door chain	Yes	–	–
Door viewer	Yes	–	–

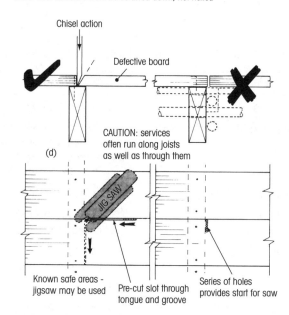

Figure 13.12   Renewing defective floor boards and forming access traps

Warning: if a board has to be cut out of the floor and its underside is not visible, there is always the possibility that services, i.e. electric cables, water or gas pipes, may run beneath it.

The whereabouts of many services can be detected by using a battery-operated detector (Section 2.1.7). However, it is always advisable to switch off the service supplies at the mains before any cutting operations are carried out – it is always better to be safe than sorry – and always proceed with caution.

Figure 13.12(d) shows how board ends can be cross cut. Ends of damaged boards can be chopped out onto a joist by using a chisel, or cut alongside it. A jigsaw is useful here, but only if the underside of the board is visible and clear of services. Alternatively a series of small holes can be bored to help the point of the flooring saw or tenon saw to get started on piercing the board. A pad saw can then be used to finish the cut close up against the adjoining board.

Special flooring saws can differ in their shape but all are designed to start a cut-off on the surface of a board and to cut into corners. They are a very useful addition to the tool kit.

A similar procedure can be used to rip down the tongue of a laid board. Equally useful is a small portable circular saw with its blade depth set so that it only cuts the tongue without protruding below the underside of the board.

To lift a single board from between two joists, possibly to be used as a small access trap, first, check as previously stated that there are no services beneath the board or indeed passing through adjacent joists. Choose a suitable heading joint then cross-cut the other end of the length to be removed – now, if the gap between the boards is wide rip down one tongue of the tongued and grooved joint between the boards, or both tongues if the gap is tight. At the heading end of the board, punch down the nails to at least two-thirds of the board's depth. Then using two wide flooring chisels (sometimes called tonguing tools – *not bricklayer bolsters*), from one edge lever the board up and out.

(g) *Staircases.* Probably the most common defects found are:
- Gaps between the wall-string and wall plaster-work.
- Creaking treads.
- Damaged nosings.
  (i) *Gaps between the wall-string and wall plaster-work.* This could simply be the result of initial drying out of the fabric, incurring differential shrinkage between dissimilar materials. On the other hand it could be that the string is receiving inadequate support or fixings, resulting in a small amount of vertical deflection. Sideways (lateral) movement could be due to

missing or defective packings between the wall-string and wall. If access to the underside of the staircase is possible checks can be made to ensure that all joints are secure, and that the general construction of the staircase is not at fault – see Section 11.6.10.

(ii) *Creaking treads.* Not an uncommon fault with older staircases, particularly if they have been subjected to prolonged activity. Assuming that the staircase was originally sound (see Section 11.6.10) this could be the result of the joint between tread and riser parting (Figure 13.13a) allowing small amounts of movement when stepped on. Access to the underside (soffit) is not usually possible having been plastered over. However, this problem can often be resolved by gluing and screwing the tread down into the top of riser of the offending step – as shown in Figure 13.13(b). If treads are carpet covered screw heads won't show, otherwise they can be slightly sunk and pelleted. Shrinkage can always be a large contributing factor in the breakdown of joints, therefore small gaps in the tread to string housing may also be apparent. In which case, provided the treads are fully located in their housing, gluing thin wood veneers into the gap(s) may be a solution. If these methods does not work,

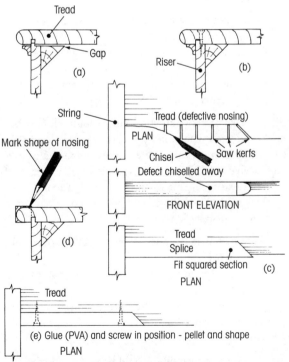

Note: all splicings etc. should be screwed as well as glued - screws not only hold the joint, but also serve as cramps

Figure 13.13   Staircase step repairs

then underside access will be required to check the condition of the staircase as a whole – such as tread housings, glue blocking, and mid-carriage support (if provided). Then modify accordingly – adding extra brackets (possibly steel brackets) may be a possible remedy.

(iii) *Damaged nosings*. Because tread nosings take the blunt of the traffic they are often the most vulnerable to damage. This is often partly due to an overdeep plough groove cut into the underside of the tread designed to receive the tongue of the riser board below it.

Defective nosings can be repaired by carefully cutting away the damaged portion and letting in a new piece of wood of the same thickness as that shown in Figure 13.13(c). Cut the spliced end at 45° to extend the glue line and help stabilise the joint. Leave the spliced portion in the square (unless the housed portion within the string is detached), the remaining portion of the tread nosing can then be used as a template to scribe its profile onto the splice (Figure 13.13d). Glue, screw, and pellet the counter bore holes blend into tread by sanding (Figure 13.13e). When a Scotia mould has been used between the underside of the nosing and riser, if possible renew the whole length with glue and pins.

Exterior work:

(h) *Eaves gutters*. See Section 13.5.2(h) (Removal, renewal or replacement).
(i) *Gates*. Wooden garden and driveway gates are permanently subjected to all-round weather exposure making them the most vulnerable of all joinery items to decay and damage.

Off-the-shelf gates (standard items) should all now be pressure treated with wood preservative against fungal decay. But many of the ornate purpose-made gates may only receive a surface treatment, therefore, like many of the other external items of joinery, if they are not regularly maintained they could be liable to fungal decay (wet rot) in places such as:

• Where members lap one another – palings over rails, etc.
• End grain exposure – tops and bottoms of stiles, paling, etc.
• Open joints (due to moisture movement) – shoulders of tenons, mortice holes, etc.
• Behind fittings and fixtures.

Hinges, bolts, and catches should be protected from corrosion by being galvanised (zinc coated), and their fixings (screws, coach bolts) should all be rust-proof types.

Most of the expensive ornate gates, particularly those made from hardwood such as English oak are nearly always worth repairing. Fortunately in the past many of these gates were dry jointed (with the possible exception of pre-painting the joint before assembly). In which case dismantling may simply be a matter of removing the draw pins (Figure 6.27d) or dowels with a hammer and steel drifter (also see 13.4.1(11)) or boring them out. Then with a soft mallet carefully dismantle the joints as required.

Rather than replacing a whole component you may wish to consider renewing a section by splicing or inserting a false tenon as shown in Figure 13.8.

13.5.2 Removal, Renewal or Replacement

The assumption here is that a repair it is not practicable. Although to renew or replace an item would usually cost more than a repair the chances are its service length would be increased and therefore become most cost effective.

It is not always possible to renew an item by size or pattern, it may obsolete as an off-the-shelf item. In which case it would have to be purpose made – with the added option of being able to modify the designed to suit the client's requirements.

(a) *Doors*. With a possible exception of purpose-built house, garage, and outbuilding doors, 6′8″ by 2′8″ house doors, other sizes can be found in most trade catalogues.

The process of re-hanging the door will be the same as a new one, this is explained in Section 6. If the hinge pattern is different from the original, their position may have to altered in which case the old hinge housing should be made good with a wood fillet (see Section 13.5.2e) – **not by using a wood filler**. The same principle would apply to the position of the lock or latch, etc.

(b) *Door frames*. Before endeavouring to remove a door frame, always check to see if there is a lintel above the opening (believe it or not this is not always the case). If there isn't, no work must be undertaken until provision can be made for one to be installed.

Providing the opening is correctly supported and a lintel is in place, then the door frame can, depending whether it is an open or closed frame, be removed in sections as shown in Figure 13.14. Once removed clean the surrounding reveal of any spent mortar and make good any holes previously occupied by the horns.

Fix the new doorframe as described in Chapter 6.

(c) *Window casements and sashes*. Because these are components of the window as a whole and not purchased separately, replacements will be purpose made.

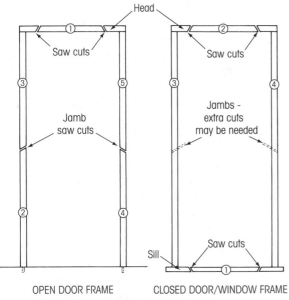

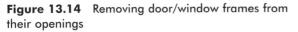

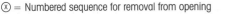

OPEN DOOR FRAME   CLOSED DOOR/WINDOW FRAME

Ⓧ = Numbered sequence for removal from opening

Figure 13.14 Removing door/window frames from their openings

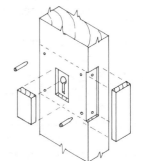

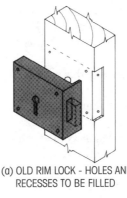

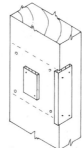

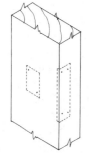

(a) OLD RIM LOCK - HOLES AND RECESSES TO BE FILLED (b) HOLES AND RECESSES ENLARGED TO RECEIVE FILLING PIECES

(c) OVERTHICK FILLING PIECE GLUED AND PINNED IN PLACE (d) DRESSED FLUSH WITH SMOOTHING PLANE, FILL NAIL HOLES AND SMALL GAPS WITH WOOD FILLER - SAND SMOOTH

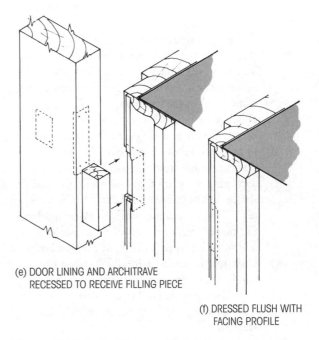

(e) DOOR LINING AND ARCHITRAVE RECESSED TO RECEIVE FILLING PIECE

(f) DRESSED FLUSH WITH FACING PROFILE

Figure 13.15 Making good holes and recesses

The process of re-hanging will be the same as when originally installed. If the hinge pattern is different from the original, the position may have to be altered, in which case any housing should be made good with a wood fillet (see Section 13.5.2e below) – **not by using a wood filler**. The same principle would apply to the position of stays, latch and fasteners, etc.

(d) *Window frames.* Before endeavouring to remove a window frame, always check to see if there is a lintel above the opening (believe it or not this is not always the case). If there isn't, no work must be undertaken until provision can be made for one to be installed.

Providing the opening is correctly supported and a lintel is in place, then the window frame can, like the closed door frame, be removed in sections as shown in Figure 13.14. Once removed, clean the surrounding reveal of any spent mortar and make good any holes previously occupied by the head and sill horns.

(e) *Door and window hardware (making-good holes and recesses).* Unless the door or window is relatively new it is highly unlikely that replacement hardware will fit onto or into the same position (especially if it is to be upgraded – see Section 13.5.1(e) and Table 13.6). Or for example a more drastic step may be taken to replace a rim dead lock for a mortice lock. In which case, as shown in Figure 13.15, possibly both the keyhole and door and jamb recesses would have to be made good. Notice how the keyhole has been enlarged to accommodate the wood fillet – this would also apply to a redundant spindle hole.

Note. If because of stile damage or reverse hanging, etc., all three hinge recesses are to be filled, then rather than letting in individual pieces, consider reducing the whole width of the door by not less than 10 mm (or removing an existing lipping piece).

A lipping piece is then glued and fixed to the edge to bring it back to its original size. Figure 13.4 shows a method of widening a door. See Chapter 6 (Door hardware).

(f) *Renewing sash cords.* The reason why this is dealt with under renewal rather than repairs is because it is false economy not to renew all four cords if only one or two are broken.

There are many schools of thought as to which is the best way to carry out this operation, but the method chosen should prove to be both quick and effective. All the work can be carried out from inside the building.

Before starting you should always protect the surrounding with dust sheets – they should not however be placed on polished floors where you might stand or walk as they will very likely slip from under you.

The following should be read in-conjunction with Figure 13.16.

(i) Close and securely fasten the sashes to prevent sliding or falling over when beading is removed.

(ii) Cut remaining cords to the lowered sash and gently lower the sash weight to the bottom of the case (never allow a weight to drop under its own weight, as this could damage the case and its fixing).

(iii) Leaving the head and sill staff beads in place, remove the right- and left-hand staff beads by first using a sharp trimming knife to cut through any layers of paint covering the joint. Then one at a time (after removing any screws) ease the beads out from the centre by using an old wide wood chisel or flooring chisel (see Figure 13.12a) until bow shaped, they can then be disengaged from their corner mitre. Nails retained in the bead should be pulled through from the back to avoid damaging the paint film (round-headed nails should not have been used, but, if they were, cut off the protruding portion). After marking (T) on the tops of the beads they should be laid to their respective sides of the window.

(iv) Unfasten the lower sash – lift it out of the window and remove old cords and sash nails from the stiles, and put to one side.

(v) Lower the top sash and cut its cords as at stage (ii).

(vi) Holding the sash steady to prevent it from falling, remove the parting beads (after cutting through any paint work as previously described) by starting at the sill end and easing them out of the grooves with a chisel. They may or may not have been nailed – if they are, remove with great care.

(vii) Lift out the top sash, remove old cords and sash nails from the stiles, and put it to one side.

(viii) Remove the pockets (one screw to each side) and take out the weights from each of their cases. Each weight should have its old cords removed and a chalk marked front (F) or back (B), then laid on to the floor at the appropriate side of the window.

Note. Front and back weights may differ in size and weight.

(ix) Marking sash cord length is as shown in Figure 13.16(a), with suitable distances 'X' and 'Y' to allow for cord fixing and pulley clearances (Figure 13.16e). These are marked with chalk onto the stiles of both sashes. These distances are then transferred to the face of the pulley stiles.

(x) Figure 13.16(b) shows how a cording 'mouse' can be made from either a small rolled piece of sheet lead or preferably a short length of sash chain (sometimes used on heavy sashes in place of cord). The mouse string (not less than 2 m long) is attached to the end of a knot (Hank) of good quality waxed sash cord as shown, and the mouse is made ready to be threaded over the pulley wheel (ensure that the pulley wheel is running freely and not broken – lubricate if necessary).

(xi) Figure 13.16(c) shows two alternative methods of cording the window. At the end of the run, the cord is attached to the appropriate back weight (top sash weight, TSW).

(xii) Working from the right-hand or left-hand side (depending on the method of cording), feed the corded weight (making sure it is on the correct side of the weight parting slip (wagtail)) for the top sash through the pocket into the box, pull it up to just below the pulley wheel – lightly wedge it in that position with one of the doweled pegs. Then pulling the cord down the pulley stile cut it off 50 mm above the chalk line. Tie the bottom sash weight (BSW) to the loose end of cord, pull it up to the pulley and wedge it in the position, then this time cut the cord 30 mm below the chalk line. Repeat the process on the opposite side (Figure 13.16d). Fix the pockets back into position.

(xiii) Attach the cords to the sashes by nailing with sash nails (some attachments are by knotting). As shown in Figure 13.16(e) use at least four nails per stile – each nail should be slightly sloping away from the rebate to avoid touching the glass. Nails should not interfere with the pulley.

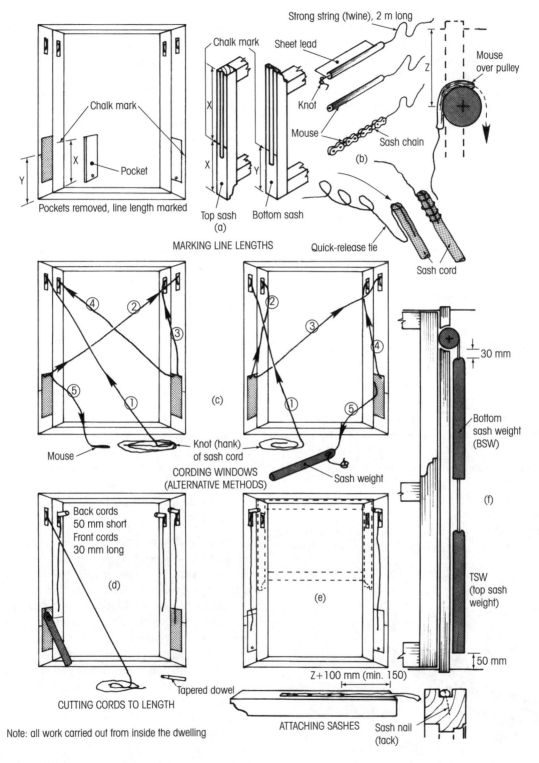

Figure 13.16 Sash cording a double hung sash window

(xiv) Cord the top sash first, then release its pulley wedges and reposition the parting beads. Then cord the bottom sash. Replaced the staff beads – if nails are to be used (screws in cups are preferred), use oval nails.

(xv) When the window is re-corded in this way, it should be noticed from Figure 13.16(f) that, when the top sash is closed, its weights will be suspended 50 mm above the sill, thus allowing for any stretch in the sash cord, and for any build-up of mortar dust, etc. in the case bottom, which could possibly interfere with the distance the weight can travel to allow sash to close properly. The bottom sash weights should be a minimum of 30 mm below the pulley when the bottom sash is closed. Ensure that the sashes run freely – rubbing candle wax over the sliding parts of the sashes can ease their movement.

(xvi) Remove dust from the sill and window board with a dust brush. Collect up spent nails and old sash cord, together with dustsheets and remove them from site.

(g) *Removal of a broken glass pane.* Work involving reglazing windows with large panes of glass should only be undertaken by specialist glaziers who should be fully equipped to deal with this type of work safely.

Where practicable opening lights can be unscrewed from their frames and the broken glass removed at ground level. Broken glass within fixed frames must be treated with extreme caution – operatives should position themselves above and away from the window, in case any broken pieces of glass become dislodged in preparing the work. The area below the window **must** be cordoned off to ensure no one enters this potentially dangerous zone.

Before starting any work, operatives must be wearing either shrouded safety glasses or goggles, which must remain in place during the whole operation as protection from glass splinters which will inevitably be shed in varying degrees as the broken glass is being removed from its frame. Stout leather gauntlet-type gloves with rubber finger grips should be worn to protect hands and wrists whenever handling glass – particularly when removing broken pieces (Figure 13.17a).

Removing the glass – start at the head or top rail, work around the casement/sash or frame removing all the old putty fronting (pointing) with a hacking knife and hammer (loose pieces should be removed as you progress downward). Once all the edges of the glass have been exposed, remove all the glazing sprigs (small squared-sectioned tacks) or panel pins with pliers. Then as shown in Figure 13.17(b), using an old wide wood chisel held flat against the back of the glass, gently tap at the putty bedding, this action will dislodge the glass from the rebate. It is advisable to lay detached casements/sashes on the ground during the latter part of this operation – a suitable protective cover should be placed between the casements/sashes and the ground. *Note.* Reglazing a window – see Section 8.13.

(h) *Eaves gutters.* By and large the plastic gutter has superseded both wood and cast iron. Early problems with brittleness and distortion have more or less been overcome. Problems of thermal movement in hot weather (distortion, buckling, and creaking) were and still remain where fixers have failed to follow manufacturers fixing guidelines. All joint and junction fittings should have moulded into then an overlap distance indicator, so that a gap is left between abutting lengths of gutter. In this way provision is made to allow for thermal movement to take place, but take note, that a **greater gaps will be required when the gutter is fixed in a colder climates than hot.**

Wood gutter – (traditionally known as 'spouting') mid-range sizes of 3″ × 5½″ (75 mm × 140 mm) and 4″ × 6″ (100 mm × 150 mm) are still available, and still popular in some areas. One of the factors being that on the basis of renewing short defective sections it can be very cost effective. Also its service life span can now, with modern preservative pressure treatment be increased considerably from its predecessor.

Before replacing a wood eaves gutter, because of its weight always check on its means of support. It may be sat on a series of brick or stone corbels with supplementary strap fixings back to the wall or rafters. Or more commonly, steel bracket attached to the foot of the rafters – possibly over a fascia board. You will also find that in many instances that the gutter has been nailed back to the spar (rafter) feet. Before taking down a length of defective gutter these nails should be cut by running a hacksaw blade (held in a pad saw handle) between the back of the gutter and the spar feet.

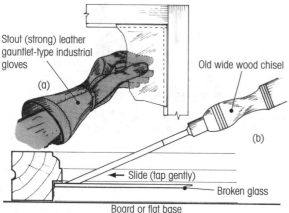

Stout (strong) leather gauntlet-type industrial gloves

(a)

Old wide wood chisel

(b)

← Slide (tap gently)

Broken glass

Board or flat base

Figure 13.17 Detaching and removing broken glass

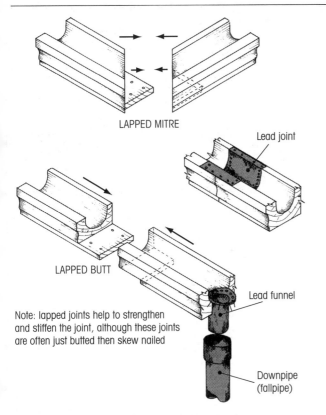

Figure 13.18 Wood eaves gutter – joints and connections

Figure 13.18 shows a lapped mitre and butt joint that will be screwed together from the underside. Smaller sectioned gutter is often simply butted and held together by skew nailing (driving nails in at an angle) the joint. Traditionally all the joints are covered with a strip of lead not less than 100 mm wide. The procedure for making these joints is as follows:

- Gently beat the lead to the channel profile (either straight or mitred).
- With a chisel recess the lead into the channel until it is flush with it.
- Coat the joint with bituminous mastic (jointing compound).
- Using copper tacks (not nails) close nail the lead to the gutter.
- Smear over the nail heads with the same bituminous compound.
- Paint over the joint with the same component used to treat the inner surface of the gutter – usually bituminous paint or tar.

Lead funnel water outlets are set into holes cut through the channel with an expansive bit [BENCH AND SITE SKILLS, FIGURE 5.59]. The seal between the funnel and the gutter channel is made much the same as the joints. A scribing gouge [BENCH AND SITE SKILLS, FIGURE 5.80] is used to recess the funnel flange.

Note. Leadwork is often the job of the plumber.

Cast iron gutters are usually replaced or reinstated by the plumber.

Metal gutters other than cast iron are usually supplied and fixed by specialist contractors.

Fall – all eaves gutters should have sufficient slope to discharge the roof water to the appropriate down pipe (fall pipe) without retaining any puddles. An acceptable minimum fall would be 1 in 350 or say about 10 mm for every 3 m of gutter.

13.5.3 Making Good Small Areas of Brickwork

For whatever purpose when brickwork has been removed or become damaged the likelihood is that for small amounts you would be expected to make it good. The first problem you could be faced with if dealing with face brickwork would be matching the pattern and possibly the make of the original.

Figure 13.19 shows an example of where three bricks have had to be removed. There are many reasons for cutting slots in brickwork, for example:

- To receive a needle (beam) as part of a temporary shore.
- Gain access to a wall cavity.
- Inspect joist ends.
- Provide a bearing for a joist or beam.

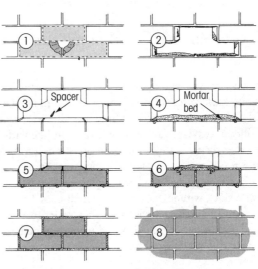

1: Forming the hole

2: Brickwork removed

3: Old mortar removed

4: Spacing blocks and mortar bed layed

5: First and second bricks layed

6: Spacing blocks and mortar bed layed

7: Third brick layed – spacers removed ready for making good

8: Pointing over to match existing brickwork – brickwork made good

Figure 13.19 Making good a hole in brickwork

- Start by cleaning away all the old mortar and particles of dust – ensure pieces don't enter any cavity. Then, after preparing a mortar mix of 1:1:6 (1 part Portland cement, 1 part lime, 6 parts yellow sand) proportions, or by using a pre-packed mortar mix, the following method can be used.
- Cut six spacers (packing) 125 mm long and square to the thickness of adjacent mortar seams.
- Lay a bed of mortar as shown.
- Insert spacer blocks (leave a 25mm protrusion) – level mortar until just slightly above them.
- Apply a bevel of mortar to the end and leading top edge of the first brick and place in position – remove surplus mortar.
- Repeat with the second brick.
- Fill the perp end (perpendicular joint) with mortar.
- Lay the second bed, but this time as shown in Figure 13.19(6) the spacers have a nail stop to prevent them being pushed forward.
- Apply a bevel of mortar to the ends and very thin bevel of mortar to the top leading edge of the brick.
- Without dislodging the spacers carefully position the brick onto the bedding.
- Carefully fill any open joints, then rake them back below the surface of the brickwork.
- Remove spacers when mortar is partially set – make good the holes – leave to set.
- Clean or rake out a few mortar joints adjacent to the new brickwork.
- Repoint the area with as near a mix as possible to the original.
- This same method can also be used with blockwork.

13.5.4 Making Good Small Areas of Plasterwork

A few years ago making good plasterwork meant hacking back to brick or blockwork and then applying a sand and cement or sand and lime base or backing coat, possibly in two layers, screeding it to be in line with the wall, then as it was drying back in preparation to receive a thin finishing coat of plaster. Today you can use a one coat patching plaster capable of being applied up to 50 mm thick without slumping, shrinking, or cracking – this you can purchase either as a dry powder to be mixed with water and used as and when required, or pre-mixed in a tub ready for use.

Pre-treatment to base materials such as brick or blockwork after any loose particles have been removed, usually involves damping the dry wall down with water – in some cases where suction has to be controlled a PVAc bonding agent can be used.

Figure 13.20 shows a common plasterwork failure after a replacement window operation. In this case, because adjacent plasterwork was found to be loose it was decided to renew all the plasterwork on one side of

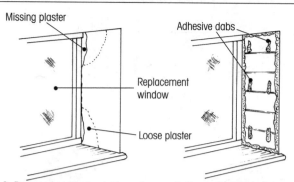
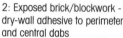
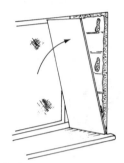
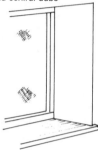

1: Damaged and loose plasterwork

2: Exposed brick/blockwork - dry-wall adhesive to perimeter and central dabs

3: Apply plasterboard

4: Use jointing plaster and/or skim over with wallboard finish

Figure 13.20 Making good defective plasterwork to a window reveal

the reveal. But instead of using a wet plaster backing, a strip of plasterboard was been used in its place. The plasterboard was first stuck to the wall with dabs of wallboard adhesive, joints and gaps were then filled with 'Dry-wall' filler, then finally the surface was skimmed over with a wallboard finish plaster.

13.5.5 Making Good Ceramic Wall and Floor Tiles

The need to replace damaged ceramic wall tiles can be as a result of:

- Cracking due to structural movement of the wall.
- Holes – previously occupied by wall plugs for redundant fittings or fixtures.
- Accidental damage.
- Damage to ceramic floor tiles would more likely be due to accidental damage (dropping a hammer doesn't help!).

Finding a good match is often very difficult – you could try taking a sample piece (Figure 13.21, stage 4) to your supplier.

If a true match is not possible you could consider removing more than one tile, and replacing them with contrasting or decorative patterned tiles.

The following procedure for removing and replacing the tile should be read in conjunction with Figure 13.21.

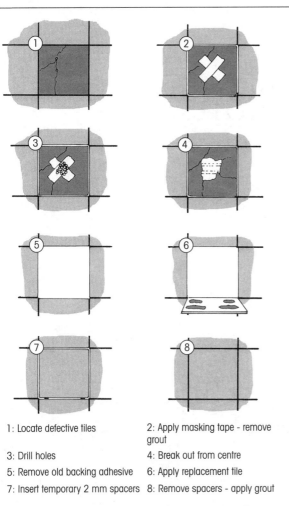

1: Locate defective tiles

2: Apply masking tape - remove grout

3: Drill holes

4: Break out from centre

5: Remove old backing adhesive

6: Apply replacement tile

7: Insert temporary 2 mm spacers

8: Remove spacers - apply grout

Figure 13.21 Renewing defective ceramic wall tiles

- Prepare the work area – cover the floor and unit areas, etc. with dustsheets (small particles of sharp ceramic tiles moved across a surface will scratch and damage it).
- Wear **full** eye protection (goggles) and facemask.
- Release perimeter stress by carefully scraping away the grout from around the tile (*a purpose-made tool is available for this purpose*). Never prise a tile up from the edge (joint) as this will inevitably crack the adjacent tile.
- Try not to damage the base (backing) material scrape or chip away old adhesive.
- Modified tiles – if the tile requires cutting to size. Use either a proprietary tile cutter or tile scorer with snap-off tool. Indents and scribes can be cut out with a tile cutting frame saw.
- Apply four dabs of adhesive to the back of the replacement tile. As the tile is pressed home there should be enough room for the adhesive to spread out without restricting it to lie flush with adjacent tiles.
- Position two 2 mm spacers into the bottom joint, and leave until the backing adhesive is set (24 hours).
- Remove spacers.

With a rubber squeegee apply grout to the joints of the replacement tile and those adjacent to it. Use the appropriate grout (cement based filler – special types are available to suit different conditions), point the joints with a slither of wood with a rounded point. When dry polish the surface with a cloth.

Note. Ceramic floor tiles can be similarly replaced although be larger spacers with needed.

Appendix: Floor Plans

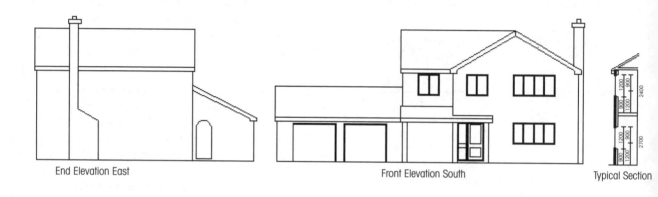

End Elevation East

Front Elevation South

Typical Section

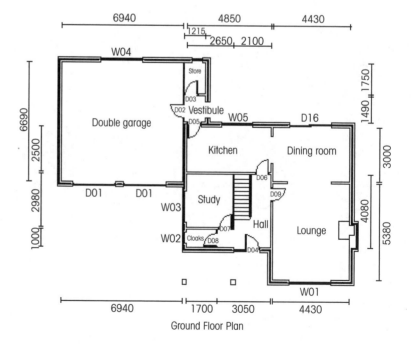

Ground Floor Plan

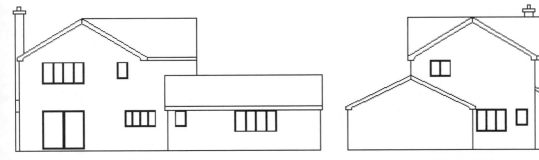

Rear Elevation North   End Elevation West

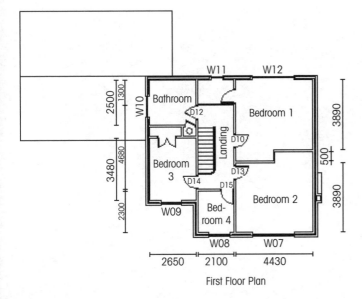

First Floor Plan

Index